D1727025

NUCLEOTIDE ANALOGS

NUCLEOTIDE ANALOGS

SYNTHESIS AND BIOLOGICAL FUNCTION

Prof. Dr. Karl Heinz Scheit
Max-Planck-Institut für Biophysikalische Chemie
Göttingen, Germany

A WILEY-INTERSCIENCE PUBLICATION
JOHN WILEY & SONS, New York • Chichester • Brisbane • Toronto

Library of Congress Cataloging in Publication Data:

Scheit, Karl Heinz, 1934–
Nucleotide analogs.

"A Wiley-Interscience publication."
Includes index.
1. Nucleotides—Synthesis. I. Title.

QD436.N85S3 574.87′328 79-25445
ISBN 0-471-04854-2

Printed in the United States of America

10 9 8 7 6 5 4 3 2 1

To F.

Preface

When the ingenious organic chemist Emil Fischer published the synthesis of a phosphorylated theophylline glucoside, the first nucleotide analog of its kind, he made the following visionary statement:

Diese Erfahrungen mit der Theophyllinglucosidphosphorsäure scheinen mir geeignet, neue Gesichtspunkte für die Beurteilung der natürlichen Nucleotide und Nucleinsäuren zu gewinnen. . . . Mit der synthetischen Erschließung der Gruppe ist die Möglichkeit gegeben, zahlreiche Stoffe zu gewinnen, die den natürlichen Nucleinsäuren mehr oder weniger nahestehen. Wie werden sie auf verschiedene Lebewesen reagieren? Werden sie zurückgewiesen oder zertrümmert oder werden sie am Aufbau des Zellkerns teilnehmen? Die Antwort darauf kann nur der Versuch geben. Ich bin kuhn genug, zu hoffen, daß unter besonders gunstigen Bedingungen der letzte Fall, die Assimilation künstlicher Nucleinsäuren ohne Spaltung des Molekuls eintreten kann. Das müßte aber zu tiefgreifenden Änderungen des Organismus führen, die vielleicht den in der Natur beobachteten dauernden Veranderungen, den Mutationen, ähnlich sind [EMIL FISCHER, *Chem. Ber.* **47**, 3196 (1914)].

(It seems to me that further advances in understanding the natural nucleotides and nucleic acids can follow from these experiences with theophyllineglucoside phosphoric acid. . . . The possibility of synthesizing this class of compounds allows many substances to be made that are more or less closely related to the natural nucleic acids. How are these substances going to react with various living organisms? Are they going to be rejected, or degraded or are they going to be incorporated into the cell nucleus? I am bold enough to hope that under especially favourable conditions the last case, the assimilation of synthetic nucleic acids without cleavage of the individual molecule can occur. This would however lead to fundamental changes of the organism which perhaps resemble mutations, the permanent changes observed in nature.)

Nothing needs to be added to this eloquent expression of the importance of nucleotide analogs. Since then a large body of information on this class of compounds has accumulated. A number of reviews have competently dealt with specific topics of nucleotide analog research but a comprehensive treatment of this subject was yet missing. This book, *Nucleotide Analogs*, is my attempt to describe the present knowledge on the synthesis, chemistry, biochemistry, and application of nucleotide analogs. To keep the size of this book within reasonable limits, I have presented information on the biochemistry of nucleotide analogs that either contributes to the understanding of structure-function relationships or that has prompted the chemical synthesis of an analog.

The book is written for biochemists as a stimulating introduction to the application of nucleotide analogs in the study of biological systems. It informs them of the present state of chemical modification of nucleotides and of the available nucleotide analogs. The book is written for the organic chemists who are fascinated by the chemistry of natural products. It introduces them to an interesting and important aspect of nucleotide chemistry. And last but not least, I hope that colleagues, working in the field of chemistry or biochemistry of nucleotide analogs, find this book a useful synopsis which will guide their research to new objectives.

I have to thank many for their help during the preparation of the manuscript—too many, to name them all. I am deeply indebted to Dr. Vic Armstrong for his constructive and competent criticism. Thanks are due to my colleagues Drs. N. N. Leonard, F. Eckstein, and E. Schlimme for supplying me with unpublished information. Dr. M. Osborne-Weber helped me with the translation of the statement by Emil Fischer cited above. The support of Dr. T. M. Jovin is gratefully acknowledged. Miss S. Schröder is thanked for her help in the preparation of drawings and Mrs. A. Bast, for her expert typing of the final manuscript. Warmest thanks go to my family for their great patience in enduring my obsession.

KARL HEINZ SCHEIT

Göttingen, Germany
January 1980

Contents

1 CHEMICAL STRUCTURE OF NUCLEOTIDES 1

1.1 Electronic Structures of Heterocyclic Bases, 1
1.2 Conformation of Nucleotides, 3

2 NUCLEOTIDES WITH MODIFIED HETEROCYCLIC SUBSTITUENTS 13

2.1 Substitution at Ring Nitrogen Atoms, 14

2.1.1 Alkylation, 14
2.1.2 *N*-Oxidation, 18

2.2 Substitution at Exocyclic Groups, 19

2.2.1 Alkylation, 19
2.2.2 Reactions with Aldehydes, 21
2.2.3 Substitution of Keto by Thioketo Groups, 24
2.2.4 Substitution of Keto by Selenoxo Groups, 33

2.3 Introduction of Exocyclic Substituents, 35

2.3.1 8-Substituted Purine Nucleotides, 35
2.3.2 2-Substituted Purine Nucleotides, 42
2.3.3 5-Substituted Pyrimidine Nucleotides, 46
2.3.4 6-Substituted Pyrimidine Nucleotides, 57
2.3.5 Miscellaneous, 59

2.4 Ring Analogs of Purine Nucleotides, 61

2.4.1 Azapurine Nucleotides, 62
2.4.2 Deazapurine and Pyrazolo (4,5-d)pyrimidine Nucleotides, 64
2.4.3 Miscellaneous, 68

2.5 Ring Analogs of Pyrimidine Nucleotides, 72

2.5.1 Azapyrimidine Nucleotides, 72
2.5.2 Deazapyrimidine, Pyridon, and Pyridazon Nucleotides, 77

3 NUCLEOTIDES WITH UNCOMMON GLYCOSIDIC BONDS 90

3.1 Nucleotides with Uncommon *N*-Glycosidic Bonds, 90
3.2 Pseudouridylic Acid, 92

4 NUCLEOTIDES WITH MODIFIED PHOSPHATE GROUPS 96

4.1 Nucleotide Analogs with Altered P—O—P Bonds, 97
4.2 Thiophosphate Analogs of Nucleotides, 101
4.3 Nucleotides with Altered P—O—C Bonds, 113
4.4 Nucleoside Phosphites, 120
4.5 Nucleoside Phosphonates, 125
4.6 Miscellaneous, 134

5 NUCLEOTIDES WITH ALTERED SUGAR PARTS 142

5.1 Arabino-, Xylo-, and Lyxo-Nucleotide Analogs, 143
5.2 Unnatural Enantiomeric and Anomeric Forms of Nucleotides, 153
5.3 Substitution of Ribose Moieties in Nucleotides, 161
5.4 Aliphatic Analogs of Nucleotides, 179
5.5 Miscellaenous, 185

6 METHODS OF PHOSPHORYLATION 195

6.1 Synthesis of Phosphate Esters, 196

6.1.1 Phosphoesterchloridates, 197
6.1.2 Phosphoric Acid Anhydrides, 202
6.1.3 Phosphorylation by Transesterification, 204
6.1.4 Miscellaneous, 207

6.2 Synthesis of Nucleoside Polyphosphates, 211

6.2.1 Displacement Reactions with Triester Pyrophosphates, 211
6.2.2 Nucleoside Phosphoramidates, 212
6.2.3 Reactive Phosphate Esters, 214

7 REACTIVE DERIVATIVES OF NUCLEOTIDES 219

7.1 Reactive Derivatives of Nucleotides, 221

7.1.1 Phosphate-Modified Nucleotides, 221
7.1.2 Base-Modified Nucleotides, 227
7.1.3 Sugar-Modified Nucleotides, 233

7.2 Photogenerated Reactive Derivatives of Nucleotides, 237

7.2.1 Phosphate-Modified Nucleotides, 237
7.2.2 Base-Modified Nucleotides, 238
7.2.3 Sugar-Modified Nucleotides, 246

AUTHOR INDEX 255

SUBJECT INDEX 271

NUCLEOTIDE ANALOGS

ONE

Chemical Structure of Nucleotides

The term nucleotide was first introduced by Levene and Mandel (1) to apply to the products isolated from acidic digests of nucleic acids. Nucleotides are the phosphate esters of nucleosides, which in turn are β-*N*-pentofuranosides of heterocyclic bases. The five major naturally occurring heterocyclic bases are adenine, guanine, cytosine, thymine, and uracil, and the pentofuranose component may be either D-ribose or 2-deoxy-D-ribose.

The general organic chemistry of nucleosides and nucleotides as well as the fascinating history of the early work on their isolation and the elucidation of their structure is the subject of a limited number of monographies (2–9). They cover the literature up to 1974. This chapter summarizes the present state of our knowledge of the structural features of nucleotides.

1.1 ELECTRONIC STRUCTURES OF HETEROCYCLIC BASES

Although the heterocyclic bases of nucleotides can theoretically exist in several tautomeric forms, the number of forms depends on the number of exocyclic OH or NH_2-substituents on the bases, and one form generally predominates. Thus for in uridine and thymidine the corresponding bases occur mainly as the diketo tautomers (10–13). The keto-amino tautomeric forms prevail in cytidine (14, 15) and guanosine (11). The adenine substituent in adenosine or adenosine 5′-phosphate adopts the amino form (9). However, the spectroscopic methods employed in those studies, namely UV, IR, and NMR are not sensitive enough to rule out the possibility that enol-imino structures exist as minor forms.

The bases of nucleotides or nucleosides can have the properties of weak acids or bases—adenine and cytosine are weak bases, uracil and thymine are acids, and guanine has the properties of both. Evidence for the site of protonation has emerged from X-ray and spectroscopic studies. In the case of adenine nucleotides (**1, 2**), the proton is attached to the N(1) atom (14, 15). Protonation of guanine nucleotides (**3, 4**) occurs at nitrogen atom N(7) (15, 16, 17, 18), and in cytosine nucleotides (**5, 6**) protons add to nitrogen

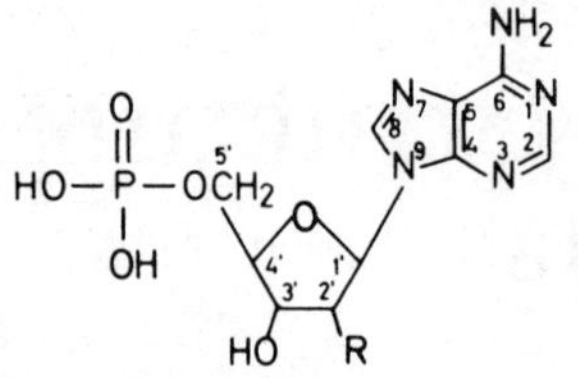

1,2

1	R = H	2′-Deoxyadenosine 5′-phosphate
2	R = OH	Adenosine 5′-phosphate

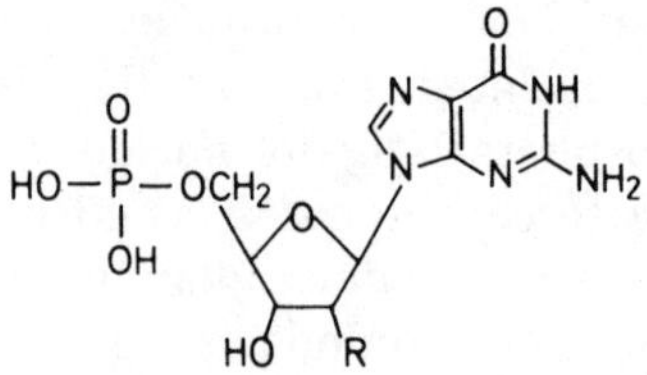

3,4

3	R = H	2′-Deoxyguanosine 5′-phosphate
4	R = OH	Guanosine 5′-phosphate

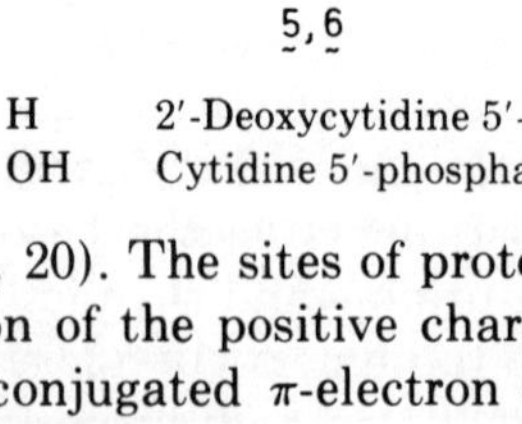

5,6

5	R = H	2′-Deoxycytidine 5′-phosphate
6	R = OH	Cytidine 5′-phosphate

atom N(3) (15, 17, 18, 19, 20). The sites of protonation are not necessarily identical with the location of the positive charges, since the heterocyclic substituents all contain conjugated π-electron systems, allowing delocalization of the charge. Thus in adenine nucleotides the positive charge is localized to a large degree on the exocyclic amino group. This also applies to the localization of negative charges formed by abstraction of protons from the base in uridine 5′-phosphate (**8**) or guanosine 5′-phosphate (**3**). In the latter case the negative charge is predominantly localized at the oxygen atom of the 6-keto group (21). The negative charge obtained after abstraction of a proton from nitrogen N(3) in **8** or uridine is located at O(4) (10, 22). However, proton abstraction from thymine in **7** or thymidine is supposed to lead to an anion with the charge distributed between atoms O(2) and O(4) (10, 22, 23). The pK_a values for the main heterocyclic bases in nucleosides and nucleotides are given in Table 1.1. These values vary with temperature and ionic strength. It is also expected that the chemical modification of the bases leading to an alteration in their electronic properties will cause a large variation in their respective pK_a values. The phosphate group of a nucleotide markedly influences the pK_a value of the corresponding heterocyclic substituent indicating interactions between these two moieties (23).

Nucleotides are esters of phosphoric acid and as such (with the exception of the cyclic phosphates) possess two or more ionization constants, the

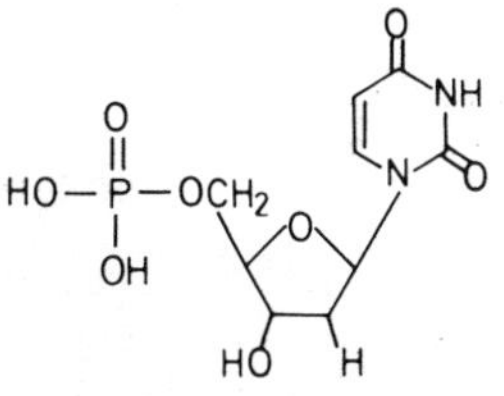

7

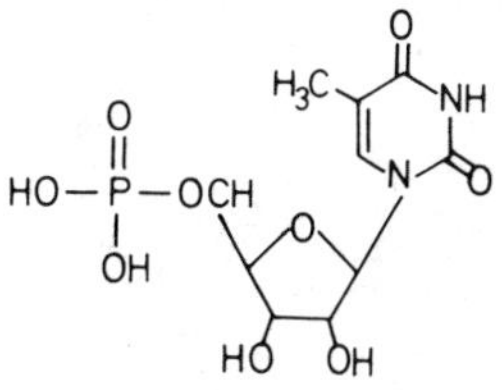

7 Thymidine 5′-phosphate

8

8 Uridine 5′-phosphate

first having a pK_a value of 1, whereas the values of the secondary ionization constants vary between 6 and 7. As shown in Table 1.2, the nature of the heterocyclic substituent has little influence on the secondary ionization constants. An invaluable physical property of nucleotides is their ultraviolet absorption spectrum of the heterocyclic substituents (Table 1.3). The relatively high molar extinction coefficients make absorption spectroscopy the method of choice in nucleotide analysis. Furthermore, chemical modification of nucleotides at heterocyclic substituents are often accompanied by changes in the absorption spectra.

1.2 CONFORMATION OF NUCLEOTIDES

The DNA of a cell contains its genetic information in the form of a linear sequence derived from four deoxyribonucleotides. In replication and transcription, this sequence is read as a sequence of functional groups, the heterocyclic bases of the four nucleotides. The chemical basis of recognition in these processes is probably a complementary-base pairing between adenine-thymine and guanine-cytosine. In addition, nucleic acids contain structural signals that are recognized by regulatory proteins and nucleic acid polymerases. The structural features of these signals are possibly not only a defined sequence of heterocyclic bases, but also a unique conforma-

Table 1.1 pK_a Values for Heterocyclic Bases in Nucleosides and Nucleotides[a]

Compound	pK_a
Cytidine	4.17
Cytidine 5′-phosphate	4.54
Adenosine	3.52
Adenosine 5′-phosphate	3.88
Uridine	9.38
Uridine 5′-phosphate	10.96
2′-Deoxythymidine	9.93
2′-Deoxythymidine 5′-phosphate	10.47
Guanosine	9.42
Guanosine 5′-phosphate	10.0

Source. Reference 24.

[a] pK_a values measured spectrophotometrically at 20°C and zero ionic strength.

tion, which in turn is determined by the conformation of the respective nucleotides in a particular sequence. Furthermore, nucleotides are substrates or regulatory ligands for many enzymes, not only those involved in nucleic acid metabolism. ATP (**9**) and to some extent GTP are the

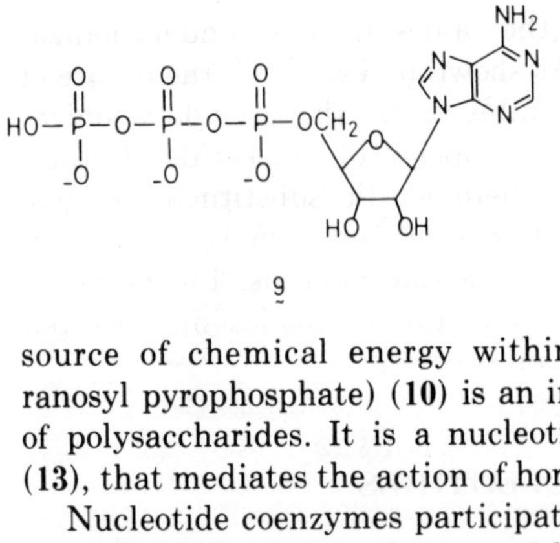

9

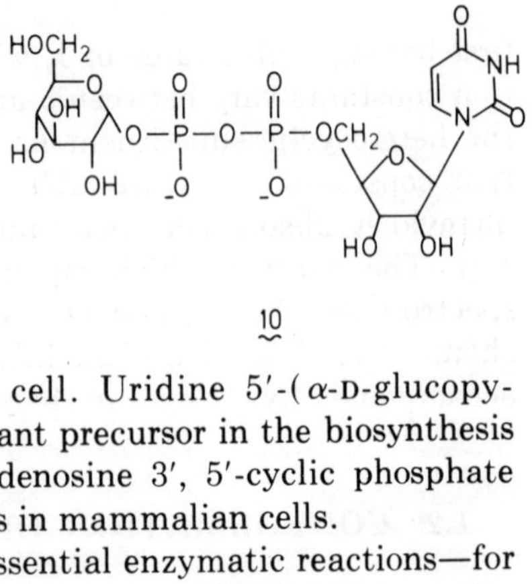

10

source of chemical energy within the cell. Uridine 5′-(α-D-glucopyranosyl pyrophosphate) (**10**) is an important precursor in the biosynthesis of polysaccharides. It is a nucleotide, adenosine 3′, 5′-cyclic phosphate (**13**), that mediates the action of hormones in mammalian cells.

Nucleotide coenzymes participate in essential enzymatic reactions—for example NAD^+ (**11**) in the case of dehydrogenases, and FAD (**12**) in oxidative phosphorylation. The numerous functions of nucleotides within a cell must comprise a large spectrum of interactions, ranging from high to low specificity, which in turn requires structural flexibility and variability.

The structure of a nucleotide is determined by the following structural parameters:

1. Bond lengths between atoms and bond angles within a molecule (Figs. 1.1 and 1.2).

Table 1.2 pK_a Values for Secondary Ionization of Nucleoside 5′-mono-, -di-, and -triphosphates

	pK_a		
Derivative	5′-mono-phosphate	5′-di-phosphate	5′-tri-phosphate
Adenosine	6.57	7.20	7.68
Guanosine	6.66	7.19	7.65
Uridine	6.63	7.16	7.58
Cytidine	6.62	7.18	7.65

Source. Reference 25.

2. Conformation about the glycosidic bond, that is the orientation of the heterocyclic substituent relative to the carbohydrate moiety.
3. Conformation of the phosphodiester bond.

Any chemical modification of a nucleotide undoubtly alters these parameters. One important aim of the chemical synthesis of nucleotide

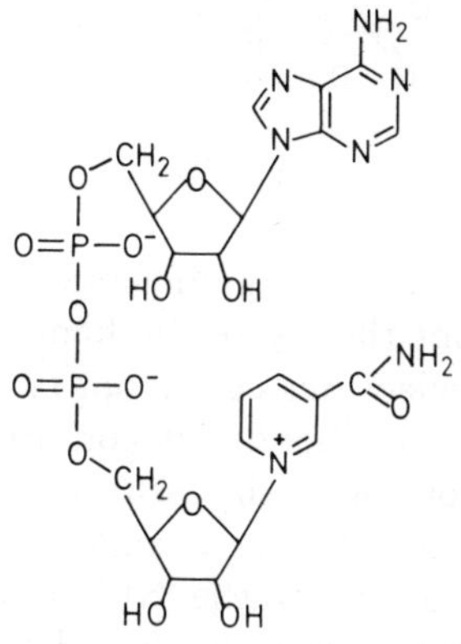

11

analogs is therefore a correlation of the structural differences with the differences in biological behavior when compared to the natural congener. This approach has often allowed valuable contributions toward a better understanding of complex biological processes. The five atoms of ribose or 2′-deoxyribose adopt a half-chair conformation in nucleotides. Stable conformations are those in which atoms C(2′) and C(3′) deviate from a plane through atoms C(4′)—O(1′)—C(1′). Atoms that are on the same side of the plane as atom C(5′) are called *endo*; those on the opposite side are termed *exo* (26) (Fig. 1.3). Energy barriers between the different forms are low. The half-chair conformations C(2′) *endo* and C(3′) *endo* appear with

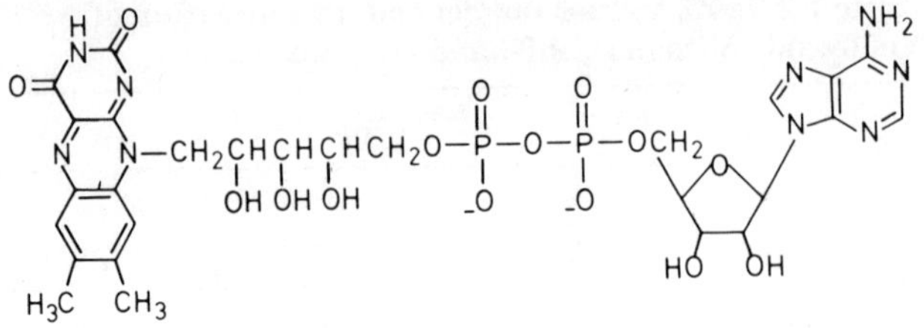

12

similar frequencies whereas conformation C(3′) *exo* is rather rare and conformation C(2′) *exo* has not yet been observed. (27).

As shown in Fig. 1.4, the C(5′)—O(5′) bond can adopt three different conformations. In contrast to energy calculations (28), crystallographic and spectroscopic data indicate that the *gauche-gauche* conformation is

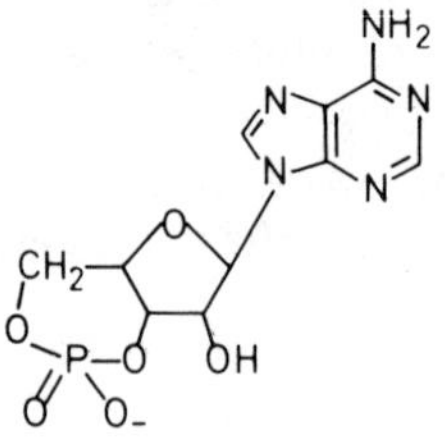

13

the favored form in natural occurring nucleotides (29, 30). Free rotation of the heterocyclic bases about the glycosidic bond, C(1′)—N(1), is hindered on steric grounds (31). Nucleosides can adopt two principle conformations with respect to the C(1′)—N(1) bond: *syn* conformation, in which the O(2) atom in pyrimidine nucleotides or the N(3) atom in purine nucleotides is located above the ribose moiety; *anti* conformation, in which the O(2) atom in pyrimidine nucleotides or the N(3) atom in purine nucleotides points away from the ribose moiety (31, 32). The exact definitions of *syn*-

Table 1.3 Ultraviolet Absorption Spectra of Nucleotides[a]

Compound	λ_{max} (nm)	E_{max} (10^{-3})	λ_{min} (nm)	A_{250}/A_{260}	A_{260}/A_{280}
5′UMP	262	10	230	0.73	0.35
5′TMP	267	10.2	—	0.65	0.73
5′GMP	252	13.7	224	1.16	0.66
5′CMP	271	9.1	249	0.84	0.98
5′AMP	259	15.4	227	0.78	0.16

Source. Handbook of Biochemistry, 2nd ed., H. A. Sober, Ed., The Chemical Rubber Co., Cleveland, 1970.

[a] All spectral data obtained at pH 7.

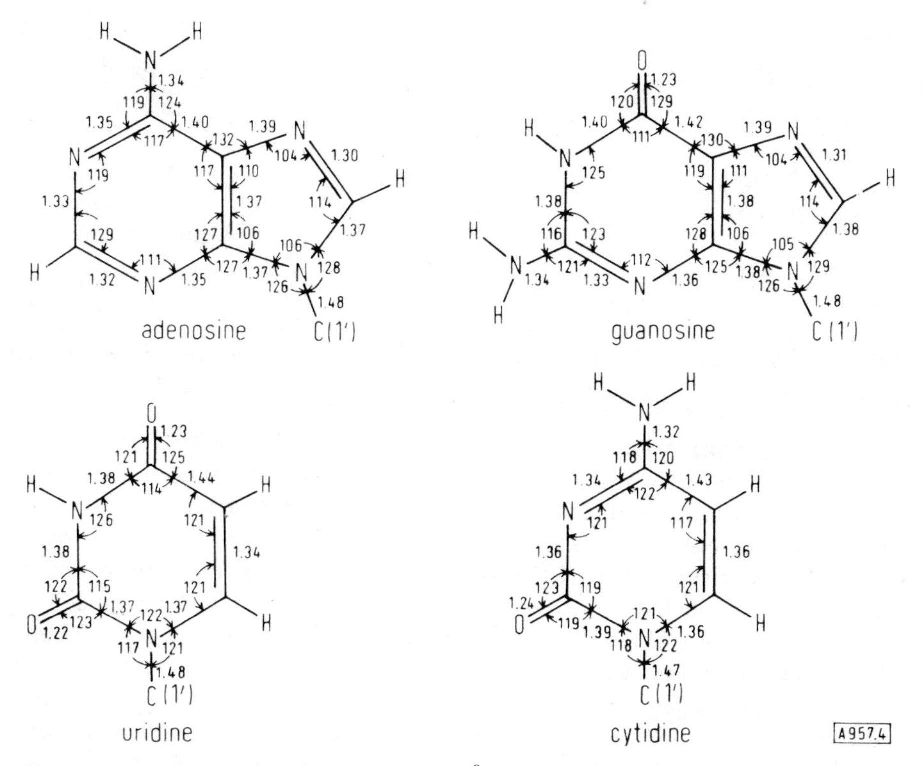

Figure 1.1 Atomic bonding distances (Å) and angles (°) of heterocyclic bases. *Source*. Reference 26.

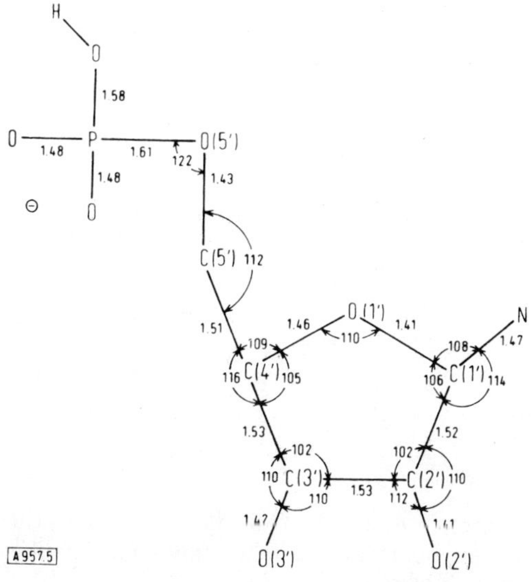

Figure 1.2 Atomic bonding distances (Å) and angles (°) of D-ribose 5-phosphate. *Source*. Reference 26.

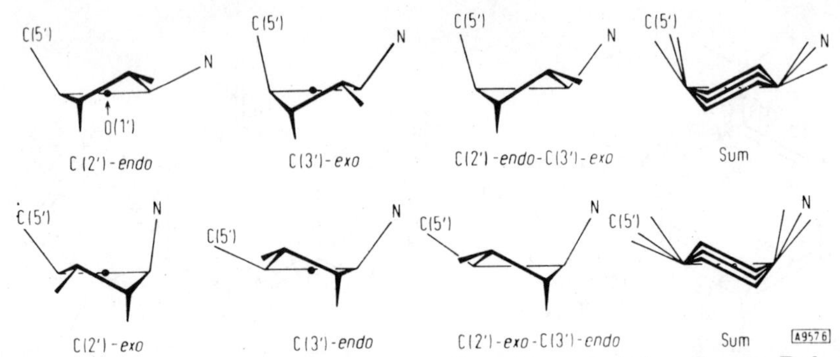

Figure 1.3 Possible conformation of D-ribose in nucleotides. *Source*. Reference 26.

and *anti*-conformations are given in Fig. 1.5. The conformations of nucleotides, listed in Tables 1.4 and 1.5 have been obtained from X-ray-diffraction studies and are relevant to the crystal structures of these molecules. These results must of course be applied to the structure of nucleotides in solution with caution. For example, ^{1}H-nmr spectroscopy indicates that nucleotides possess flexible structures in solution and that they are in a state of rapid equilibrium between the anti- and the syn-conformations with preference for either the anti- or the syn-form (34).

What stimulates interest in the synthesis of chemically modified nucleotides? As already pointed out above, nucleotides play an important

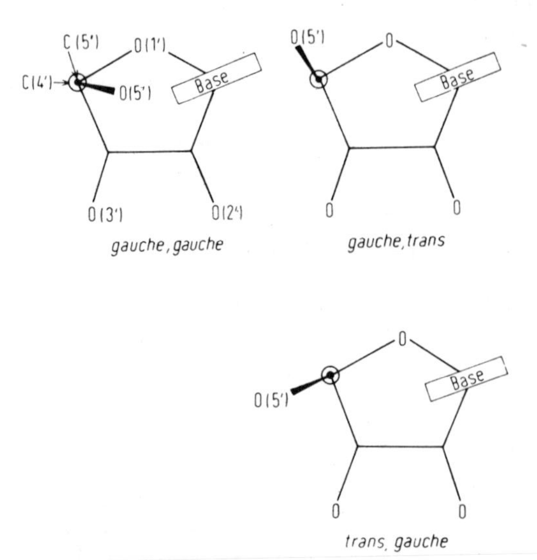

Figure 1.4 Conformation at C(5′)—O(5′) bonds in nucleotides. Also employed is the nomenclature gauche^{-}(trans,gauche), gauche^{+}(gauche,gauche) and trans (gauche,trans). *Source*. Reference 26.

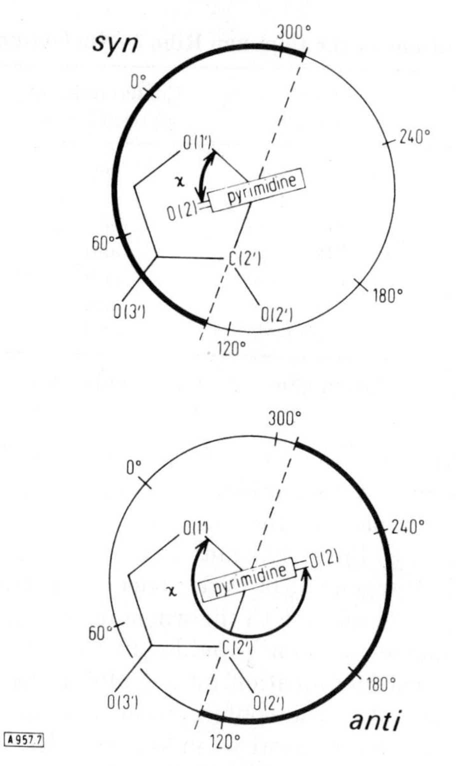

Figure 1.5 Conformation at the C(1′)—N(1) bond in pyrimidine nucleotides. *Source.* Reference 26.

role in cellular metabolism. Hence, there are many enzymes and regulatory proteins that specifically bind nucleotides either as substrates or effectors. A nucleotide molecule can formally be divided into three structural parts: a heterocyclic substituent, a sugar moiety, and a phosphate residue. Binding of a nucleotide to an enzyme presumably involves

Table 1.4 Conformations of the Common 2′-Deoxy 5′-Nucleotides

Nucleotide	Conformation of sugar	Conformation of glycosidic bond	Conformation of C(5′)—O(5′) bond
2′-Deoxyadenosine 5′-phosphate	C(2′)-endo	anti	gauche$^+$
2′-Deoxyguanosine 5′-phosphate	O(1′)-endo	anti	trans
2′-Deoxycytidine 5′-phosphate	C(3′)-exo	anti	gauche$^+$
2′-Deoxythymidine 5′-phosphate	C(3′)-endo	anti	gauche$^+$

Source. Data from M. Sundaralingam (1975). Reference 33.

Table 1.5 Conformations of the Common Ribo 5′-Nucleotides

Nucleotide	Conformation of sugar	Conformation of glycosidic bond	Conformation of C(5′)—O(5′) bond
Adenosine 5′-phosphate	C(3′)-endo	anti	gauche$^+$
Adenosine 5′-triphosphate			
Molecule 1	C(3′)-endo	anti	gauche$^+$
Molecule 2	C(2′)-endo	anti	gauche$^+$
Guanosine 5′-triphosphate	C(3′)-endo	anti	gauche$^+$
Uridine 5′-phosphate	C(2′)-endo	anti	gauche$^+$
Cytidine 5′-phosphate	C(2′)-endo	anti	gauche$^+$

Source. Data from M. Sundaralingam (1975). Reference (33).

recognition and correct positioning of the part of the molecule that is to be chemically transformed. Among others, three major questions arise: which structural parts participate in the recognition process, which structural parts merely function as links or means for positioning the molecule, and what is the spatial arrangement of that nucleotide portion involved in the chemical transformation relative to the whole molecule. Specific chemical alteration of a nucleotide can provide answers to these questions. However, any chemical modification inflicts alterations on the nucleotide structure as a whole. Thus the conformation in solution of a nucleotide analog has to be regarded a priori as unknown unless adequate data are present. Simple analogy conclusions with respect to the parent natural nucleotide can be misleading. The behavior of a nucleotide analog as a substrate for an enzyme when compared to its natural congener yields valuable information with respect to the conformation of the enzyme bound nucleotide, provided sufficient knowledge of the conformational parameters of the analog are available. It can be, for example, very dangerous to conclude from the significant change in substrate properties induced by chemical substitution of the heterocyclic base of a nucleotide that the latter exclusively participates in recognition by a protein, if this chemical modification drastically alters the conformation of the sugar moiety of the nucleotide analog.

A number of synthetic and natural nucleoside analogs with interesting biological properties are known, some of which have already been successfully employed as therapeutic agents. These nucleoside analogs all share the common feature that their biological function requires their intracellular conversion to the corresponding nucleotides. Investigations of the mechanism by which these nucleoside analogs interfere with the cellular metabolism therefore require the chemical synthesis of phosphorylated derivatives.

Further motivation for the synthesis of nucleotide analogs comes from the need for modified nucleotides that have desirable physical properties, such as fluorescence and absorption in the visible wavelength range, or that possess reactive functional groups that can covalently substitute a protein on specific interaction.

Biological studies with synthetic DNA's and RNA's containing nucleotide analogs of various kinds and in defined positions have demanded the enzymatic synthesis of those polynucleotides from chemically modified precursors. It may be sufficient to mention two examples for activities of this kind, namely the search for polyribonucleotide analogs able to stimulate the induction of interferon, and the effect on their template function of modified nucleotides in polynucleotides.

In conclusion, nucleotide analogs are extremely useful tools in studies directed toward the understanding of the function of enzymes and regulatory proteins. Biochemical problems are therefore the motivation for their preparation and application. The chemical synthesis of a nucleotide analog should always be accompanied by attempts to acquire as much data possible to define the analog's structure and conformation.

REFERENCES

1. P. A. Levene and J. A. Mandel, *Chem. Ber.* **41,** 1905–1909 (1908).
2. D. O. Jordan, *The Chemistry of Nucleic Acids*, Butterworths, London, 1960.
3. P. A. Levene and L. W. Bass, *Nucleic Acids*, The Chemical Catalog, Co., Inc., New York, 1931.
4. A. M. Michelson, *The Chemistry of Nucleosides and Nucleotides*, Academic Press, New York, 1963.
5. G. R. Pettit, *Synthetic Nucleotides*, Van Nostrand Reinhold Co., New York, 1972.
6. *Synthetic Procedures in Nucleic Acid Chemistry*, Vol. 1, W. W. Zorbach and R. S. Tipson, Eds., Wiley, New York, 1968.
7. *Organic Chemistry of Nucleic Acids*, N. K. Kochetkov and E. I. Budovskij, Eds., Plenum Press, London, 1972.
8. L. Goodman, Chemical Syntheses and Transformations of Nucleosides, in *Basic Principles in Nucleic Acid Chemistry*, Vol. 1, P. O. P. Ts'o, Ed., pp. 94–194, Academic Press, New York, 1974.
9. D. M. Brown, Chemical Reactions of Polynucleotides and Nucleic Acids, in *Basic Principles in Nucleic Acid Chemistry*, Vol. 2, P. O. P. Ts'o, Ed., pp. 2–78, Academic Press, New York, 1974.
10. D. Shugar and J. J. Fox, *Biochim. Biophys. Acta* **9,** 199–218 (1952).
11. G. L. Angell, *J. Chem. Soc.* **1961,** 504–515.
12. J. P. Kokko, J. H. Goldstein, and L. Mandell, *J. Am. Chem. Soc.* **83,** 2909–2911 (1961).
13. L. Gatlin and J. C. Davis, *J. Am. Chem. Soc.* **84,** 4464–4470 (1962).
14. H. T. Miles, *Biochim. Biophys. Acta* **27,** 46–52 (1958).
15. H. T. Miles, *J. Am. Chem. Soc.* **85,** 1007–1008 (1963).
16. M. Sundaralingam, *Acta. Cryst.* **21,** 495–506 (1966).
17. C. A. Dekker, *Ann. Rev. Biochem.* **29,** 453–474 (1960).

18. J. M. Broomhead, *Acta. Cryst.* **4,** 92–100 (1951).
19. C. E. Bugg and R. E. Marsh, *J. Mol. Biol.* **25,** 67–82 (1967).
20. M. Sundaralingam and L. H. Jensen, *J. Mol. Biol.* **13,** 930–943 (1965).
21. H. T. Miles, F. B. Howard, and J. Farier, *Science* **142,** 1458–1463 (1963).
22. J. J. Fox and D. Shugar, *Biochim. Biophys. Acta* **9,** 369–384 (1952).
23. R. L. Sinsheimer, R. L. Nutter, and G. R. Hopkins, *Biochim. Biophys. Acta* **18,** 13–27 (1955).
24. J. Clauwaert and J. Stockx, *Zeitschrift für Naturforsch.* **23b,** 25–30 (1968).
25. R. Phillips, P. Eisenberg, P. George, and R. J. Rutman, *J. Biol. Chem.* **240,** 4393–4397 (1965).
26. W. Saenger, *Angew. Chem.* **85,** 680–690 (1973).
27. S. Arnott in *The Jerusalem Symposia on Quantum Chemistry and Biochemistry IV*, The Israel Academy of Sciences and Humanities, Jerusalem, 1972.
28. H. R. Wilson and A. Rahman, *J. Mol. Biol.* **56,** 129–142 (1971).
29. E. Shefter and K. N. Trueblood, *Acta Cryst.* **18,** 1067–1077 (1965).
30. A. E. V. Haschemeyer and A. Rich, *J. Mol. Biol.* **27,** 369–384 (1967).
31. J. Donohue and K. N. Trueblood, *J. Mol. Biol.* **2,** 363–371 (1960).
32. W. Saenger and K. H. Scheit, *J. Mol. Biol.* **50,** 153–169 (1970).
33. M. Sundaralingam, *Ann. N.Y. Acad. Sci.* **255,** 3–42 (1975).
34. M. Gueron, C. Chachaty, and T. D. Son, *Ann. N.Y. Acad. Sci.* **222,** 306–322 (1973).

TWO

Nucleotides with Modified Heterocyclic Substituents

The Chemical reactivity of purine and pyrimidine substituents of nucleotides allows a variety of chemical reactions to take place, thus enabling a large number of modified nucleotides to be synthesized. The chemical reactions that are employed in the chemical modification of nucleotides can be grouped in (1) electrophilic substitution and addition reactions at the ring nitrogen atoms, (2) substitutions at exocyclic amino groups and oxygen atoms, and (3) electrophilic substitution and addition reactions at carbon atoms. The general aspects of these reactions with respect to their application to the synthesis of nucleotide analogs are shortly discussed in the pages that follow.

1. Electrophilic Substitution and Addition Reactions at the Ring Nitrogen Atoms. The heterocyclic bases of nucleotides contain two types of ring nitrogen atoms. Those with two substituents which, according to qualitative considerations of conjugation of atomic orbitals as well as quantum chemical calculations, carry partial negative charges. Nitrogen atoms N(3) and N(7) of purines, as well as N(1) of adenine and N(3) of cytosine fall into this category. They are readily attacked by electrophilic reagents. Ring nitrogen atoms with three substituents carry partial positive charges. This class comprises N(1) of guanine, N(3) of uracil, and N(3) of thymine. Because one of the substituents is a hydrogen atom that can be abstracted by base, the resulting anions are the targets for electrophilic substitution.

2. Substitutions at Exocyclic Amino Groups and Exocyclic Oxygen Atoms. Although qualitative orbital bonding considerations postulate partial positive charges at exocyclic amino groups, quantum chemical calculations of σ-electron densities, however, demonstrate negative net charges localized at these groups. The latter is in agreement with the experimental observations that electrophilic reagents attack exocyclic amino groups. Similar considerations predict reactions of electrophiles with carbonyl oxygen atoms of heterocyclic bases in nucleosides and nucleotides.

3. Electrophilic Substitution and Addition Reactions at Carbon Atoms. Addition reactions at the carbon—carbon double bond can be either nucleophilic or electrophilic in nature. Hitherto only the C(5)—C(6) double bonds of pyrimidine bases have been shown to undergo addition reactions—the C(4)—C(5) double bonds of purines being seemingly inert. Substitution of the C(5)—C(6) double bonds in pyrimidine nucleosides proceeds readily. Double bonds in purine nucleosides seem to be inert against addition reactions. On the contrary, substitution at C(8) of purine bases by various electrophilic reagents is feasible.

2.1 SUBSTITUTION AT RING NITROGEN ATOMS

2.1.1 Alkylation

A detailed account of the reactions of heterocyclic bases in nucleotides and nucleic acids with alkylating agents has been given by Singer (1). Many *N*-alkylated nucleotides are minor constituents of nucleic acids (2), and in a variety of mammalian mRNA molecules N^7-methylguanosine 5′-phosphate is attached to the 5′-termini via a P—O—P linkage (3–6). N^7-Methylguanosine 5′-phosphate and its derivatives have therefore gained considerable interest, since they appear to function as analogs of the 5′-terminal part of mammalian mRNA molecules in translation (7–9). Interest in the alkylation of nucleotides was initiated by the observations that the alkylating agent mustard gas was mutagenic (10) and in addition inactivated DNA viruses (11). The principal target of alkylation of DNA by many alkylating reagents seemed to be deoxyguanylic residues (1). Alkylation of guanosine 5′-phosphate by dimethylsulfate leads to N^7-methylguanosine 5′-phosphate (**1**) (12). The glycosidic bond in compound **1** is extremely labile and hydrolyzes even at pH 7 (12, 13).

At alkaline pH, however, the N^7-methylguanine moiety of **1** is attacked by hydroxyl ions yielding the pyrimidine nucleotide **3** (12, 13). The formation of nucleotide phosphate esters by dimethylsulfate becomes a

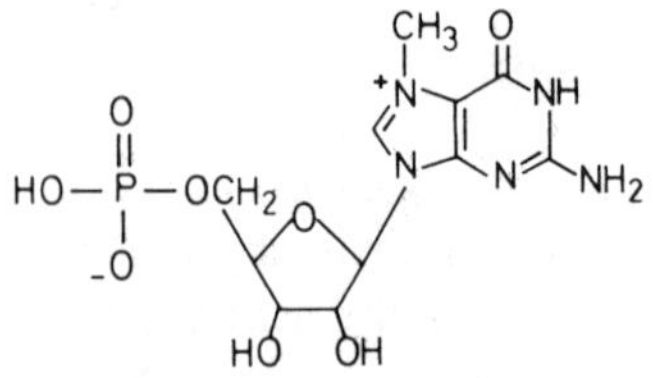

1

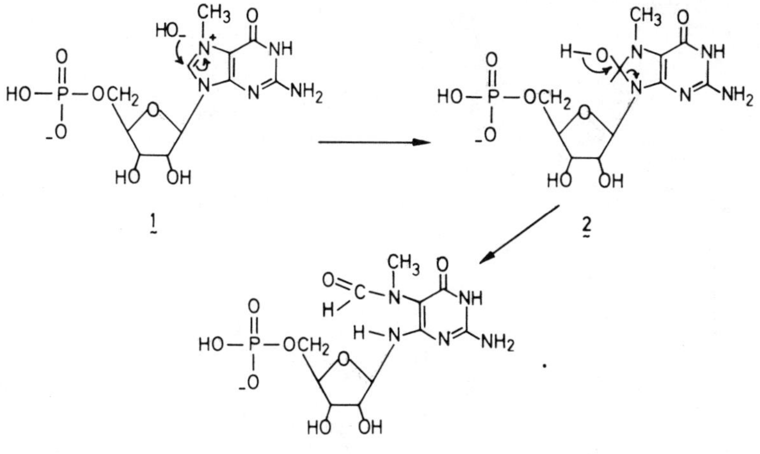

prominent side reaction if alkylation is carried out at pH > 6, but is completely avoided at pH 4 or lower (6). Alkylation of guanosine 5′-phosphate by diazomethane in heterogeneous phase also yields **1** besides the corresponding nucleotide phosphate methyl ester (13, 15). Inosine 5′-phosphate is similar to guanosine 5′-phosphate in its behaviour toward alkylation by alkyl sulfates, alkyl halides and diazoalkanes (16–18). As judged from the alkylation of inosine with alkyl sulfates, alkyl halides and diazo alkanes (16–18), inosine 5′-phosphate probably behaves in a manner similar to guanosine 5′-phosphate on alkylation. In N^7-methylinosine the glycosidic bond is labile and the imidazole ring is readily opened by attack of hydroxyl ions at carbon atom C(8).

On reaction of adenosine 5′-phosphate with dimethylsulfate at pH 7, the N(1) and N(3) methylated derivatives are obtained, the former being favored (19). When alkylation of adenosine 5′-phosphate, adenosine 5′-diphosphate (14) and adenosine 5′-(α-D-glucopyranosyl pyrophosphate) (20) by dimethylsulfate was performed at pH 5, the main reaction was alkylation of N(1). Methylation of the adenine moiety at N(7) by dimethylsulfate has been reported to occur to a small extent (21). Using diazomethane in an aqueous heterogeneous phase, esterification of the phosphate residue predominates over methylation of the adenine moiety. The main target for methylation of the heterocycle in adenosine 5′-phosphate under these conditions is N(1) (15). A Dimroth type rearrangement (22) was exploited for the synthesis of N^6-methyladenosine 5′-phosphate (**6**) and N^6-methyladenosine 5′-diphosphate (**7**) (14) from the corresponding N(1) methylated derivatives **4** and **5**. This can be applied to the synthesis of N(6) alkylated analogs in general as explained in Section 2.2.1. The only product formed in the reaction of cytidine 5′-phosphate with dimethylsulfate at pH 4 to 5 is N^3-methylcytidine 5′-phosphate (23, 24). N^3-Methylcytidine 5′-diphosphate was obtained in a similar way from

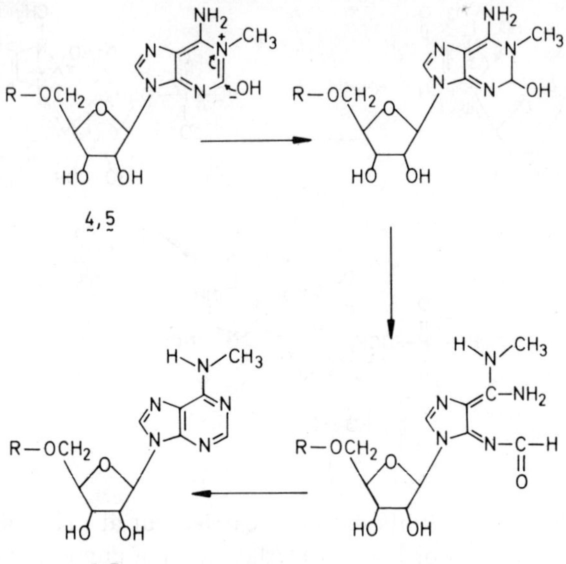

4, 6 $R = P(O)(OH)O^-$

5, 7 $R = P(O)O{-}O{-}P(O)(OH)O^{2-}$

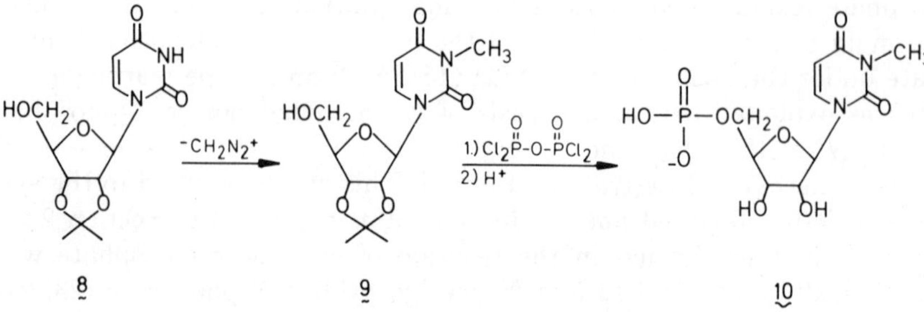

cytidine 5′-diphosphate (25). The reaction of dialkylsulfates with cytidine 5′-phosphate is very sensitive to experimental conditions as shown by Shugar (26). In alkaline medium and under controlled conditions there is virtually no alkylation of either nitrogen atom N(3) or the exocyclic amino group, but instead methylation of the 2′- and 3′-OH functions occurs. Uridylic and thymidylic acids seem to be inert to methylation by dimethylsulfate, although Michelson et al. (27) reported methylation of uracil moieties in poly (rU) at N(3) by the latter reagent in the presence of tri-*n*-butylamine. These reaction conditions have not yet been applied to either uridine 5′-phosphate or thymidine 5′-phosphate. The main target of methylation in pyrimidine nucleotides by diazomethane is N(3), accompanied by excessive esterification of the phosphate residues (15). Therefore to obtain N^3-methyluridine 5′-phosphate (**10**) it was necessary to phosphorylate the N(3)-methylated nucleoside **9**, prepared by methylation of 2′,3′-*O*-isopropylideneuridine (**8**) (28). The monophosphate **10** served as a precursor for N^3-methyluridine 5′-(D-glucopyranosyl pyrophosphate) (29) and N^3-methyluridine 5′-diphosphate (30).

An alternative approach to the synthesis of N^3-methylpyrimidinenucleoside 5′-polyphosphates, which is generally applicable, is demonstrated by the synthesis of N^3-methyl-2′-deoxythymidine 5′-triphosphate (**13**) (31).

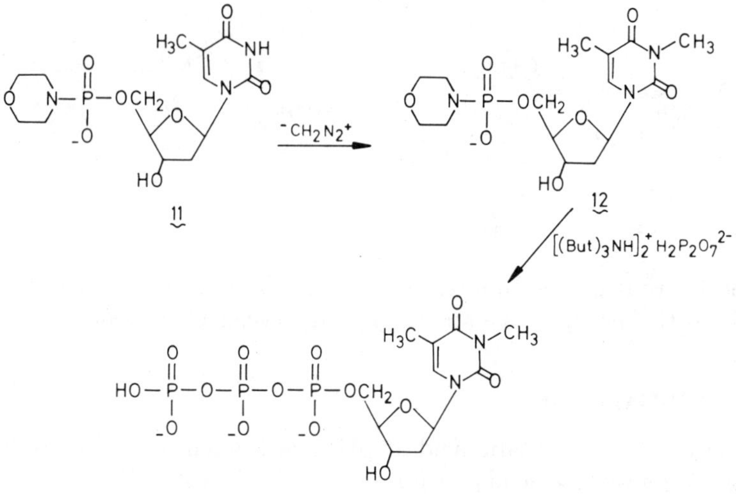

This synthesis exploits the fact that phosphodiester anions are virtually inert to diazomethane. The nucleotide morpholidates **11** and **12** were found to behave like phosphodiester anions. N^3-Methyluridine 3′-phosphate was prepared by phosphorylation of 5′,2′-*O*-ditrityl-N^3-methyluridine with 2-cyanoethylphosphate using dicyclohexylcarbodiimide as condensating reagent. After removal of the blocking groups under standard conditions, N^3-methyluridine 3′-phosphate was isolated in high yield (32).

The total synthesis of N^3-(3-L-amino-3-carboxypropyl)uridine-5′-phosphate (**16**), a minor constituent of transfer RNA^{Phe} from *E. coli*, has been reported by Seela and Cramer (33). N^3-(3-L-Amino-3-carboxypropyl)-uridine (**15**) was obtained by alkylation of 2′,3′-*O*-isopropylideneuridine (**8**) with α-L-benzamido-γ-bromobutyrate and subsequent removal of protecting groups. Phosphorylation of **15** with phosphorus oxychloride afforded **16**. Finally it should be noted that alkylation of N(7) in guanylic acids, of N(1) in adenylic acids and of N(3) in cytidylic acids leads to quaternization of the respective nitrogen atom. A common feature of nucleotides that possess quaternary ring nitrogen atoms, and hence carry formal positive charges, is a strong blue fluorescence emission.

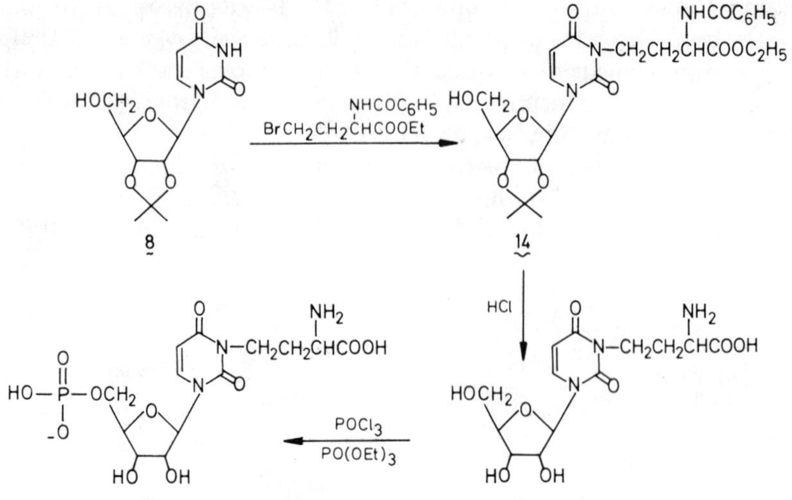

2.1.2 *N*-Oxidation

Reaction of monoperphtalic acid at pH 7 with adenosine 5′-phosphate or cytidine 5′-phosphate leads to formation of adenosine-N^1-oxide 5′-phosphate (**17**) or cytidine-N^3-oxide 5′-phosphate (**18**) (34, 35), respectively. Similarly the N^1-oxides of adenosine 5′-diphosphate and adenosine 5′-triphosphate were prepared analogously (34). Inosine N^1-oxide 5′-phosphate is available by deamination of adenosine-N^1-oxide to inosine-N^1-oxide and subsequent phosphorylation with phosphorus oxychloride in triethylphosphate (36). Treatment of cytidine-N^3-oxide 5′-phosphate with mild alkali yields an intermediate with an opened pyrimidine ring, whose glycosidic bond is readily cleaved by dilute acetic acid. *N*-Oxidation and treatment with alkali followed by mild hydrolysis with acetic acid has been employed to specifically cleave deoxyoligonucleotides at deoxycytidine residues (37). In a weak alkaline medium, 2′-deoxyadenosine 5′-phosphate reacts slowly with hydrogen peroxide to yield the corresponding

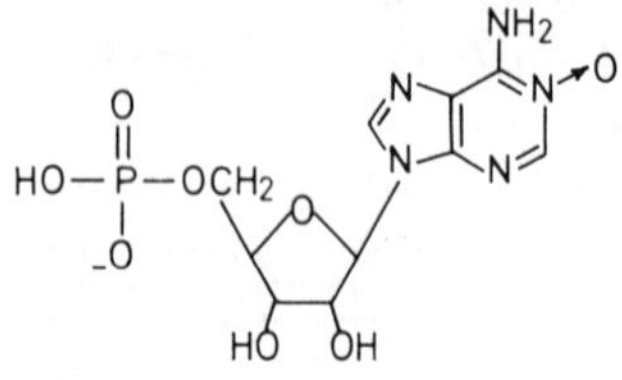

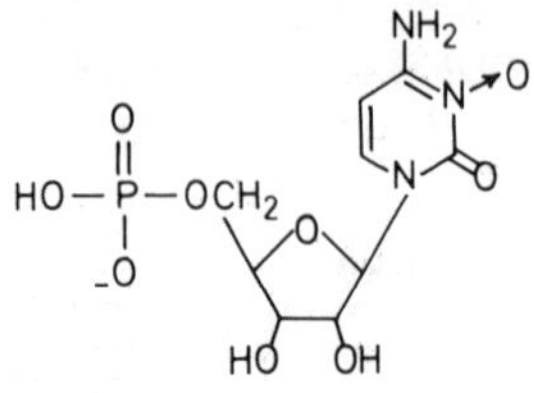

N^7-oxide (**19**) (38). Because of many side reactions, the yield of **19** is rather low. Whereas **19** is stable to alkali, adenosine-N^1-oxide undergoes fission of the pyrimidine ring with concomitant loss of atom C(2) to give an imidazol-*N*-1-(*β*-D-ribofuranosyl) derivative (**20**) (39, 40).

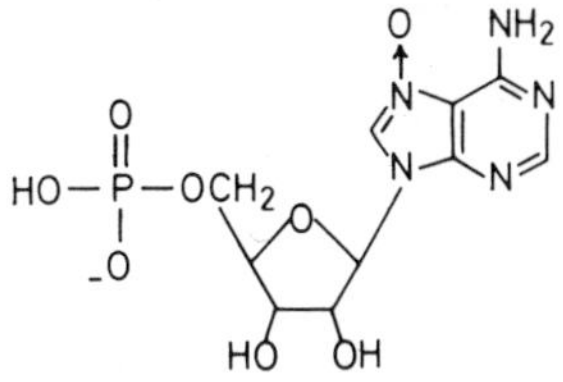

19

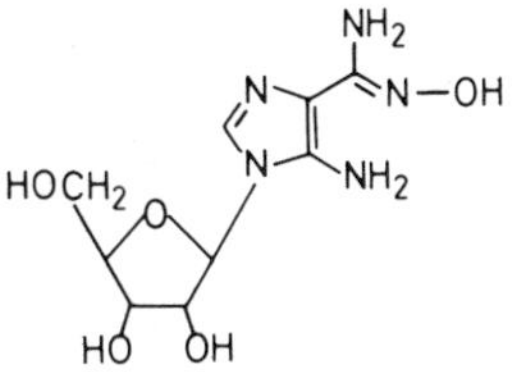

20

2.2 SUBSTITUTION AT EXOCYCLIC GROUPS

2.2.1 Alkylation

Although methylation of guanosine in a homogeneous solution by diazomethane yields a fair proportion of O^6-methylguanosine besides the *N*-methyl derivatives (41), this does not hold for methylation of guanylic acids in a heterogeneous aqueous solution (15). Similar observations were made in the case of inosine (8). Arguments that the different behavior of nucleosides and nucleotides toward alkylation merely results from differences in the experimental conditions does not seem to be plausible (41) since alkylation of nucleic acids or oligonucleotides by diazoalkanes or *N*-alkyl-*N*-nitroso compounds leads to formation of O^6-methylguanine (3). Thus O^6-alkylguanine nucleotides therefore had to be prepared by phosphorylation of the corresponding nucleosides. Phosphorylation of O^6-methylguanosine (**22**) by phosphotransferase from carrots to the 5′-phosphate (**23**) and subsequent chemical conversion of **23** to the 5′-diphosphate (**25**) has been reported by Gerchman et al. (42). Nucleosides and nucleotides with O^6-methylguanine moieties possess strong blue fluorescence emission. It has not yet been possible to alkylate directly the exocyclic amino groups of guanine or adenine nucleotides. N^6-Monoalkyl derivatives of adenine nucleotides are principally available from the Dimroth rearrangement of the respective N^1-alkylated compounds (14, 19, 43, 44). A series of N^6-monoalkyladenosine 5′-phosphates, for example, those with alkyl substituents Δ^2-isopentenyl, benzyl, furfuryl, and adamantyl were synthesized by phosphorylation of the respective nucleoside with phosphorous oxychloride in triethyl phosphate (45). The rearrangement of N^1-alkylated adenine nucleotides to N^6-monoalkyl derivatives was exploited by Leonard et al., for an elegant synthesis of N^6-(Δ^2-isopentenyl) adenosine 5′-phosphate. Adenosine 5′-phosphate was reacted with Δ^2-isopentenylbromide to give N^1-(Δ^2-isopentenyl)adenosine 5′-phosphate, which on treatment with dilute ammonium hydroxide at elevated temperature, yielded the N^6-(Δ^2-isopentenyl) derivative (43, 44). The Dimroth rearrangement of N^1-alkylated adenine derivatives is of particular interest because of the cytokinine activities of N^6-alkyladenine compounds (45). Because of the relative stability of 6-chloropurine ribosides, the corresponding nucleotides can be prepared by standard phosphorylation procedures. This was demonstrated by the syntheses of the 5′-monophosphates and 5′-diphosphates of 6-chloro-9-(β-D-ribofuranosyl)purine (46) and 2-amino-6-chloro-9-(β-D-ribofuranosyl)purine (47) respectively. However, the 6-chloro substituent is reactive enough to undergo nucleophilic displacement reactions by primary and secondary amines or hydrogen sulfide ions, respectively.

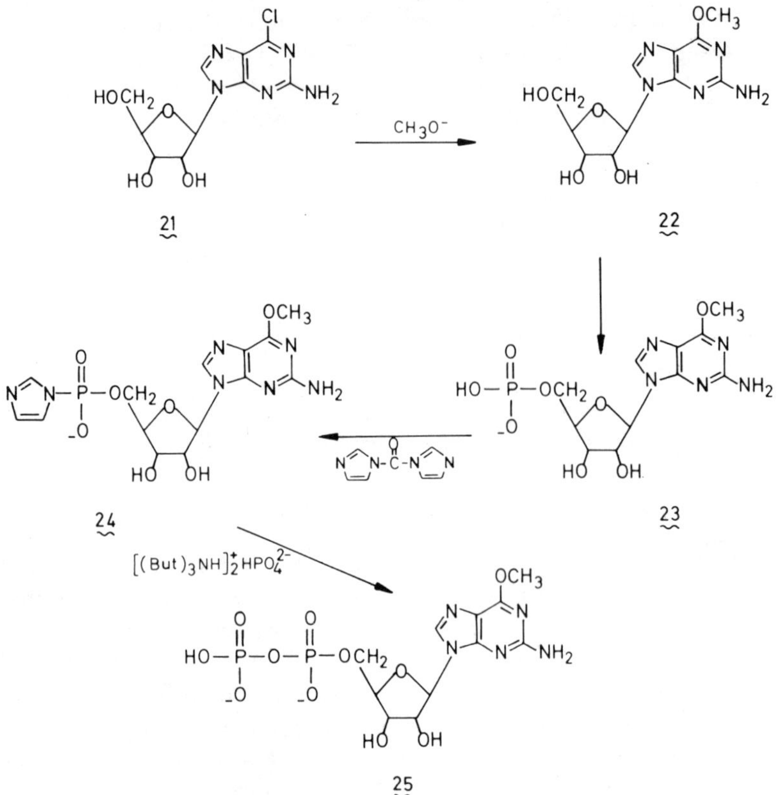

Hong and Chedda achieved the synthesis of a number of N-purin-6-ylcarbamoyl-9-β-D-riboside 5′-phosphates (**29**), among them the L-threonine derivative (**29**), R = $CH(COOH)CH(OH)CH_3$, which is a minor constituent of hypermodified transfer RNA's (48). The reaction sequence involved acylation of the protected adenine nucleotide **26** with ethyl chloroformate and aminolysis of the resulting N^6-ethylcarbamoyl intermediate **27** to give a series of N-purin-6-ylcarbamoyl derivatives (**28**). Removal of blocking groups from **28** by conventional procedures furnished

26 → (Cl-C(=O)-O-Et) → 27 → (RNH_2) → 28 → 29

28, 29 R = $CH(COOH)CH(OH)CH_3$
$CH(CH_3)_2$
$CH_2CH{=}CH_2$
$CH_2CH_2CH(CH_3)_2$

29. The alkylation of the keto groups in uridine- and thymidine 5′-phosphate by diazoalkanes or *N*-alkyl-*N*-nitroso compounds could not be demonstrated, but does occur in DNA, RNA, and oligonucleotides (1). Direct alkylation of the exocyclic amino group in cytidine 5′-phosphate could not be achieved (26).

2.2.2 Reactions with Aldehydes

Because of low nucleophilicity, the exocyclic amino groups are only attacked by rather reactive carbonyl compounds. Thus the amino groups of the heterocyclic bases adenine, guanine, and cytosine react with aqueous formaldehyde to form *N*-hydroxymethyl derivatives (49–54). These reactions are readily reversible and hence satisfying chemical characterization of the reaction products has not been possible. α-Dicarbonyl compounds such as glyoxal or 3-ethoxybutanon-2-al-1 (ketoxal) react specifically with guanine nucleotides forming a relatively stable tricyclic ring system (**30**) (55, 56). The structure of **30** was proven by nmr spectroscopy (57). This reaction has been extensively used for the modification of guanine residues in nucleic acids. Reaction of adenosine 5′-phosphate with chloroacetaldehyde in aqueous solution at room temperature leads to the formation of 1,N^6-ethenoadenosine 5′-phosphate (**31**), a nucleotide containing a tricyclic imidazo(2,1-*i*)purine moiety (58, 59). It

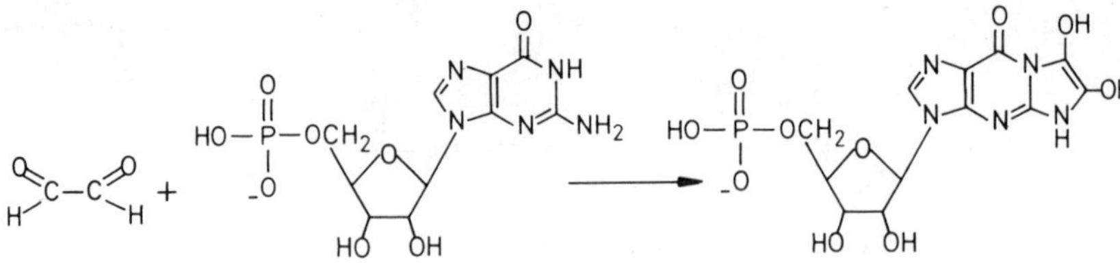

30

could be demonstrated by the use of the lanthanide-shift ^{1}H-nmr technique that the conformation of 1,N^6-ethenoadenosine 5′-phosphate is very similar to that of adenosine 5′-phosphate (60).

X-Ray analysis of the reaction product between α-chloro-*n*-butyraldehyde and adenosine established that reaction of α-haloaldehydes with

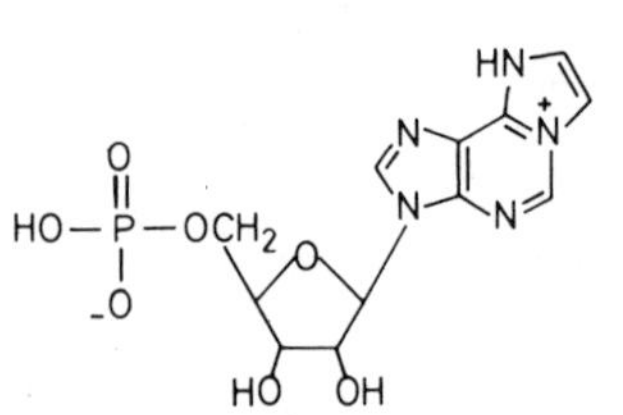

31

the adenine ring proceeds by initial alkylation of N(1) followed by ring closure between the aldehyde and the 6-amino group with elimination of water (61). An invaluable property of the 1,N^6-ethenoadenine chromophore is its characteristic fluorescence (62). Furthermore the introduction of the 1,N^6-etheno substituent often does not abolish the biological function of adenine nucleotides. Leonard and Tolman (67) have reviewed the large body of information on the use of these derivatives in biochemical studies. The mild reaction conditions required for the modification by chloroacetaldehyde allows the facile synthesis of the 1,N^6-etheno derivatives of adenosine 5′-diphosphate, adenosine 5′-triphosphate (58), flavine adenine dinucleotide (63), triphosphopyridine nucleotide (64), and nicotinamide adenine dinucleotide (58, 65). Measurements of fluorescence life times and quantum yields have provided insight into the structures of flavine 1,N^6-ethenoadenine dinucleotide and nicotineamide 1,N^6-ethenoadenine dinucleotide in solution. The former was found to exist to approximately 90% and the latter to approximately 45% in the form of an intramolecularly stacked conformation (65, 66).

In contrast to adenosine, guanosine does not react at pH 4 with chloroacetaldehyde (68). However, slow reaction does occur at pH 6.3, the reaction product, 1,N^2-ethenoguanosine, being nonfluorescent (68).

Reaction of chloroacetaldehyde with cytidine derivatives at pH 3.5 yields fluorescent 5,6-dihydro-5-oxo-6-β-D-ribofuranosylimidazo(1,2-*c*) pyrimidine derivatives (**32**) (69, 70). By analogy to 1,*N*⁶-ethenoadenine, and to demonstrate the relation of the modified derivatives to the parent compounds, the designation 3,*N*⁴-ethenocytosine is preferred. The crystal and molecular structure of 3,*N*⁴-ethenocytidine has been determined by X-ray diffraction (71). A thorough study of its fluorescence properties

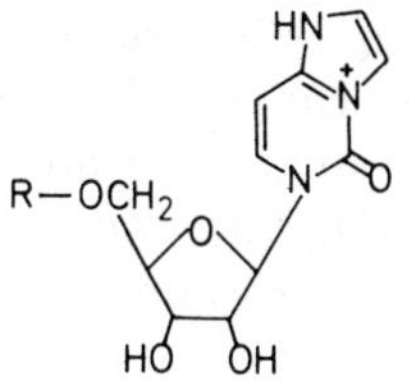

32

R = $P(O)(OH)O^-$, $P(O)O\text{-}O\text{-}P(O)(OH)_2^{2-}$,
$P(O)O\text{-}O\text{-}P(O)O\text{-}O\text{-}P(O)(OH)O^{3-}$

revealed that only the protonated form fluoresces; its fluorescence quantum yield of 0.01 is much lower than that of 1,*N*⁶-ethenoadenine derivatives (72). Nicotinamide 3,*N*⁴-ethenocytosine dinucleotide (**33**) (73), prepared from 3,*N*⁴-ethenocytidine 5′-phosphate and nicotinamide mononucleotide by reaction with dicyclohexylcarbodiimide, behaved as an analog of nicotinamide adenine dinucleotide in many enzymatic reactions. The reason is presumably the close spatial similarity between 3,*N*⁴-ethe-

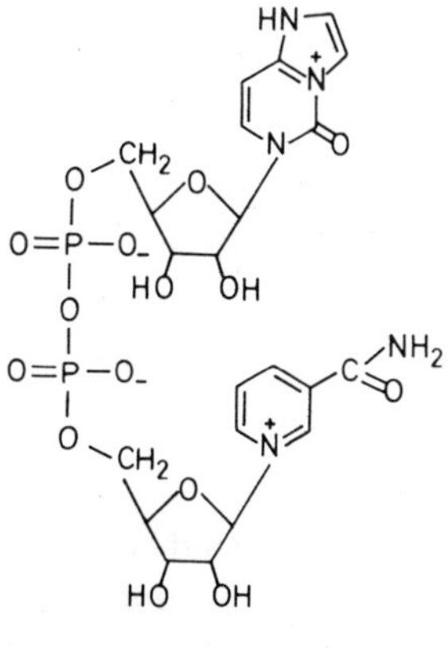

33

nocytosine and adenine in the dinucleotide (73). Divalent mercury cations specifically bind to 1,*N*⁶-ethenoadenine and 3,*N*⁴-cytosine residues. Thus, after reaction with chloroacetaldehyde, DNA from herring sperm, bound Hg^{2+} cations to 1,*N*⁶-ethenoadenine and 3,*N*⁴-ethenocytosine bases (74).

This may prove to be a useful procedure for the identification of adenine and cytosine bases in nucleic acids by means of electron microscopy.

2.2.3 Substitution of Keto by Thioketo Groups

Purine Nucleotides. Purine-6(1H)-thion (75) and 2-aminopurine-6(1H)-thion (6-thioguanine) (76) are extremely effective anticancer agents, particular against human cancer (77). Since these thioketo purines have to be converted to the corresponding ribonucleotides within the cell to exert their biological activity (78, 79), considerable attention has been devoted to the synthesis of nucleosides and nucleotides of 6-mercaptopurine or 6-thioguanine. 6-Mercaptopurine-9-β-D-riboside (6-thioinosine) and 2-amino-6-mercaptopurine-9-β-D-riboside (6-thioguanosine) are accessible by thiation with P_2S_5 in pyridine of suitably blocked nucleosides (80). The synthesis of the 5′-phosphates of 6-thioinosine and 6-thioguanosine by routine procedures has been reported (81, 82). A more direct approach involves thiation of *O*-acylated nucleotides **34** and **35** (83). Formation of the undesired phosphorothioates (**38, 39**) seemed to be a serious drawback at first, but treatment with dicyclohexylcarbodiimide converts the phosphorothioates **38** and **39** into the corresponding phosphates (**40, 41**) (84).

34, 36, 38, 40 R = H

35, 37, 39, 41 R = NH_2

Perini and Hampton (85) observed that on phosphorylation of 6-thioinosine (**42**) with phosphorus oxychloride in trimethyl phosphate, a mixture of 6-thioinosine 5′-phosphate (**43**) and 6-methylthioinosine 5′-phosphate (**44**) is formed (85). *S*-Alkylation had occurred during the subsequent hydrolysis step at pH 9 of the original reaction mixture. Prolongation of this treatment with excess trimethyl phosphate at pH 9 yielded almost exclusively the *S*-methylated product **44**. Perini et al., point out that this *S*-alkylation might prevent the application of phosphorylation with phosphorus oxychloride in the presence of trialkyl phosphate to thio-

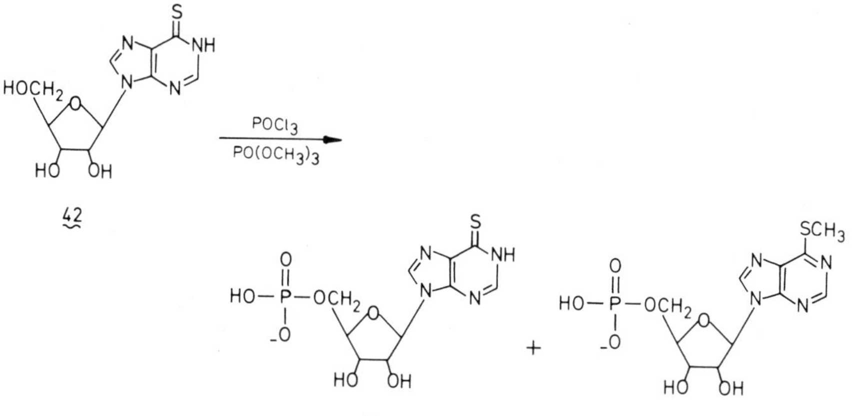

42 $\xrightarrow[PO(OCH_3)_3]{POCl_3}$ 43 + 44

keto nucleosides. However, if the formed thioketonucleoside 5′-phosphate is not subjected to prolonged treatment at pH 9, the phosphorylation of 6-thioinosine with phosphorus oxychloride in triethyl phosphate can be successfully employed (86). 6-Thioinosine 5′-phosphate survives the rather rigorous conditions for conversion to the corresponding 5′-morpholidate, which then can be reacted with tri-*n*-butylammonium phosphate to give 6-thioinosine 5′-diphosphate (87). Reaction of 6-thioinosine 5′-phosphate with *N*,*N*′-carbonyldiimidazole at room temperature yields the nucleotide imidazolidate. The latter is a suitably reactive intermediate for the preparation of 6-thioinosine 5′-triphosphate (86, 88). The synthesis of 6-thioguanosine 5′-triphosphate by conventional procedures was reported by Darlix et al. (89). An alternative approach to the synthesis of 6-thioguanosine 5′-phosphate is the selective phosphorylation of the unprotected nucleoside (**45**) by bis-(2,2,2-trichloroethyl)phosphoryl chloride (90). The resulting triester (**46**) can be conveniently isolated in crystalline form. The 2,2,2-trichloroethyl groups are removed by treatment with zinc dust in refluxing pyridine and pure **47** is obtained as the crystalline sodium salt.

An interesting synthesis route to 6-thioguanine nucleotides has been developed by Ueda et al. (91). They discovered that adenosine 5′-phosphate-N^1-oxide (**48**) reacts with cyanogen bromide to form 2-imino-6-β-D-ribofuranosyl-(1,2,4-oxadiazolo(3,2-f)purine)-5′-phosphate (**49**). Treatment of **49** with ammonia yields the N^6-cyano derivative **50**. Methylation of **50** furnished the N^1-methoxy compound **51** which was converted to an iminoimidazole carboxamidoxime intermediate **52** upon treatment with alkali. The latter cyclized in a Dimroth-type rearrangement to the 2-amino-6-methoxyaminopurine derivative **53**, which gave 6-thioguanosine 5′-phosphate **47** on reaction with hydrogen sulfide. This reaction sequence was also applied to 2′-deoxyadenosine 5′-phosphate-N^1-oxide and arabinoadenosine 5′-phosphate-N^1-oxide. Evidence that 6-thioguanine derivatives predominantly exist in the thioketo tautomer form comes from the crystal structure of 6-thioguanosine (92) as well as spectroscopic studies (80).

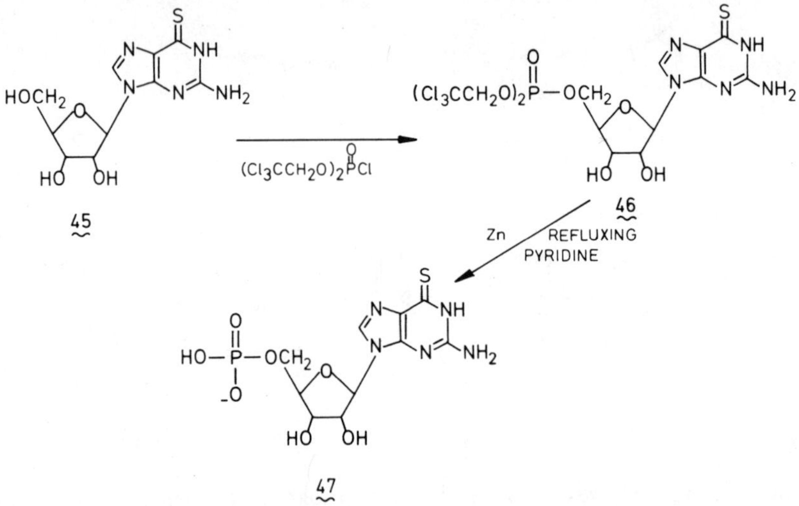

The 6-thioketopurine derivatives are not only of interest because of their pharmacological importance. The reactivity of the thioketo group allows a variety of specific chemical reactions to be applied, opening new synthetic routes to otherwise relatively inaccessible purine derivatives. Desulfurization of 6-thioinosine or 6-thioguanosine derivatives with Raney

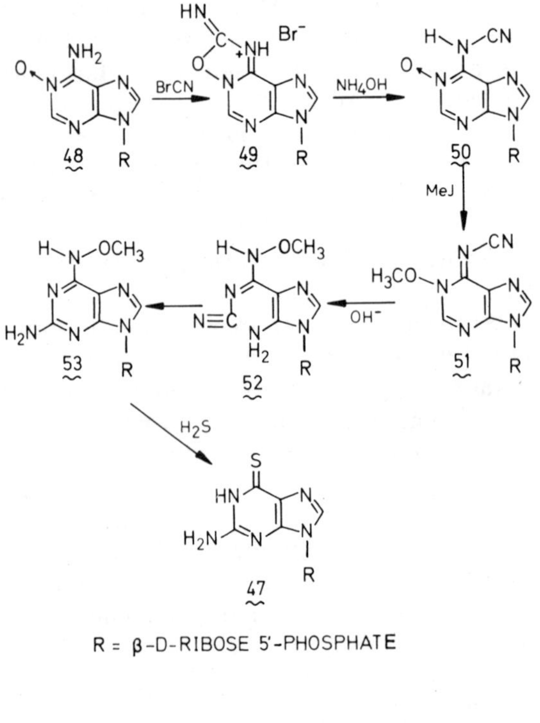

R = β-D-RIBOSE 5′-PHOSPHATE

nickel gives purine-9-β-D-ribofuranoside (nebularine) or 2-amino-purine-9-β-D-ribofuranoside derivatives, respectively (80). Derivatives of 6-thioinosine are selectively methylated at the thioketo group. The resulting 6-methylthioinosine derivatives readily undergo nucleophilic displacement reaction with primary and secondary amines. The 6-methylthioguanosine derivatives, however, only react with such nucleophiles under more vigorous conditions. Oxidation of thioketopurine compounds by iodine leads to formation of disulfides **54** (93). Using permanganate in alkaline

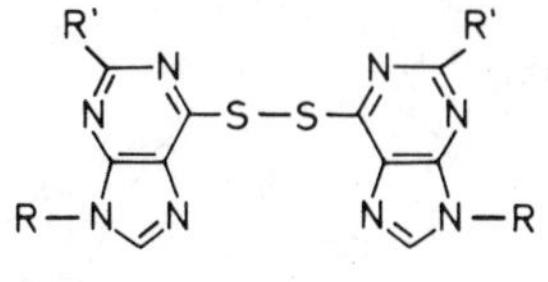

R' R'
N N
N S—S N
R—N N N N—R

54

medium 6-thioketopurines are oxidized to the stable purine 6-sulfonates (80). In analogy to 4-thioketopyrimidine nucleosides, 6-thioinosine and 6-thioguanosine derivatives react with sulfite ions in the presence of oxygen forming the corresponding 6-sulfonates **58**, **59**, and **60** (86). From experiments with ^{35}S-6-thioinosine the following mechanism for the reaction of

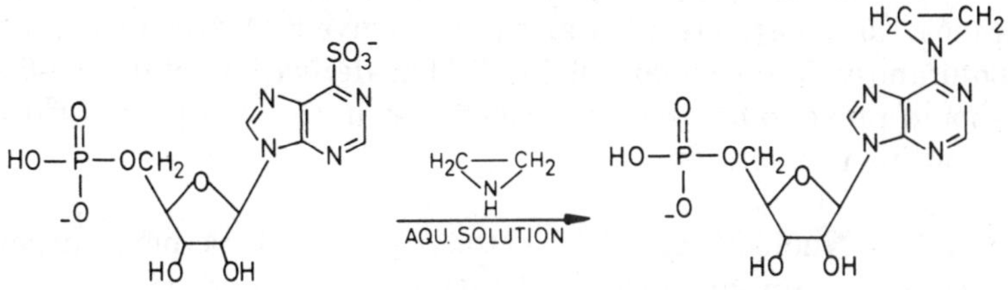

SO_3^- $H_2C—CH_2$
HO—P—OCH₂ $H_2C—CH_2$ / N / H HO—P—OCH₂
AQU. SOLUTION
HO OH HO OH

58 64

sulfite ions with thioketopurine derivatives in the presence of oxygen has been proposed (**61–63**) (86). Purine-6-sulfonates are also formed on photooxidation of 6-thioketopurine derivatives (86).

Acid hydrolysis of derivatives of purine-6-sulfonates at room temperature leads to 6-ketopurine derivatives (86). Purine-6-sulfonates are suffi-

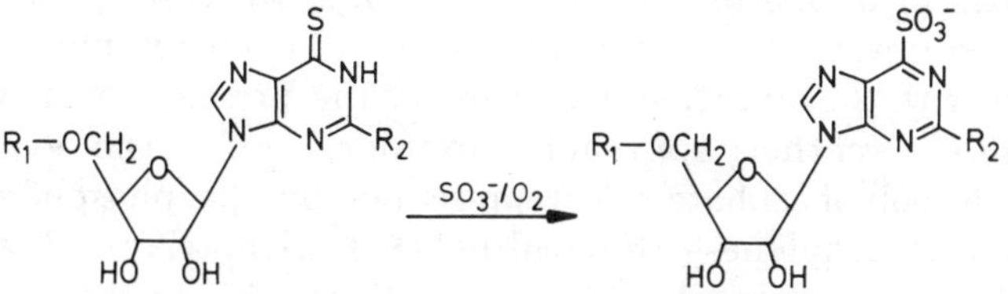

S SO_3^-
NH
R_1—OCH₂ R_2 SO_3^-/O_2 R_1—OCH₂ R_2
HO OH HO OH

55,56,57 58, 59, 60

55, 58 R_1 = P(O)(OH)O⁻ , R_2 = H
56, 59 R_1 = P(O)(OH)O⁻ , R_2 = NH_2
57, 60 R_1 = P(O)O–O–P(O)O–O–P(O)(OH)O^{3-}, R_2 = H

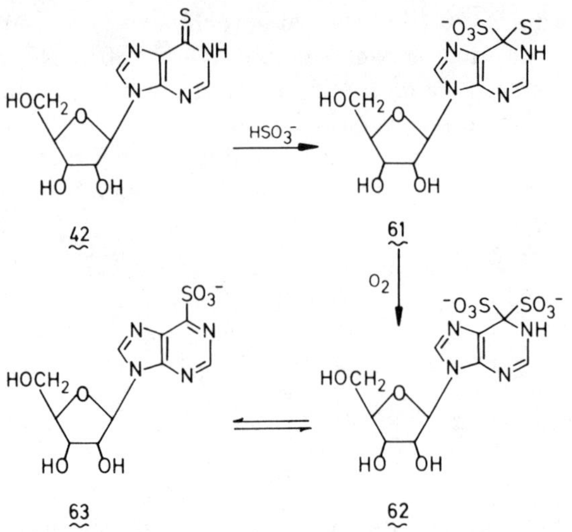

ciently stable between pH 6 and 9 to allow their isolation by anion exchange chromatography. However, they react smoothly at room temperature with nucleophilic primary and secondary amines in aqueous solution. This is exemplified by the reaction of purine-6-sulfonate-9-β-D-ribofuranosyl 5′-phosphate with aziridine to give 6-(1-aziridinyl)purine-9-β-D-ribofuranosyl 5′-phosphate (**64**) (86). The derivatives of purine-6-sulfonates should prove to be useful intermediates in the synthesis of *C*(6)-substituted purine nucleotides.

Pyrimidine Nucleotides. Thioketopyrimidines have not achieved the pharmacological importance of the thioketopurines. However, some interest in their nucleosides and nucleotides has resulted because of their useful spectroscopic properties and the chemical reactivity of the thioketo groups. Thus 4-thioketopyrimidine derivatives are extensively used as stable intermediates in the chemical transformations of natural or synthetic pyrimidine compounds (94). Furthermore, in contrast to thioketopurine nucleotides, thioketopyrimidine nucleotides are present in transfer RNA's as minor constituents (27).

4-Thiouridine 5′-phosphate (95, 96) as well as the corresponding 2′,(3′)-phosphate (96), were obtained by phosphorylation of suitably protected 4-thiouridine with 2-cyanoethylphosphate in the presence of dicyclohexylcarbodiimide. Another approach, appropriate for the synthesis of nucleoside 5′-polyphosphates, is represented by the phosphorylation of 2′,3′-*O*-ethoxymethylene-4-thiouridine (**64**) with tris-imidazolylphosphinoxide. The stable nucleotide imidazolidate **65** can be employed in subsequent reactions with phosphate or pyrophosphate to give 4-thiouridine 5′-diphosphate (**66**) and 4-thiouridine 5′-triphosphate (**67**), respectively (97). By the same procedure 2′-deoxy-4-thiothymidine 5′-triphosphate was prepared starting from 2′-deoxy-3′-*O*-acetyl-4-thiothymidine (97). Direct thiation of 2′,3′-*O*-diacetyluridine 5′-phosphate by P_2S_5 in pyridine yields after deacylation a mixture of 4-thiouridine 5′-phosphate and 4-thiouridine 5′-phosphorothioate (98). Treatment of the phosphorothioate with dicyclohexylcarbodiimide according to Eckstein (84)

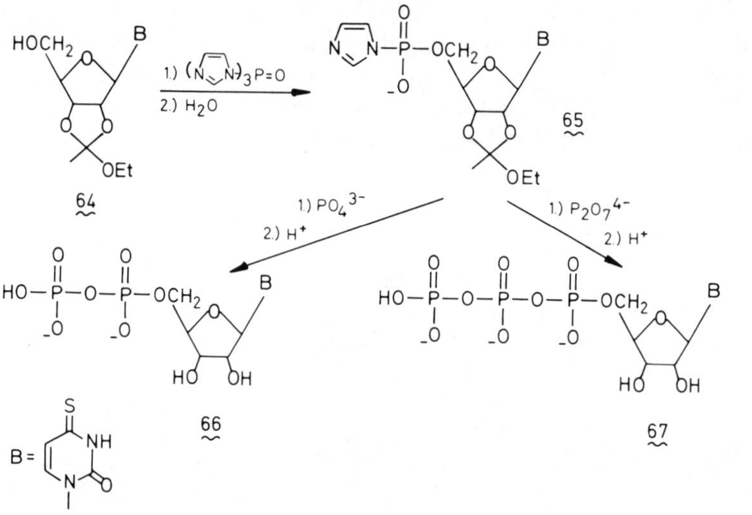

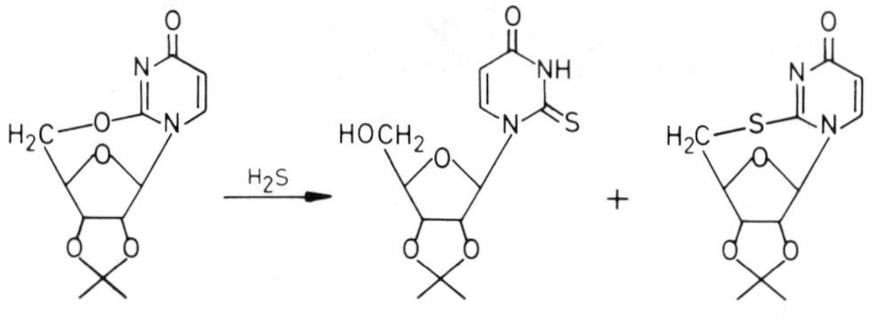

converts it to the corresponding 5′-phosphate (84). An even more economical approach to the synthesis of 4-thiouridine derivatives is reaction of cytidine derivatives with hydrogen sulfide in pyridine at elevated temperature. Cytidine 5′-mono,-di, and -triphosphate have thereby been converted to 4-thiouridine 5′-mono, -di, and -triphosphate respectively (99).

2,4-Dithiouridine is available by thiation of protected uridine with P_2S_5 under rigorous conditions (100), and phosphorylation with phosphorus oxychloride in triethyl phosphate gave 2,4-dithiouridine 5′-phosphate. Conversion of the nucleotide to the nucleotide imidazolidate by means of *N*,*N*′-carbonyldiimidazole and subsequent reaction with either phosphate or pyrophosphate furnished the 2,4-dithiouridine 5′ di and triphosphate respectively (101).

The first synthesis of 2-thiouridine 5′-phosphate (**71**) involved the tedious chemical transformation of uridine into 2-thiouridine. This was achieved by reaction of 2′,3′-*O*-isopropylidene-5′,O^2-anhydrouridine (**68**) with hydrogen sulfide yielding 2′,3′-*O*-isopropylidene-2-thiouridine (**69**) besides a major side product 2′,3′-*O*-isopropylidine-5′,S^2-anhydrouridine

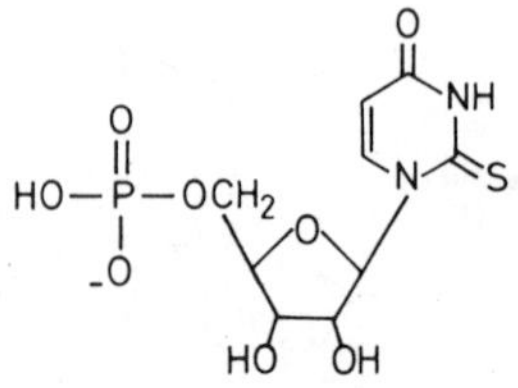

71

(**70**). Phosphorylation of **69** by 2-cyanoethylphosphate and dicyclohexylcarbodiimide gave **71** after removal of blocking groups (102). 2-Thiouridine can now be conveniently prepared by nucleoside synthesis according to Vorbrüggen et al. (103), employing 5,3,2-tri-*O*-benzoyl-1-*O*-acetyl-D-ribose and the trimethylsilyl derivative of 2-thiouracil. Phosphorylation of

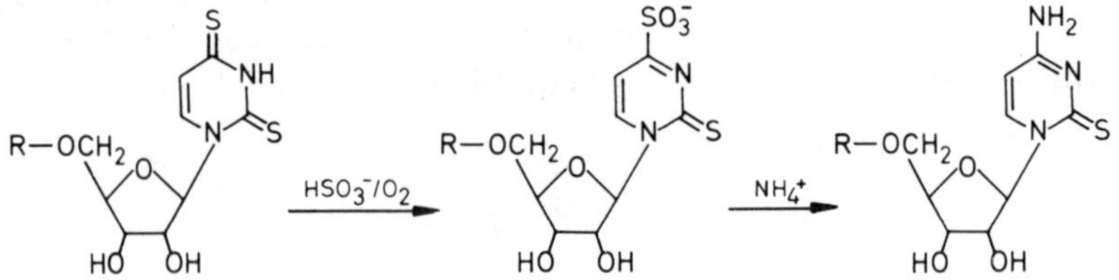

HSO_3^-/O_2 → NH_4^+ →

72, 73, 74 — 75, 76, 77 — 78, 79, 80

72, 75, 78 R = H
73, 76, 79 R = P(O)(OH)O$^-$
74, 77, 80 R = P(O)O–O–P(O)(OH)O^{2-}

unprotected 2-thiouridine with phosphorus oxychloride in triethyl phosphate gave **71**, which was converted via the nucleotide imidazolidate to the corresponding 5′-di (104, 106) and -triphosphate (105). By analogy to 4-thiouridine derivatives (107), 2,4-dithiouridine derivatives react with sulfite ions in the presence of oxygen to form the 2-thiouridine-4-sulfonates **75**, **76**, and **77**, which can then react with ammonium ions to give the 2-thiocytidine derivatives **78**, **79**, and **80** (108). More suitable for the large scale preparations of **79** and **80** is the phosphorylation of unprotected 2-thiocytidine, available by nucleoside synthesis (103), with phosphorus oxychloride in triethyl phosphate (109). The imidazolidate of **79** was reacted *in situ* with phosphate or pyrophosphate to give **80** and 2-thiocytidine 5′-triphosphate respectively (101, 109–111).

2′-Deoxy-thioketothymidine derivatives were obtained by chemical transformation of thymidine (101, 112). Reaction of 5′,O^2-anhydro-3′-*O*-acetylthymidine (**81**) with hydrogen sulfide led to 3′-*O*-acetyl-2-thiothymidine (**82**), which was converted to 2,4-dithiothymidine (**83**) by treatment with phosphorus pentasulfide. The phosphorylation procedure employed in the synthesis of 2′-deoxy-2-thiothymidine 5′-phosphate (**85**) could not be applied to 2′-deoxy-2,4-dithiothymidine (**83**) because of concomitant reduction of the 4-thioketo group.

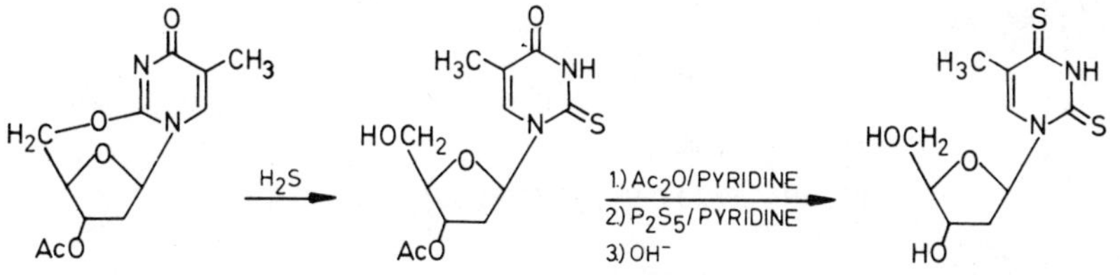

Instead 2′-deoxy-2,4-dithiothymidine 5′-phosphate (**86**) and the respective 5′-triphosphate (**87**) were obtained by phosphorylation with phosphorus oxychloride in triethyl phosphate to yield the monophosphate **86** and subsequent reaction to the 5′-imidazolidate, which in turn was treated with pyrophosphate to give **87** (101). Kochetkov et al. reported the synthesis of thioketo analogs of uridine 5′-(α-D-glucopyranosyl pyrophosphate) (113) by reaction of the respective morpholidate of 2-thioketouridine 5′-phosphate and 4-thiouridine 5′-phosphate with α-D-glucopyranosyl phosphate.

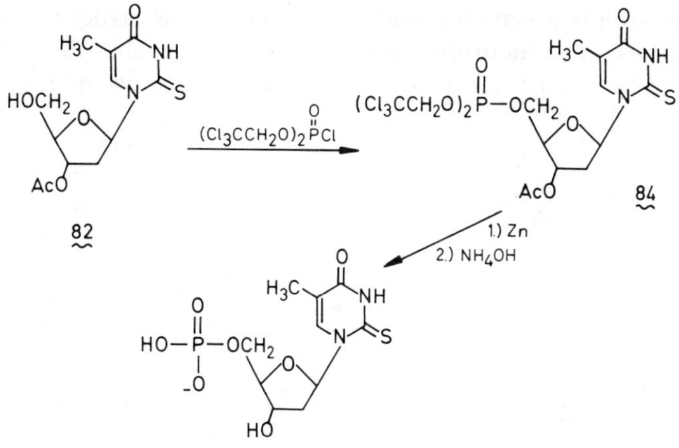

4-Thioketosubstitution of pyrimidine derivatives, contrary to 2-thioketo substitution, leads to the appearance of an absorption maximum between

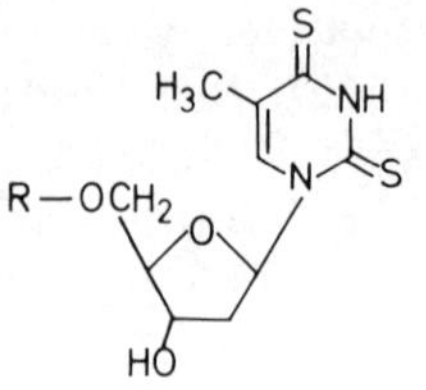

86, 87

86, R = P(O)(OH)O^-

87, R = P(O)O-O-P(O)O-O-P(O)(OH)O^{3-}

330 and 345 nm (97, 101, 112). This feature has made 4-thioketopyrimidine nucleotides attractive candidates for spectroscopic investigations of their interactions with enzymes (114, 115). 4-Thioketopyrimidine nucleotides show a weak fluorescence (quantum yield approx. 10^{-3}), which is characterized by a large Stoke shift (116). X-ray diffraction studies of 2,4-dithiouracil (117), 2,4-dithiouridine (118), 4-thiouridine (119), 2′-deoxy-4-thiothymidine (120), 2-thiouridine (121) and 2-thiocytidine (122) have shown that the thioketo tautomeric structures prevail in all cases. 4-Thiouridine adopts a syn-conformation in the crystal structure (119), but ^{1}H-nmr studies on 4-thiouridine and 4-thiouridine 5′-phosphate, however, revealed that both exist in the anti-conformation in solution (123, 124).

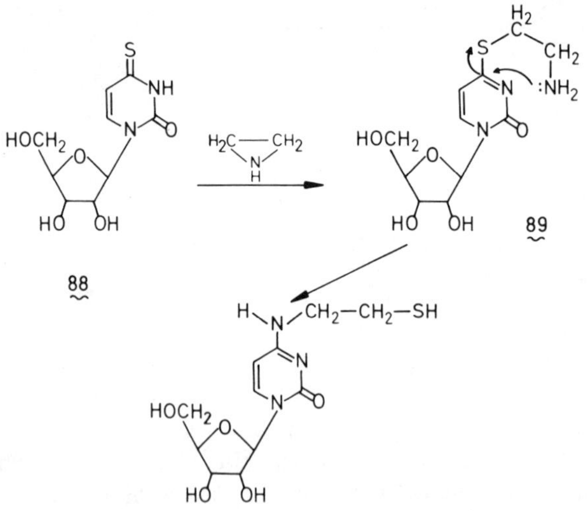

88 89

90

The thioketo groups of thiopyrimidine derivatives are the targets of specific chemical reactions. It generally holds that 4-thioketo groups are more reactive than 2-thioketo groups. They are selectively alkylated by alkyl halides in the presence of base, demonstrated in the case of 4-thiouridine 5′-diphosphate (125) and 2′-deoxy-4-thiothymidine nucleotides (126). Many alkylation studies have been carried out with thioketopyrimidine nucleosides (127–129). Methylthiopyrimidine derivatives readily undergo nucleophilic displacement reactions with primary and secondary amines. Specific alkylation at a thioketo group as well as the nucleophilic displacement of an alkylthio substituent can both be demonstrated in the case of alkylation of 4-thiopyrimidine derivatives with aziridine in aqueous solution (130). The initially formed β-aminoethylthio compound **89** rearranges to the cytidine derivative **90** by intramolecular attack of the primary amino group. The intermediate **89** cannot be isolated. If no precautions are taken, **90** is obtained as the disulfide.

The reaction of 4-thioketopyrimidines with sulfite ions in the presence of oxygen has already been treated above. The resulting 2-oxypyrimidine-

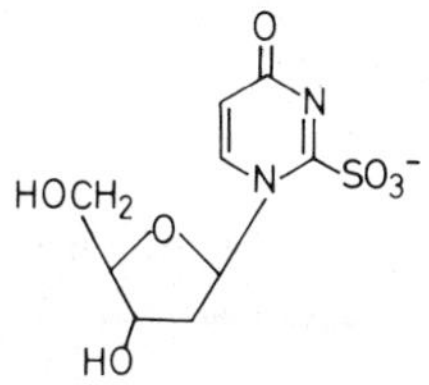

91

4-sulfonates can also be prepared by oxidation of 4-thioketopyrimidine derivatives with periodate (131) or permanganate (132, 133). 2-Thiouridine reacts analogously with permanganate or sulfite ions to form 4-oxypyrimidine-2-sulfonate-1-(β-D-riboside) (**91**) (134). Acid hydrolysis of the latter yields exclusively uridine, whereas hydrolysis at pH 10 gives a complex mixture, uridine being the major product.

4-Thiouridine and its corresponding nucleotides undergo photooxidation to intermediates that react with nucleophiles under mild conditions (135, 136). From the photooxidation studies with 6-thioinosine and 6-thioguanosine it may be safely concluded that the intermediates formed are 2-oxypyrimidine-4-sulfonates (86). Further applications of this photooxidation reaction are discussed in Chapter 7. On reaction of thioketopyrimidine derivatives with hydrogen peroxide at pH values above 7, oxidative hydrolysis occurs to give the corresponding ketopyrimidine compounds exclusively (126, 137). This reaction was successfully applied as an analytical tool (126, 137). All thioketopyrimidines bind mercurials specifically, but with varying affinities (**92–94**) (126, 137, 138).

2.2.4 Substitution of Keto by Selenoxo Groups

An appreciable antitumor activity and improved therapeutic index of 6-selenoguanine compared to 6-thioguanine prompted the first synthesis of

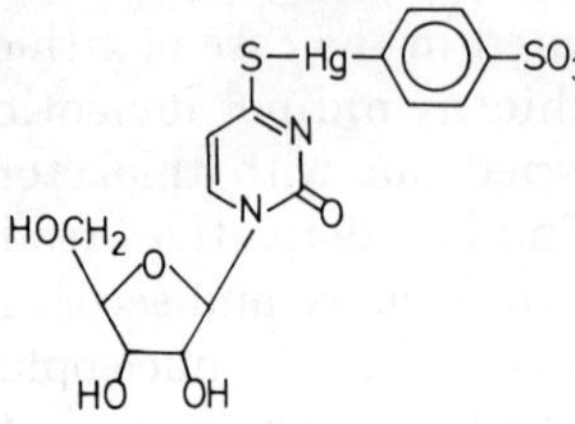

S—Hg—C6H4—SO_3^- HOCH2 HO OH
92

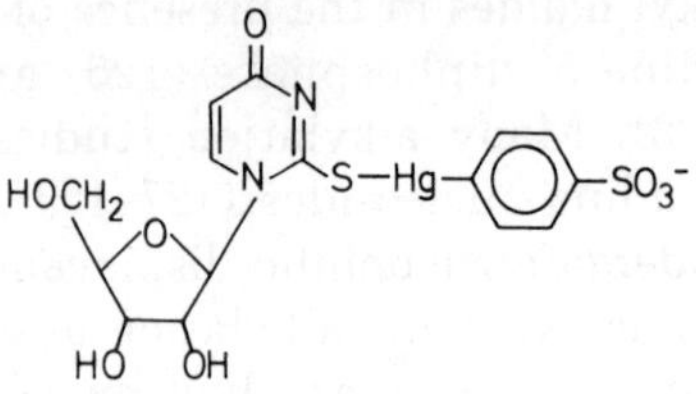

O HOCH2 S—Hg—C6H4—SO_3^- HO OH
93

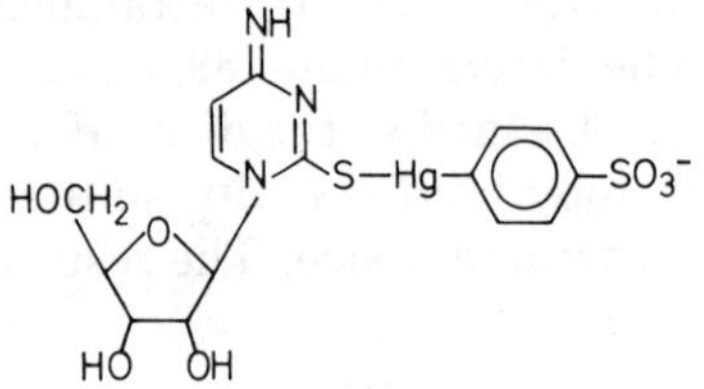

NH HOCH2 S—Hg—C6H4—SO_3^- HO OH
94

6-selenoguanine nucleosides by Townsend et al. (139, 140). Direct substitution of the 6-amino group in adenosine and adenine nucleotides by selenide ions in aqueous solution offers convenient access to 6-selenopurine nucleotides (**95**) (141, 142). This reaction is noteworthy since attempts to substitute the 6-amino group in adenosine derivatives by hydrogen sulfide

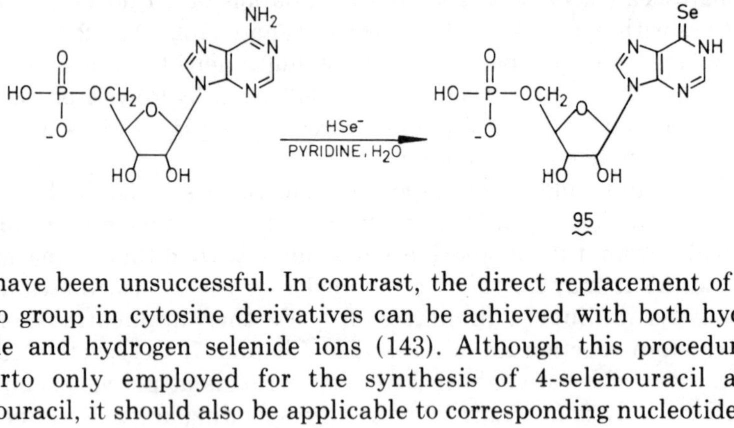

NH_2 HO—P—OCH2 O ⁻O HO OH

HSe^- / PYRIDINE, H_2O

Se NH HO—P—OCH2 O ⁻O HO OH
95

ions have been unsuccessful. In contrast, the direct replacement of the 4-amino group in cytosine derivatives can be achieved with both hydrogen sulfide and hydrogen selenide ions (143). Although this procedure was hitherto only employed for the synthesis of 4-selenouracil and 2-selenouracil, it should also be applicable to corresponding nucleotides.

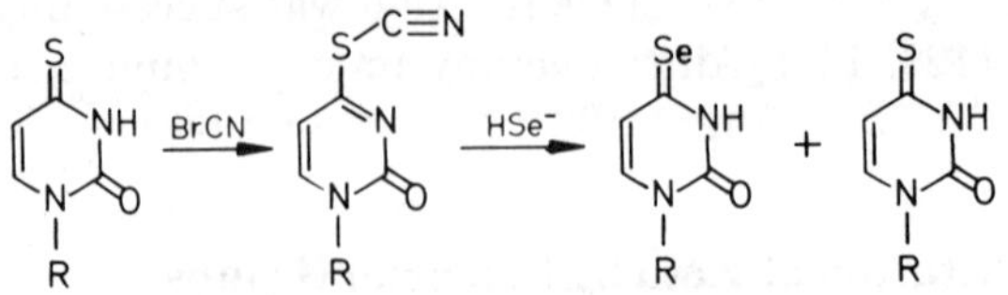

S NH R —BrCN→ S—C≡N N R —HSe^-→ Se NH R + S NH R

96 97 98 96

R = β-D-RIBOSE

Table 2.1 Effect of Thioketo and Selenoxo Substitution on Absorption Spectra and pK$_a$ Values of Purines and Pyrimidines

Compound[a]	λ_{max} (nm)	E_{max}	pK$_a$
Uracil	258	7.980	9.45
2,4-Dithiouracil	284, 361	21.570, 8.770	6.4
2,4-Diselenouracil	314, 400	16.070, 10.040	5.5
Hypoxanthine	249	8.100	8.94
6-Mercaptopurine	330	19.180	7.37
6-Selenopurine	361	14.550	7.33

Source. Reference 145. Reprinted with permission from H. G. Mautner, *J. Am. Chem. Soc.* **78,** 5293 (1956). Copyright by the American Chemical Society.

[a] All spectra measured in ethanol.

An interesting reaction for the conversion of 4-thiouridine into 4-selenouridine was discovered by Pal (144) and applied to the transformation of 4-thiouridine residues in *E. coli* transfer RNA. The simultaneous formation of **98** and **96** results from the fact that hydrogen selenide ions can either attack the C(4)—S or the S—C—N bond.

As displayed in Table 2.1, replacement of thioketo by selenoxo groups in purine or pyrimidine derivatives has a marked effect on the absorption spectra and the pK$_a$-values of the acidic protons.

The wavelength of the absorption maxima increases in going from keto to thioketo to selenoxo compounds. This is attributed to the increased contribution of the activated states $^+$C—B$^-$ (where B = O,S,Se); the greater stabilization of which in the thioketo and selenoxo compounds may be because of the availability of more than eight orbitals for accomodations of electrons. The increase in polarizability in going from keto to thioketo to selenoxo groups is responsible for the drastic enhancement in acidicity of the N—H protons of the respective heterocyclic bases.

Selenoxo analogs of pyrimidine and purine derivatives are of limited stability. The most stable seeming to be derivatives of 6-selenoguanine. The reactivity of the selenoxo groups in 6-selenopurines should be potentially useful in the chemical transformations of purine derivatives. Thus 6-selenoguanosine is converted to 2-aminopurine-9-(β-D-riboside) on treatment with Raney nickel in water at room temperature (139).

2.3. INTRODUCTION OF EXOCYCLIC SUBSTITUENTS

2.3.1 8-Substituted Purine Nucleotides

Substitution of the hydrogen atom C(8) of guanosine and guanosine 5′-phosphate by bromine occurs most readily with bromine-water at pH 3 (146, 147). The first synthesis of 8-bromoguanosine (**99**) was reported by

Holmes and Robins (148). The experimental conditions for preparation of 8-bromoguanine derivatives are critical, because the latter are rapidly oxidized by bromine with concomitant destruction of the heterocyclic ring (149). Thus the yield of 8-bromoguanosine 5′-phosphate (**104**) is

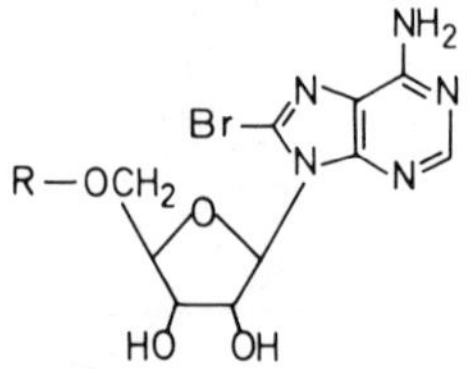

99, 100, 101, 102

considerably lower if bromination is carried out at pH 11 (150). 8-Bromoguanosine 5′-diphosphate (**105**) was synthesized via the nucleotide morpholidate (146, 151). Kapuler and Reich (152) achieved bromination of guanosine 5′-mono (**104**), -di (**105**), and -triphosphate (**106**) in formamide. Bromination by bromine-water can be applied to adenosine, adenosine 5′-

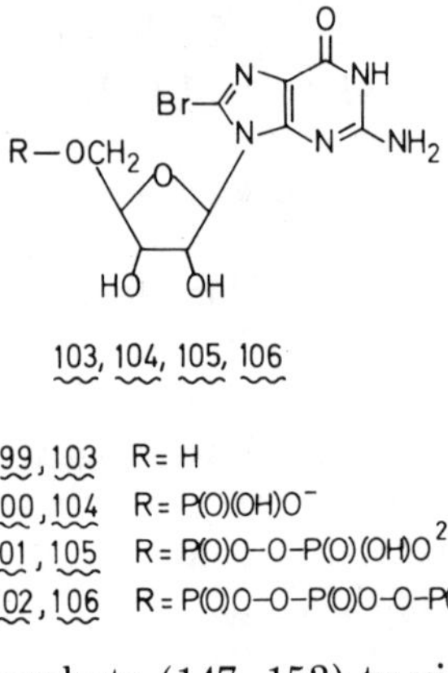

103, 104, 105, 106

99, 103 R = H
100, 104 R = P(O)(OH)O$^-$
101, 105 R = P(O)O–O–P(O)(OH)O^{2-}
102, 106 R = P(O)O–O–P(O)O–O–P(O)(OH)O^{3-}

mono, -di, and -triphosphate (147, 153) to yield the 8-bromoadenine compounds **99–102**. It is interesting to note that no attempts have been reported to prepare the guanosine derivative **105** and **106** using bromine-water. 8-Bromoadenosine 5′-(α-D-glycopyranosyl)diphosphate was obtained by bromination with bromine-water of the parent adenine compounds (154).

Introduction of bromine into position 8 of purines causes a slight red shift of the absorption maximum in the ultraviolet spectrum. Furthermore, the bulky bromine atom restricts rotation about the glycosidic bond. X-ray diffraction studies by Tavale and Sobell indicated the syn conformation for 8-bromoguanosine and 8-bromoadenosine in their crystal structures (155). This also appears to be the preferred conformation of 8-bromopurine nucleotides in solution and thus confers interesting biological properties on these purine nucleotide analogs.

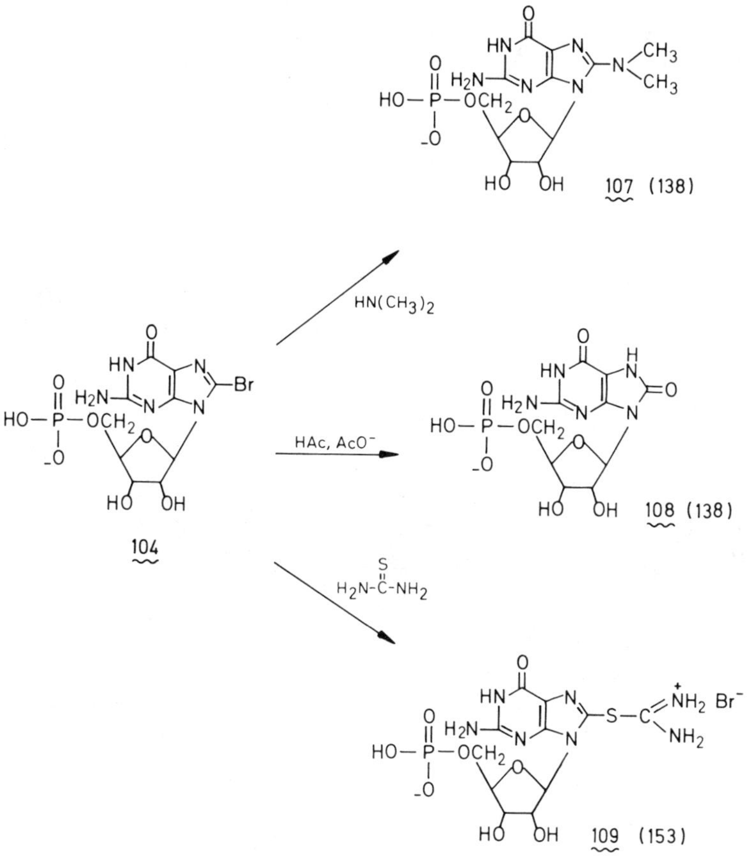

104 → $HN(CH_3)_2$ → 107 (138)

104 → HAc, AcO^- → 108 (138)

104 → $H_2N\text{-}C(=S)\text{-}NH_2$ → 109 (153)

Scheme 2.1

In addition, 8-bromopurine nucleotides are key compounds for the preparation of a great variety of 8-substituted purine nucleotides because of the relative ease with which the 8-bromo substituent can be replaced by nucleophiles. The possible reactions are outlined in the following schemes (Scheme 2.1, **107–109**; Scheme 2.2, **110–115**).

The synthesis of the 5′-diphosphates of 8-oxyadenosine (152, 157, 156), 8-azidoadenosine (165, 166), 8-oxyguanosine (169), 8-methylaminoguanosine, 8-aminoguanosine (168), as well as of the 5′-triphosphates of 8-bromoguanosine (153), 8-bromoadenosine (170–173), 8-oxyguanosine (169), 8-azidoadenosine (170–173), 8-aminoadenosine (170), and 8-mercaptoadenosine (170) involved the application of standard procedures.

Only a few nucleophilic displacement reactions of the 8-bromo substituent have been performed at the level of the triphosphate. Treatment of 8-bromoadenosine 5′ triphosphate with azide ions at elevated temperature gives the 8-azido derivative in good yield, and with aqueous sodium

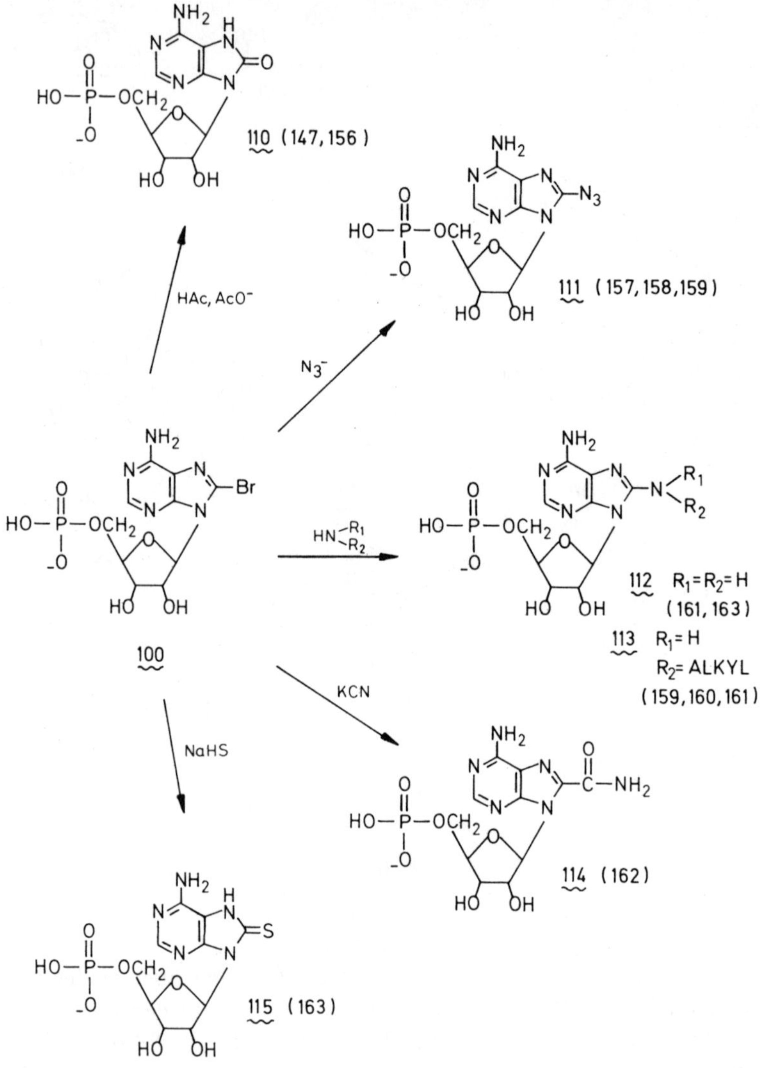

Scheme 2.2

hydrosulfide at room temperature the respective 8-mercapto compound is obtained (163). A number of nucleophilic reactions at the 8-bromo substituent have been reported for the corresponding 3′,5′-cyclic nucleotides, but have not yet been applied to nucleoside 5′-polyphosphates: for example, the reactions of 8-bromoguanosine 3′,5′-cyclic phosphate with seleno urea (164) or 8-bromoadenosine 3′,5′-cyclic phosphate with cyanide ions (162). In the latter case, however, no reaction was observed with the corresponding 5′-monophosphate. Furthermore, and this originated mainly from studies with 3′,5′-cyclic nucleotides, there are marked differences in the reactivity of the 8-bromo substituents in guanine and adenine nucleotides. 8-Bromoadenosine 3′,5′-cyclic phosphate reacts with both sodium hydrosulfide (163) and sodium hydroselenide (174), but these reactions have not yet been observed with 8-bromoguanosine 3′,5′-cyclic phosphate. A similar situation seems to exist with respect to the behavior of 8-bromoguanine and 8-bromoadenine derivatives toward azide ions.

8-Azidoadenine analogs of nicotinamide adenine dinucleotides and flavine adeninedinucleotide have been synthesized from 8-azidoadenosine 5′-phosphate and nicotinamide mononucleotide or flavine mononucleotide respectively (175).

Chlorination at position 8 of adenine nucleotides occurs with tetrabutylammonium iodotetrachloride (**116, 117**) (176). Although X-ray

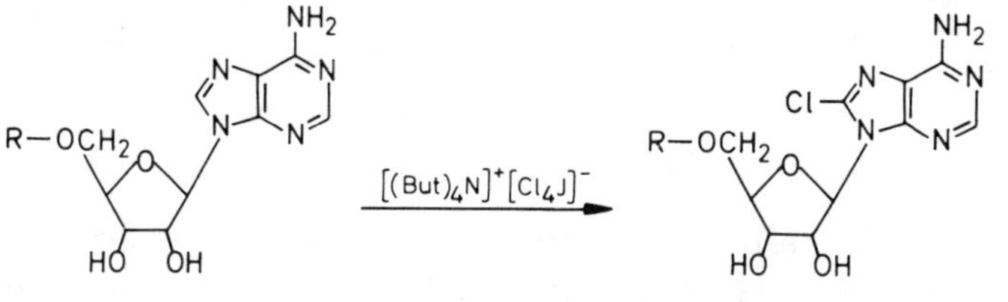

$[(But)_4N]^+[Cl_4J]^-$

116, 117

116, R = P(O)(OH)O$^-$

117, R = P(O)O-O-P(O)(OH)O^{2-}

diffraction studies showed that 8-bromoadenosine adopts the syn conformation in the crystal structure (155), and spectroscopic investigations have indicated that the syn conformation is also to be found for 8-bromoadenosine 5′-phosphate in solution (177), it is not generally true that any substituent in the 8-position of adenine nucleotides causes the syn conformation. Leng et al., studied the conformation in solution of a series of 8-monoalkylaminoadenosine 5′-phosphates by means of nmr, ultraviolet absorption, CD and their interaction with adenosine 5′-phosphate specific antibodies (160). At pH 7, 8-(2-aminoethylamino)adenosine 5′-phosphate (**118**) possesses the anti conformation about the glycosidic bond and the gauche$^+$ conformation at the C(4′)—C(5′) bond, because of a strong electrostatic interaction between the 8-ethylamino and the phosphate group. If the pH is changed to 11.5, **118** undergoes transition to the

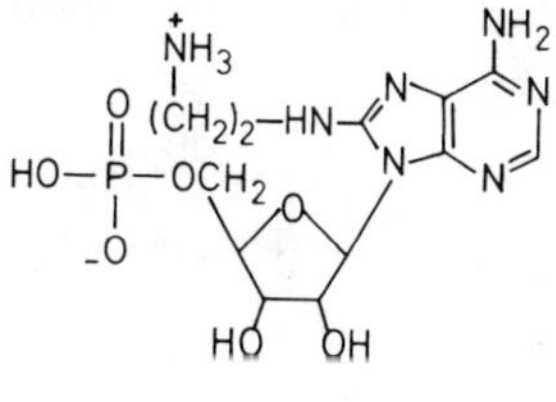

118

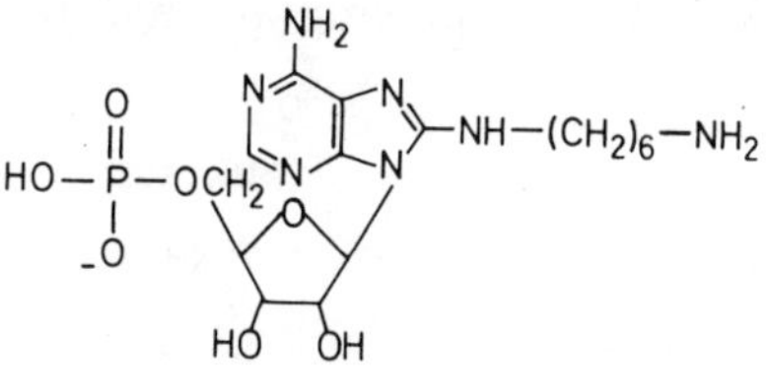

119

syn conformation. In contrast, 8-(6-aminohexylamino)adenosine 5′-phosphate (**119**) adopts the syn conformation at all pH values. It is surprising that **119** and adenosine 5′-phosphate have approximately similar affinities toward antibodies specific for the latter, whereas the 8-amino derivative **120** displays much higher affinity.

Evans and Kaplan (159) investigated the conformation in solution of 8-aminoadenosine 5′-phosphate (**120**), 8-methylaminoadenosine 5′-phosphate (**121**) and 8-dimethylaminoadenosine 5′-phosphate (**122**) by nmr

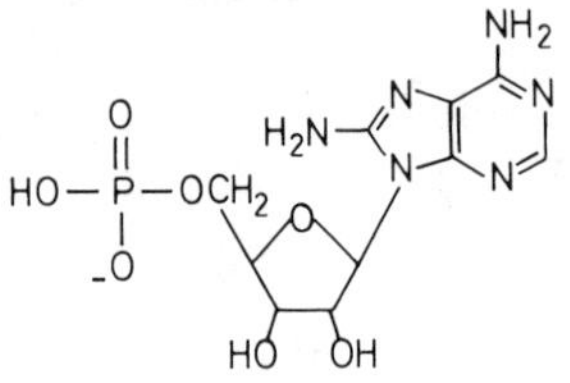

120

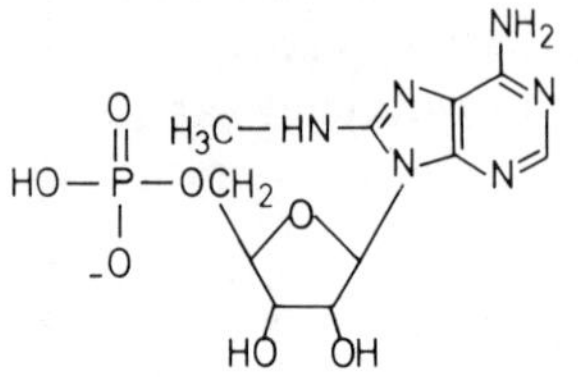

121

spectroscopy. The results indicate that **120** and **121** prefer the anti conformation, but that **122** exists predominantly in the syn conformation. Interestingly, 8-fluoro and 8-oxy substituents (167, 177) cause syn conformation despite the fact that their van der Waals radii are smaller than of the 8-amino group in **120**. On the other hand, the pK_a of the 8-amino

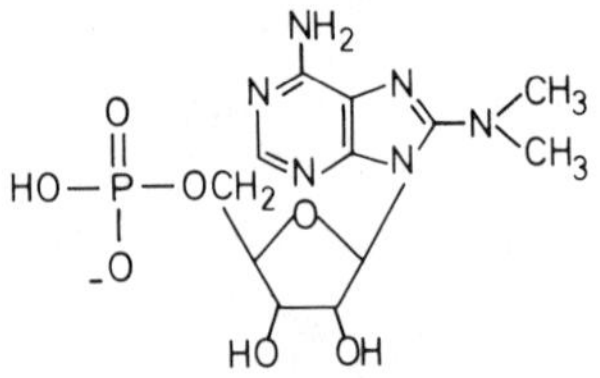

122

group is 1.5. Thus **120** will barely carry a formal positive charge at pH 7, that would favor electrostatic interactions with the 5′-phosphate group. A factor stabilizing the anti conformation in **120** and **121** could be hydrogen bonding between the 8-amino hydrogens and *O*(5′) or one of the other phosphate oxygen atoms.

8-Methylguanosine 5′-phosphate (**123**) is obtained by homolytic *C*-methylation employing methyl radicals generated from *t*-butyl hydro-

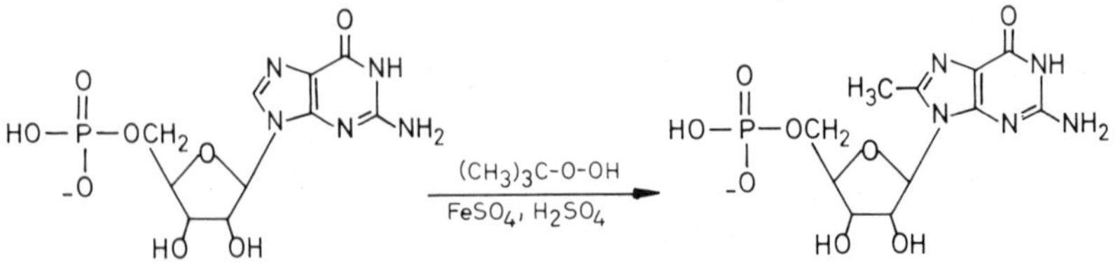

$(CH_3)_3C\text{-}O\text{-}OH$

$FeSO_4$, H_2SO_4

123

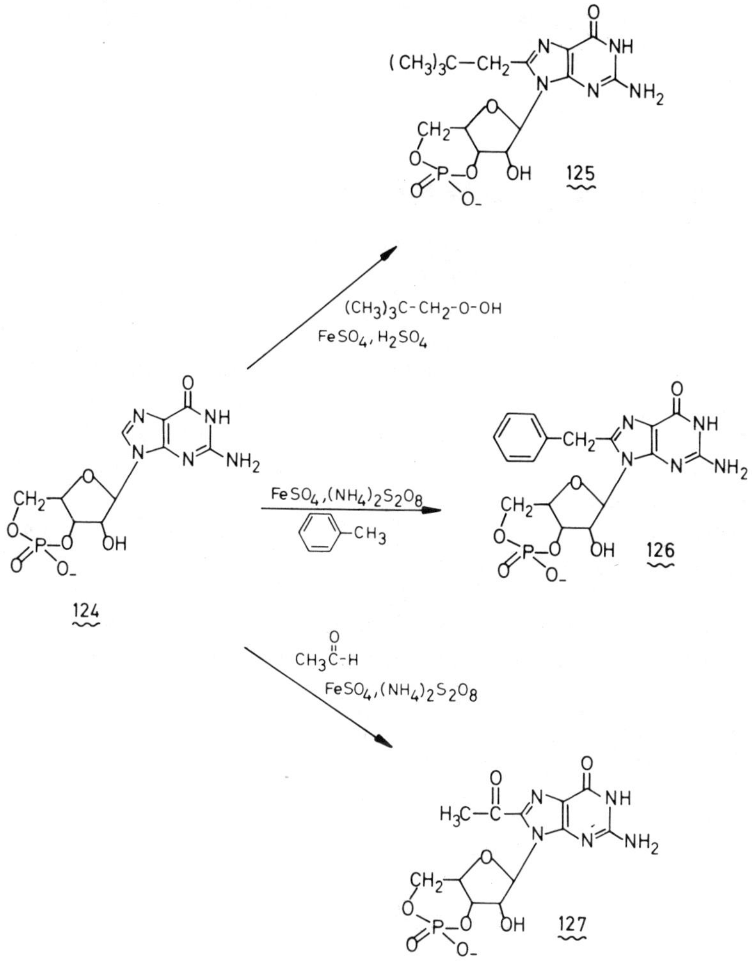

Scheme 2.3

peroxide (178). Robins et al., extended this homolytic alkylation to the synthesis of various 8-alkylguanosine 3′,5′-cyclic phosphates (179). Furthermore, these authors made use of the fact that abstraction of a hydrogen from the carbonyl atom of aldehydes by a radical source yields highly reactive acyl radicals. These species proved to be useful for the syntheses of 8-acylguanosine 3′,5′-cyclic phosphates (179).

The 8-acetyl group in **127** undergoes reactions typical of ketones. With $NaBH_4$ 8-(1-hydroxyethyl)guanosine 3′,5′-cyclic phosphate was formed. Furthermore, **127** yielded a thiosemicarbazone, semicarbazone and phenyl hydrazone. It is very likely that reactions similar to those depicted in Scheme 2.3 (**125–127**) will be general to guanine nucleotides. Adenosine 3′,5′-cyclic phosphate, however, does not undergo homolytic alkylation under similar reaction conditions. Because inosine 3′,5′-cyclic phosphate was found to react with tert.-butyl hydroperoxide-$FeSO_4$ in the presence of dilute H_2SO_4 forming the 8-methyl derivative, the possibility exists of converting 8-methylinosine derivatives into 8-methyladenosine derivatives without major difficulties.

2.3.2 2-Substituted Purine Nucleotides

2-Methyladenylic acid was identified as a minor constituent of RNA by Littlefield and Dunn (180). The chemical synthesis of 2-methyladenosine was first reported by Davoll et al. (181) and Yamazaki (182). Phosphorylation of the unprotected 2-methyladenosine led to the 5′-phosphate, which was converted to the 5′-diphosphate via the nucleotide morpholidate (183). The condensation of 5-amino-β-D-ribofuranosylimidazole 4-carboxamide with esters provides easy access to 2-alkylinosines (184). By this route 2-methylinosine was obtained, which could be phosphorylated using standard procedures to 2-methylinosine 5′-mono- and diphosphate (185).

Condensation of 5-amino-1-β-D-ribofuranosylimidazole-4-carboxamide 5′-phosphate (**128**) with phenylisothiocyanate leads to the versatile 2-mercaptoinosine 5′-phosphate (**129**) (186). 2-Methylthioinosine 5′-phosphate (**130**) was formed on methylation of **129** and this was phosphorylated to the corresponding 5′-diphosphate (187). Oxidation of the 2-methylthio group by *N*-chlorosuccinimide to the methylsulfonyl group opens up the possibility of nucleophilic substitution at position 2, thus allowing the synthesis of *N*(2)-alkylguanylic acids (**132**) (187).

128 129

The halogen substituents in 2,6-dichloro- or 2,6-dibromo-9-(2′,3′,5′-*O*-triacetyl-β-D-ribofuranosyl)-9H-purine, prepared by a fusion reaction between 2,6-dihalogenopurine and tetra-acetyl-D-ribose, exhibit different reactivities toward nucleophiles. At room temperature ammonia selectively displaces the 6-halogeno substituent to give 2-chloroadenosine (188) and 2-bromoadenosine (189, 190) respectively. The syntheses reported for 2-chloroadenosine-5′-phosphate (192), 2-bromoadenosine 5′-phosphate (192), 2-chloroadenosine 5′-diphosphate (191), and 2-chloroadenosine 5′-triphosphate (170) followed standard procedures. The unprotected nucleosides 2-chloroinosine and 2-methoxyinosine could be phosphorylated by means of phosphorus oxychloride (193).

Substitution of purine nucleotides in position 2 by alkyl groups or halogen atoms will definitely affect the conformation of the respective nucleotides about the glycosidic bonds. Experimental data concerning conformation in solution of this class of purine nucleotides is not yet available. It can, however, be anticipated that either bulky alkyl substituents or highly electronegative halogen atoms in position 2 will favor the anti conformation of such nucleotides. In the former case the reason would probably be steric hindrance, whereas in the latter case repulsion between electronegative groupings might prevail.

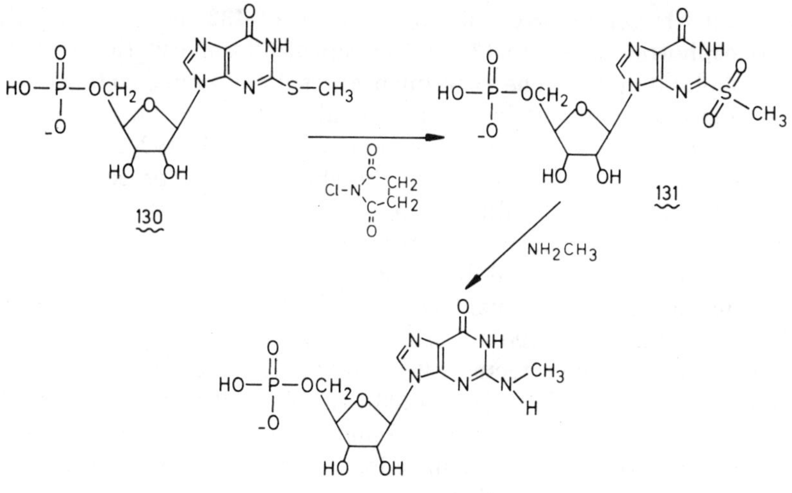

2-Aminopurine is one of the most potent mutagens among known base analogs (194). Its mutagenic action produces exclusively A·T→G·C transitions (195). This biological effect and the strong fluorescence emission of 2-aminopurine aroused considerable interest in the synthesis of 2-aminopurine nucleotides. 2-Aminopurine-9-(β-D-riboside) (**134**) is easily accessible by treatment of 6-thioguanosine with Raney nickel (80). Phosphorylation of protected 2-aminopurine-9-(β-D-riboside) by conventional methods to give the respective 5′-mono, -di, and -triphosphate (**133, 134, 135**) was first reported by Ward et al. (196). Unprotected **132** could be phosphorylated by phosphorus oxychloride in triethyl phosphate. The nucleotide imidazolidate of **133** was further converted to **134** and **135** (197).

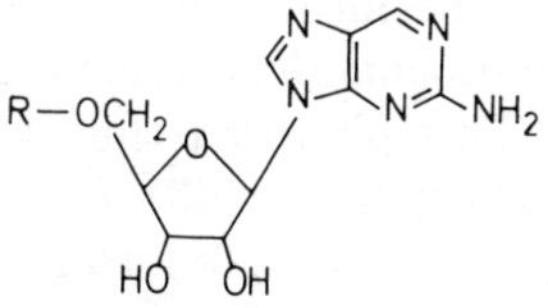

133, 134, 135

133, R = P(O)(OH)O$^-$

134, R = P(O)O-O-P(O)(OH)O^{2-}

135, R = P(O)O-O-P(O)O-O-P(O)(OH)O^{3-}

Janion and Shugar reported phosphorylation of **132** by means of wheat shoot phosphotransferase to **133** and chemical phosphorylation to **134** via the morpholidate (198). The experimental results obtained so far, indicate that 2-aminopurine nucleotides resemble adenine nucleotides more closely than guanine nucleotides in biochemical reactions (196, 199–201).

Davoll and Lowy (202) employed the chloromercuri procedure for the first synthesis of 2,6-diaminopurine-9-(β-D-riboside) from 2,6-diaminopurine and protected ribofuranosyl chloride. 2-Aminoadenosine was also obtained from 2,6-dichloropurine-9-(β-D-riboside) by reaction with azide ions and subsequent catalytic hydrogenation of the resulting 2,6-diazido compound (190). Application of standard phosphorylation procedures to suitably protected **136** led to 2-aminoadenosine 5′-phosphate (**137** and the 5′-diphosphate (**138**) (203). An alternative route to **137, 138,** and **139** is represented by the reaction of the corresponding 6-thioguanosine derivatives with sulfite ions in the presence of oxygen to the respective 6-sulfonates, followed by reaction with ammonium ions at pH 10 and room temperature (204). The synthesis of the 5′-mono and -diphosphate of 2-dimethylaminoadenosine has been reported (205).

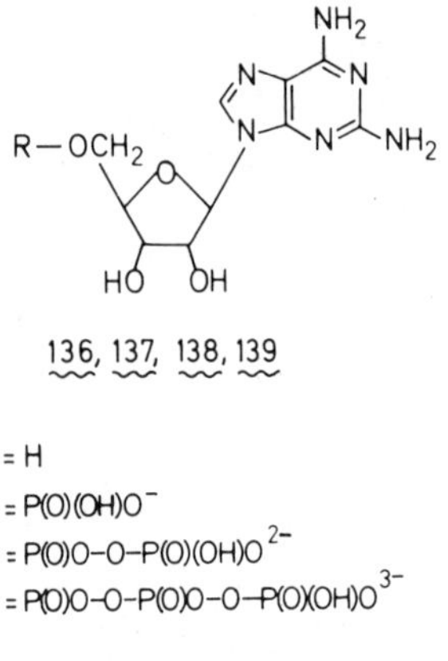

136, 137, 138, 139

136, R = H
137, R = P(O)(OH)O^-
138, R = P(O)O–O–P(O)(OH)O^{2-}
139, R = P(O)O–O–P(O)O–O–P(O)(OH)O^{3-}

Although the experimental evidence is as yet incomplete, 2,6-diaminopurine nucleotides seem to behave as analogs of adenine nucleotides in biochemical processes (199). Irradiation of adenosine-*N*(1)-oxide in dilute ammonium hydroxide gives 2-keto-6-aminopurine-9-(β-D-riboside) (isoguanosine) besides a small amount of adenosine (206). It is assumed that this photochemical rearrangement proceeds through an oxirane intermediate. This photoreaction was applied to the synthesis of isoguanosine 5′-monophosphate (**143**) (207), 5′-diphosphate (**144**) (208, 209) and 5′-triphosphate (**145**) (208) from the respective adenine-*N*(1)-oxide compound.

Treatment of guanosine 5′-phosphate with nitrosonium tetrafluoroborate in the presence of cupric chloride gives 2-chloroinosine 5′-phosphate (**146**) in high yield (210). Attempted displacement of the 2-chlorine atom

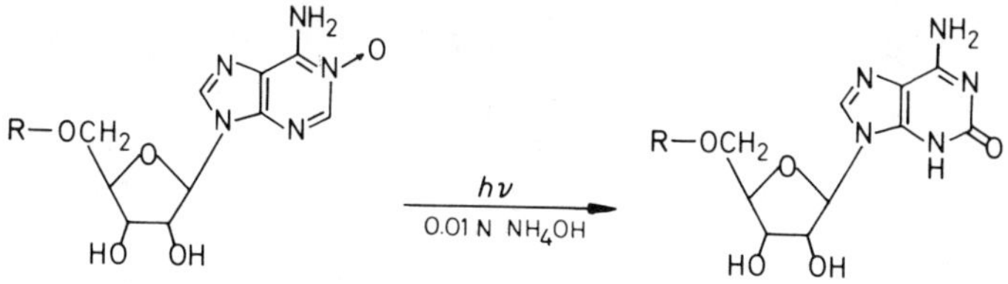

140, 141, 142 → 143, 144, 145 ($h\nu$, 0.01 N NH_4OH)

140,143 R = P(O)(OH)O⁻
141,144 R = P(O)O–O–P(O)(OH)O $^{2-}$
142,145 R = P(O)O–O–P(O)O–O–P(O)(OH)O $^{3-}$

by azide ion led to the tetrazolo(5,1-b)6-oxopurine-9-(β-D-riboside)5′-phosphate (**147**), which was shown to coexist in equilibrium with 5% of the 2-azido compound (**148**). Compound **147** fluoresces with a quantum yield of 0.26. Photolysis of **147** and **148** also occurs because of the equilibrium between both. The nucleotide **147** was phosphorylated to the corresponding 5′-triphosphate by a standard technique (210).

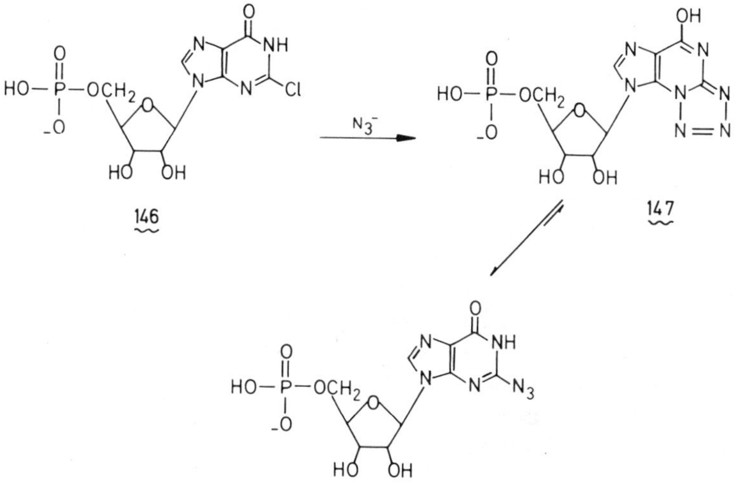

146 → (N_3^-) 147 ⇌ 148

Yamaji et al. (211) offer a promising new synthetic route to 2-substituted adenine nucleotides, which might be of general importance, although the reactions have hitherto only been performed with 3′,5′-cyclic nucleotides. Treatment of 1,N^6-ethenoadenosine 3′,5′-cyclic phosphate with alkali leads to cleavage of the six-membered ring with concomitant loss of carbon atom C(2). The resulting 3-β-D-(3′,5′-cyclic phospho)-ribofuranosyl-4-amino-5-(imidazol-2yl)-imidazole (**149**) reacts with various reagents under cyclization to form 2-substituted 1,N^6-ethenoadenosine 3′,5′-cyclic

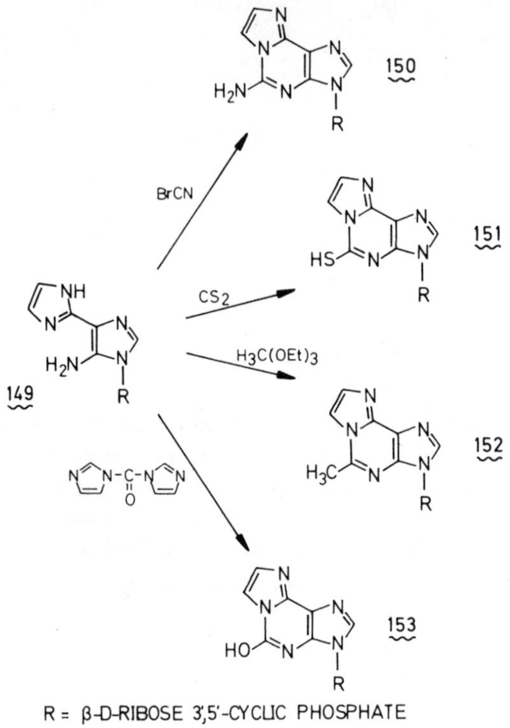

R = β-D-RIBOSE 3′,5′-CYCLIC PHOSPHATE

Scheme 2.4

phosphates (**150–153**) (Scheme 2.4). Similar ring closure reactions have been observed with 5-amino-1-β-D-ribofuranosylimidazole-4-carboxamidine 3′,5′-cyclic phosphate (212). In the latter case, however, side reactions predominate and hamper the isolation of the desired products. The importance of the cyclization reactions of **149** rests on the observation by Yamaji et al. (213) that the 1,N^6-etheno substituent can be removed by treatment with *N*-bromosuccinimide. The authors explain this reaction by the mechanism shown in Scheme 2.5.

Because of the 1,N^6-etheno substituent the reactivity of the 2-mercapto group in **151** is very high, and it readily reacts with halogens to form the corresponding 2-halogeno-1,N^6-ethenoadenosine 3′,5′-cyclic phosphates (**154**), from which the 2-halogenoadenosine 3′,5′-cyclic phosphates (**155**) could be obtained by treatment with *N*-bromosuccinimide (214).

2.3.3 5-Substituted Pyrimidine Nucleotides

The chemistry and synthesis of 5-substituted pyrimidine nucleosides and nucleotides have been competently reviewed by Bradshaw and Hutchinson (215). These authors suggest three types of reaction mechanisms for substitution reactions at C(5) of the pyrimidine ring.

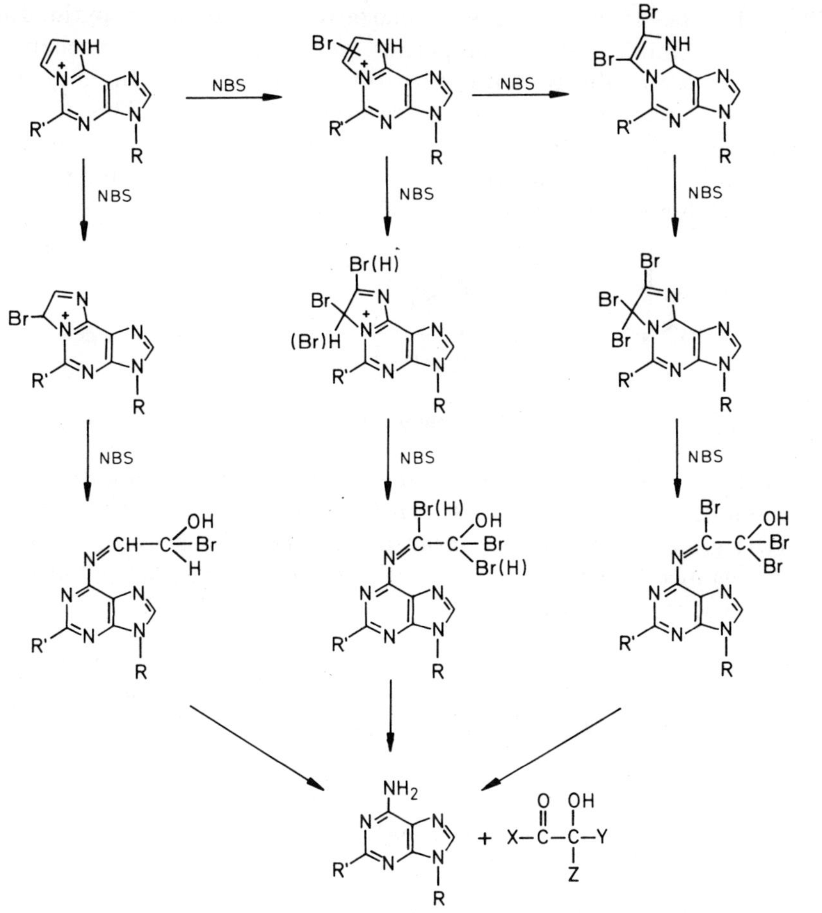

NBS = N-BROMOSUCCINIMIDE
X,Y,Z = H or Br

Scheme 2.5

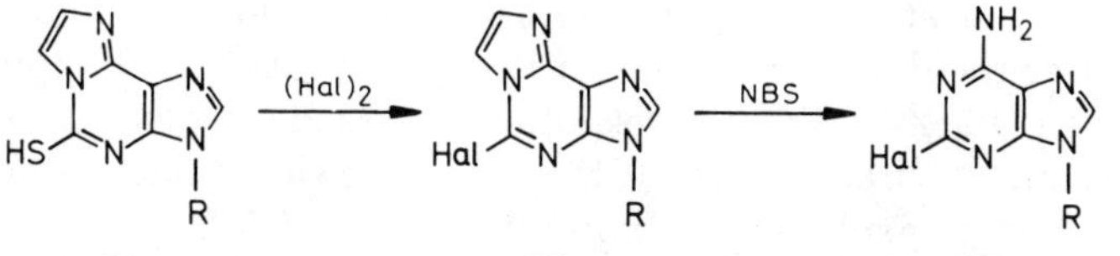

151 154 155

Hal = Cl, Br, J
NBS = N-BROMOSUCCINIMIDE
R = β-D-RIBOSE 3',5'-CYCLIC PHOSPHATE

Type 1. Reactions at the electronegative C(5) atom in pyrimidine bases, which exhibit aromatic properties, might proceed by a mechanism resembling electrophilic aromatic substitution (Scheme 2.6). In most cases cytidine derivatives do not undergo reactions of Type 1. This may be explained by the fact that the basic nitrogen atoms of the cytosine ring react initially with electrophiles. The ring thus acquires a formal positive charge which hinders further reactions at C(5).

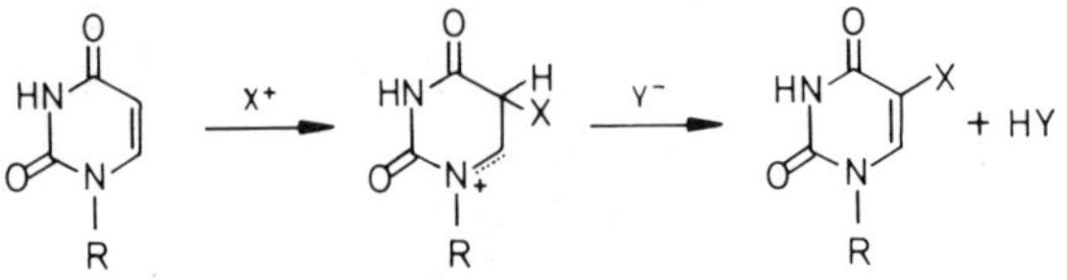

Scheme 2.6

Type 2. The second mechanism resembles the Michael addition. The nucleophile that attacks C(6) first can be water, an alcohol (particularly the 5′-OH group of nucleosides in an intramolecular reaction), or halide ions. In this mechanism a subsequent elimination of the nucleophile from C(6) must occur (Scheme 2.7).

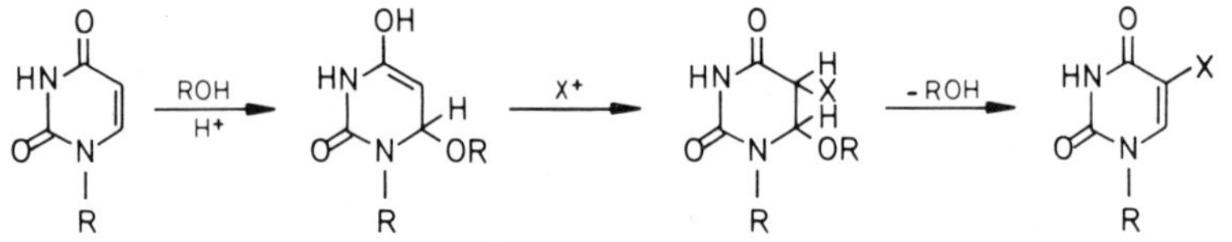

Scheme 2.7

Type 3. A third reaction mechanisms involves free radicals. Addition of a free radical to the heterocyclic ring in the excited state leads to an intermediate, from which the 5-substituted pyrimidine can be generated directly by loss of a hydrogen radical. Alternatively, addition of a second radical may lead to a dihydropyrimidine species that can undergo an elimination reaction as outlined in Scheme 2.8.

The 5-halo substituted pyrimidine nucleotides gained enormous interest after the discovery of the antitumor activity of 5-fluoropyrimidine derivatives (216, 217) and the usefulness of 5-iodo-2′-deoxyuridine in the chemotherapy of viral diseases (218). The following modes of action have been proposed to explain the inhibitory effects of 5-fluoropyrimidines on mammalian, bacterial and viral growth. (1) 5-Fluoro-2′-deoxyuridine-5′-phosphate is a strong and specific inhibitor of deoxythymidylate synthetase. This inhibition specifically blocks DNA synthesis. (2) The incorporation of 5-fluorouridylic acid into RNA species may result in the synthesis of nonfunctional enzymes, transfer RNA's or ribosomes. (3) 5-

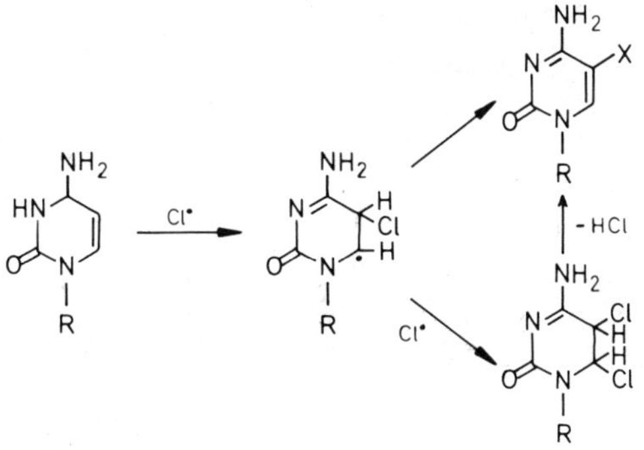

Scheme 2.8

Fluoropyrimidine nucleotides may interefere with *de novo* synthesis of pyrimidine nucleotides (70).

The first synthesis of 2′-deoxy-5-fluorouridine 5′-phosphate (219), 5-fluorouridine 5′-phosphate, and 5′-triphosphate (220) employed blocked nucleosides and conventional phosphorylation procedures. The 5-fluoropyrimidine nucleosides are available by nucleoside synthesis (221, 222). 5-Fluorocytidine 5′-phosphate (**156**) and 5′-diphosphate (**157**) were prepared

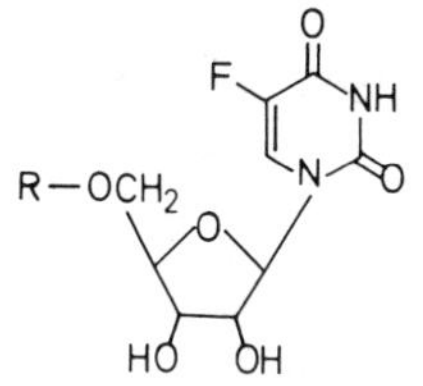

156, R = P(O)(OH)O⁻
157, R = P(O)O–O–P(O)(OH)O²⁻

from 5-fluorocytidine, which in turn was obtained by fluorinating 5′,3′,2′,-*O*-triacetyl-N^4-acetylcytidine with trifluoromethyl hypofluorite (223). The latter reaction, which was originally applied to uridine and 2-deoxyuridine (224), may proceed either via a Type 1 or Type 2 mechanism. Direct fluorination by trifluoromethyl hypofluorite of uridine 5′-phosphate has been reported (225).

Iodination of uridine (226), 2′-deoxyuridine (227), uridine 2′,3′-cyclic phosphate (228), and uridine 5′-phosphate (229) occurs in the presence of aqueous nitric acid. Therefore iodine nitrate has been suggested as the reactive agent in these iodinations. Under similar conditions bromination to 5-bromouridine 5′-phosphate and chlorination to 5-chlorouridine 5′-phosphate also occurs (228–230). 5-Bromouridine nucleotides can be obtained by treatment with bromine (231) or *N*-bromosuccinimide (232,

233) in formamide. Chlorination of pyrimidine nucleotides is achieved by tetrabutylammonium iodotetrachloride in dimethylformamide. This procedure is applicable to uridine 5′-diphosphate and cytidine 5′-diphosphate (234). Cytidine nucleotides are halogenated in aqueous solution (235), formamide (236), or acetic acid (237), and they can be iodinated with iodine and iodic acid in acetic acid as solvent (238) or iodine and iodinetrichloride in nitric acid (239). In these cases the existence of substituents in the 5′-position excludes the possibility that the reactions follow a Type 2 mechanism. They are more likely to be of Type 1.

The 5′-diphosphates of 5-chlorouridine (**164**), 5-bromouridine (**165**), and 5-iodouridine (**166**) were prepared from the corresponding 5′-phosphates

158, 159, 160 — $(C_6H_5O)_2P(O)Cl$ → 161, 162, 163 — PO_4^{3-} → 164, 165, 166

158, 161, 164 B=5-CHLOROURACIL
159, 162, 165 B=5-BROMOURACIL
160, 163, 166 B=5-IODOURACIL

158, 159, and **160** by means of the diphenyl phosphate displacement approach (230, 235). Because diphenylphosphoric acid is a stronger acid than the respective nucleotide, the anion added displaces diphenylphosphate to form the nucleoside 5′-diphosphate. This method is extremely versatile and works equally well in the synthesis of nucleoside 5′-triphosphates. The scope of this reaction is discussed in depth in Chapter 6.

Reaction of the 5′-phosphate, 5′-diphosphate, or 5′-triphosphate of uridine with bromine in aqueous solution does not yield the 5-bromo but rather the 5-hydroxy derivatives **167, 168,** and **169** (240). Similarly 5-hydroxycytidine 5′-diphosphate was prepared from cytidine 5′-diphosphate (241). Uridine and cytidine derivatives react with formaldehyde to give 5-hydroxymethyl compounds. The reaction conditions, however, vary considerably from case to case. 5-Hydroxymethyluridine 5′-phosphate is formed in low yields on acid (242) or base catalysis (243). 2′-Deoxycytidine-5′-phosphate reacts with formaldehyde only in the presence of base to give the 5-hydroxymethyl derivative (243). The reaction of nucleoside 5′-phosphate with formaldehyde cannot proceed by a Type 2 mechanism. It is interesting in this context that uridine does not react

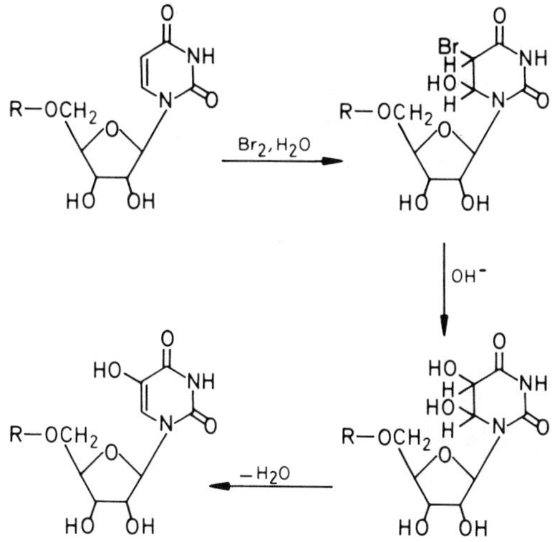

Br_2, H_2O

OH^-

$-H_2O$

167, **168**, **169**

167, R = P(O)(OH)O$^-$
168, R = P(O)O–O–P(O)(OH)O^{2-}
169, R = P(O)O–O–P(O)O–O–P(O)(OH)O^{3-}

with formaldehyde under base catalysis at all, whereas 2′,3′-*O*-isopropylideneuridine (**170**) gives a high yield of the respective 5-hydroxymethyl compound (**171**) under these conditions (244). The conformation of 2′,3′-*O*-isopropylideneuridine may be such that the 5′-OH group can attack C(6), thus allowing the reaction with formaldehyde to proceed via mechanism 2.

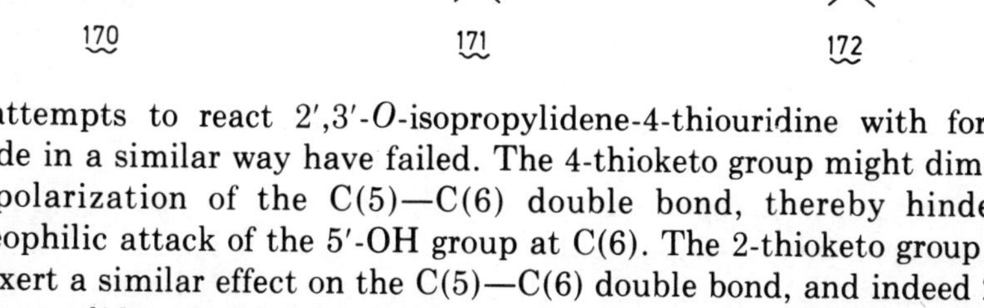

CH_2O, KOH

CH_3COOH, H^+

170 **171** **172**

All attempts to react 2′,3′-*O*-isopropylidene-4-thiouridine with formaldehyde in a similar way have failed. The 4-thioketo group might diminish the polarization of the C(5)—C(6) double bond, thereby hindering nucleophilic attack of the 5′-OH group at C(6). The 2-thioketo group cannot exert a similar effect on the C(5)—C(6) double bond, and indeed 2′,3′-*O*-isopropylidene-2-thiouridine was found to react with formaldehyde under base catalysis forming the 5-hydroxymethyl compound in high yield (245). The synthesis of 5-hydroxymethyluridine 5′-phosphate (**173**) and 5′-

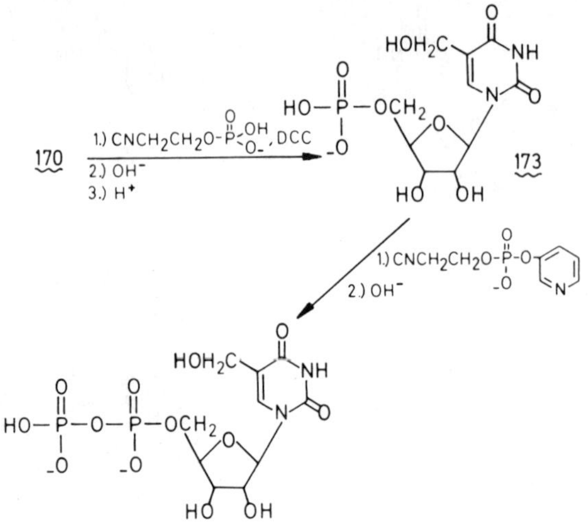

diphosphate **174** starting from 2′,3′-*O*-isopropylideneuridine (**170**) was reported by Scheit (244). The selective acetylation of **171** to **172** is feasible because of its ability to form a stable carbonium ion under acidic conditions. The 5′-diphosphate **174** is prepared by reacting **173** with the activated ester (2-cyanoethyl) 2-pyridyl phosphate (246).

Substitution of pyrimidine rings at position 5 with halogens effects the electronic structure of the heterocyclic systems to a varying degree, as indicated by bathochromic shifts of the absorption maxima in the ultraviolet. Furthermore, halogen substituents at C(5) of uracil increase the acidity of the N(3) proton of the respective uridine derivatives. Halogen substitution at C(5) of cytidine derivatives reduces the basicity of the cytosine ring.

Substitution by halogen atoms at C(5) of pyrimidine compounds offers a means for further derivatization. Bradshaw and Hutchinson (215) outlined a general scheme by which displacement reactions might take place at C(5)-halogen substituents (Scheme 2.9).

The reaction of uridine derivatives with bromine in aqueous solution to give 5-hydroxy compounds could be envisaged to follow route (a) of Scheme 2.9. 5-Bromouridine derivatives react with anhydrous ammonia via mechanism (a) to form 5-aminouridine derivatives (247–250). The synthesiss of 5-aminouridine 5′-phosphate was first described by Thiry (251). Enzymatic phosphorylation of 5-aminouridine 5′-phosphate led to the corresponding 5′-triphosphate (252). 5-Dimethylaminocytidine 5′-phosphate is obtained through displacement of bromine in 5-bromocytidine 5′-phosphate by dimethylamine (253). 2′-Deoxyuridine 5′-phosphate reacts with methyl hypobromite to form a stable compound of type **177** in Scheme 2.9. Reaction of this intermediate with sodium

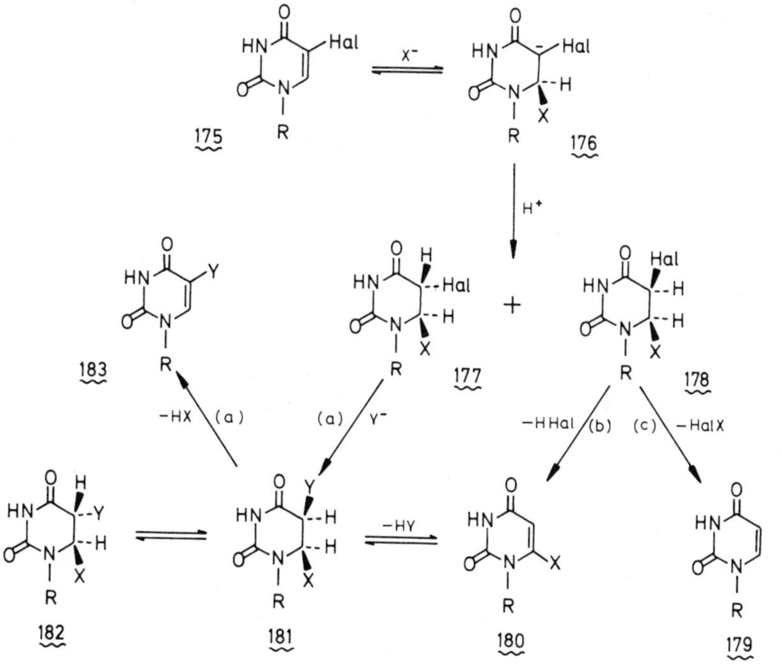

Scheme 2.9

hydrosulfide leads to formation of 2′-deoxy-5-mercaptouridine 5′-phosphate (254, 255).

Reaction of mercuric acetate with derivatives of uridine and cytidine affords covalent substitution by mercury at position 5 (256, 257). The reaction conditions are such that even sensitive derivatives of uridine and cytidine can be subjected to this mercuration. Mercurated pyrimidine nucleotides (**184–191**), provided the mercuri substituents have been complexed with thiol compounds, still possess substrate properties for enzymes like DNA polymerase, RNA-dependent RNA polymerase, and polynucleotide phosphorylase (256, 257). They furthermore offer new possibilities for the derivatization of pyrimidine nucleotides (258). Reduction of **184** with either deuterated or tritiated $NaBH_4$ could easily be

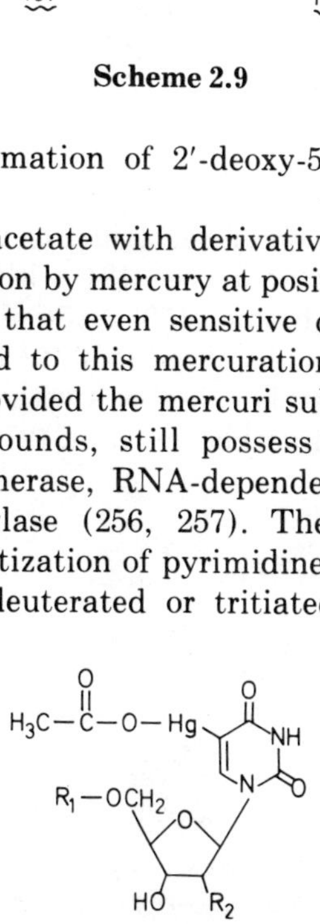

184, 185, 186, 187

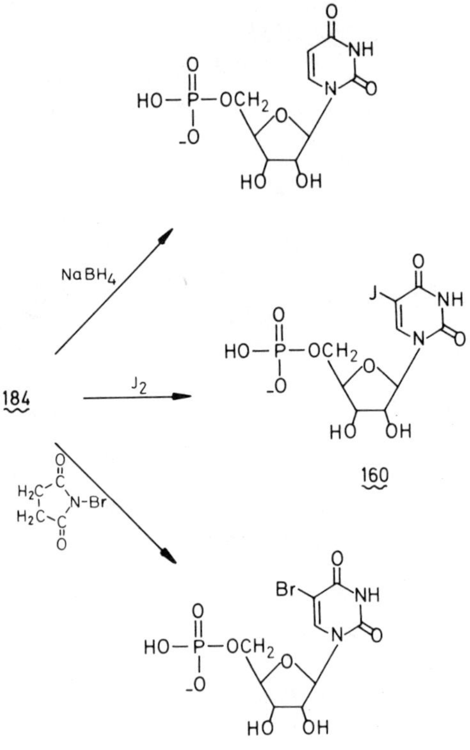

employed for the introduction of deuterium or tritium at position C(5) of pyrimidine nucleotides. Reaction of **184** with *N*-bromosuccinimide or iodine allows to prepare 5-bromo (**159**) or 5-iodopyrimidine (**160**) derivatives respectively. The report by Ward et al. (256–258) that 5-mercuripyrimidine nucleotides form stable complexes with thio compounds contrasts with observations by van Broeckhoven et al. (259, 260). The latter authors

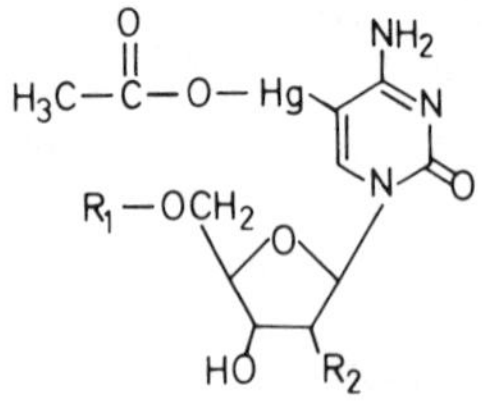

188, 189, 190, 191

184, 188 $R_1 = P(O)(OH)O^-$, $R_2 = OH$
185, 189 $R_1 = P(O)O\text{-}O\text{-}P(O)O\text{-}O\text{-}P(O)(OH)O^{3-}$, $R_2 = OH$
186, 190 $R_1 = P(O)(OH)O^-$, $R_2 = H$
187, 191 $R_1 = P(O)O\text{-}O\text{-}P(O)O\text{-}O\text{-}P(O)(OH)O^{3-}$, $R_2 = H$

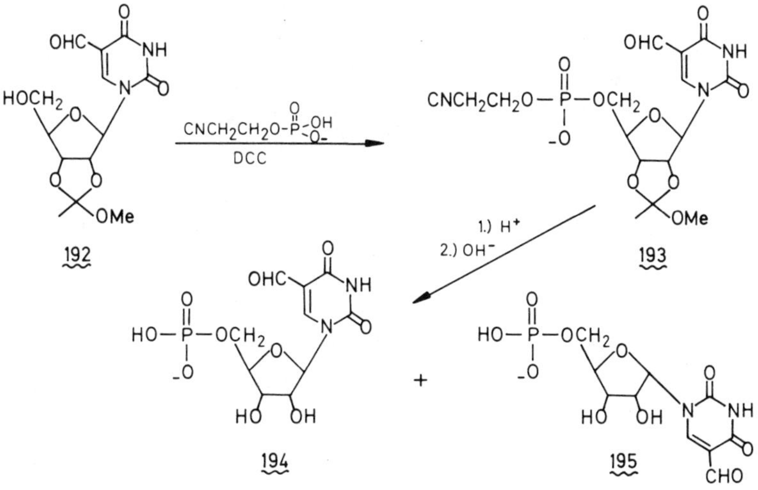

OHC, NH, O, HOCH$_2$, N, O, OMe — **192**

$\xrightarrow[DCC]{CNCH_2CH_2O\text{-}P(O)(OH)O^-}$

$CNCH_2CH_2O-P(=O)(O^-)-OCH_2$, OHC, NH, N, O, OMe — **193**

1.) H^+ 2.) OH^-

$HO-P(=O)(O^-)-OCH_2$, OHC, NH, N, HO, OH — **194**

\+

$HO-P(=O)(O^-)-OCH_2$, HO, OH, N, NH, CHO — **195**

state that the initially formed 5-mercuripyrimidine-mercaptane complexes are converted into 5-thiomercuri derivatives that decompose to the unsubstituted parent pyrimidine nucleotides.

Nitration of 5′,3′,2′-*O*-triacetyluridine 5′-phosphate with nitronium tetrafluoroborate gave the 5-nitro compound. Attempts to apply this reaction to cytidine nucleotides were unsuccessful (261). Using the same reagent in sulfolane as solvent, Huang and Torrence achieved nitration at position 5 of unprotected uridine 5′-phosphate and 2′-deoxyuridine 5′-phosphate. In the latter case, 3′-*O*-nitro-5-nitro-2′-deoxyuridine 5′-phosphate was identified as a side product of the reaction. The same authors report the phosphorylation of 5-nitrouridine to the corresponding 5′-phosphate by means of phosphorus oxychloride in triethyl phosphate (262).

5-Hydroxymethyluridine is readily oxidized to 5-formyluridine. The latter was phosphorylated by Armstrong et al., employing 2-cyanoethyl phos-

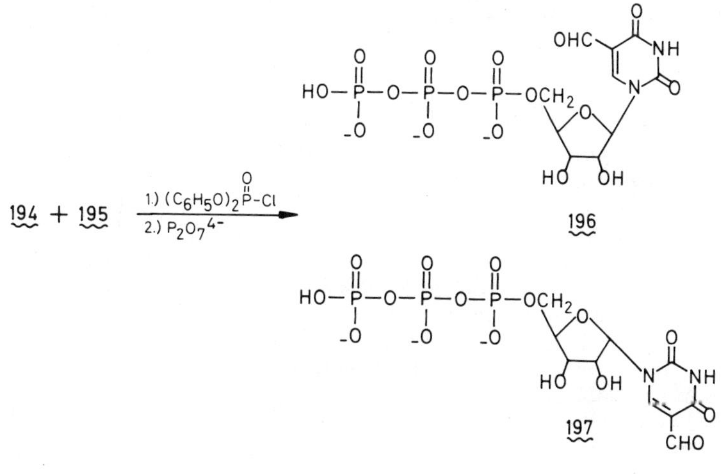

194 + **195** $\xrightarrow[2.)\ P_2O_7^{4-}]{1.)\ (C_6H_5O)_2P(O)\text{-}Cl}$

$HO-P(=O)(O^-)-O-P(=O)(O^-)-O-P(=O)(O^-)-OCH_2$, OHC, NH, N, O, HO, OH — **196**

$HO-P(=O)(O^-)-O-P(=O)(O^-)-O-P(=O)(O^-)-OCH_2$, HO, OH, N, NH, O, CHO — **197**

phate and dicyclohexylcarbodiimide (263). The conversion of the mixture of anomeric 5-formyluridine 5′-phosphates, formed on hydrolysis of **194** to the corresponding 5′-triphosphates was achieved by the diphenylphosphate displacement technique. The anomerization occurred during the base catalyzed β-elimination of the cyanoethyl blocking group in **193**. The last step of the above reaction sequence furnished β-formyluridine 5′-triphosphate (**196**) and α-formyluridine 5′-triphosphate (**197**) in a ratio of 1.2:0.86. Later Armstrong et al., demonstrated that 5-formyluridine indeed undergoes base catalyzed anomerization and suggested the following mechanism (**199–202**) (264). A series of 5-substituted pyrimidine nucleotides, listed in Table 2.2, was synthesized by conventional procedures.

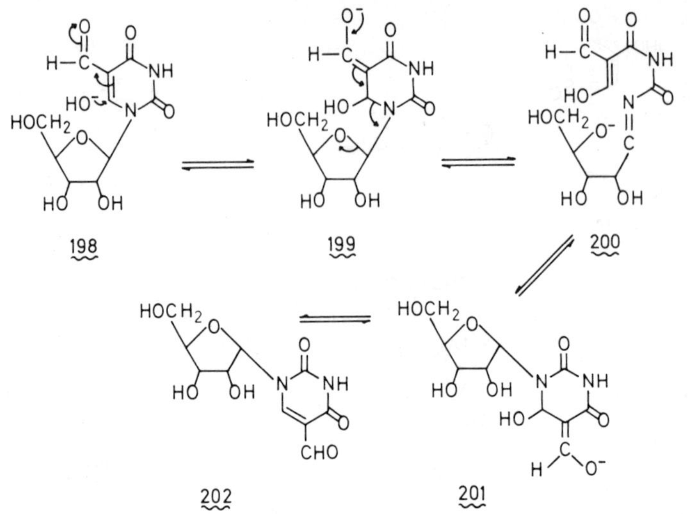

Table 2.2 5-Alkylpyrimidine Nucleotides

Nucleotide	Reference
5-Methyluridine 5′-phosphate	244, 265, 266, 267
5-Methyluridine 5′-diphosphate	244, 265, 266
5-Methylcytidine 5′-phosphate	269
5-Ethyluridine 5′-phosphate	268, 272
5-Ethylcytidine 5′-phosphate	270, 271
5-Ethyluridine 5′-diphosphate	268, 272
5-Ethylcytidine 5′-diphosphate	270, 271
β-5-Ethyl-2′-deoxycytidine 5′-phosphate	273
α-5-Ethyl-2′-deoxycytidine 5′-phosphate	273
5-Carboxyethyluridine 5′-phosphate	274
5-Ethoxycarbonyluridine 5′-phosphate	274
5-Methylsulfonyluridine 5′-phosphate	275
5-Methylsulfonyluridine 5′-diphosphate	275

Table 2.3 5-Substituted 2′-Deoxyuridine 5′-Phosphates

Substituent
Hydroxymethyl
Methoxymethyl
Benzyloxymethyl
Formyl
Acetyl
Allyl
2,3-Oxypropyl
Azidomethyl

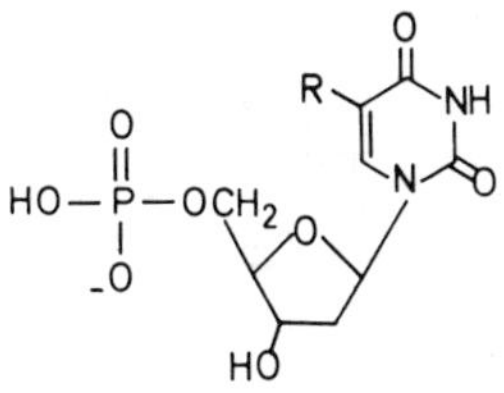

Kampf et al., synthesized various 5-substituted derivatives of 2′-deoxyuridine 5′-phosphate and studied their function as potential inhibitors of thymidylate synthetases from different sources (276). The compounds are listed in Table 2.3.

Holy and Bald (277) described the synthesis of 5-substituted uridine 2′,3′-cyclic phosphates (**203**), which were obtained in an elegant manner

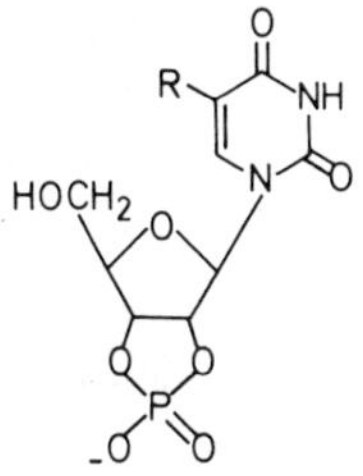

203

by oxidation of the respective 3′(2′)-uridine phosphites with hexachloro acetone. The principles of this phosphorylation procedure are presented in Chapter 6.

The fluorinated analog of thymidine, 5-trifluoromethyl-2′-deoxyuridine, exhibits strong antitumor and antiviral activities (278). The mechanism of this biological function is probably the powerful inhibition of thymidylate synthetase by 5-trifluoromethyl-2′-deoxyuridine 5′-phosphate (279, 280). Because of the base lability of the trifluoromethyl group, the synthesis of 5-trifluoromethyl-2′-deoxyuridine 5′-phosphate (**208**) and the corresponding 5′-triphosphate (**209**) required a synthetic procedure that was devoid of any treatment with strong base (281). The 5′-triphosphate **209** was prepared from **208** via the nucleotide imidazolidate.

2.3.4 6-Substituted Pyrimidine Nucleotides

Substitution of pyrimidine nucleotides in position 6, which is position α with respect to the glycosidic bond, should have similar implications on the conformation about the glycosidic bonds as 8-substitution in purine nucleotides. Indeed, X-ray diffraction studies on 6-methyluridine (282) and nmr spectroscopy of 6-carboxy-uridine (orotidine) (124) revealed that both 6-substituted uridine derivatives exist in the syn conformation. It is, however, questionable whether these results imply that the corresponding nucleotides exclusively adopt a syn conformation in solution.

The syntheses of the 5′-monophosphate, -diphosphate, and -triphosphate of 6-methylcytidine was reported by Kapuler et al. (152, 167). Phosphorylation of unprotected 6-methyluridine by phosphorus oxychloride in triethyl phosphate led to the 5′-phosphate in moderate yield (283). The abnormal behavior of 6-methylpyrimidine nucleotides in enzymic reactions does not contradict the assumption that rotation about the glycosidic bonds in these nucleotides is severely restricted.

Phosphorylation at the 5′-position of suitably blocked orotidine, a metabolite of the biosynthesis of pyrimidine nucleotides, was achieved by a conventional procedure (284): reaction of the unprotected orotidine with phosphorus oxychloride in triethyl phosphate to give the corresponding 5′-phosphate in unsatisfactory yield (285). Almost exclusive formation of 2′(3′)-phosphate was obtained on phosphorylation with pyrophosphoryl-chloride in acetonitrile (285).

An elegant synthesis of orotidine 5′-phosphate was developed by Ueda et al. (286). As outlined in Scheme 2.10, 5-bromouridine 5′-phosphate

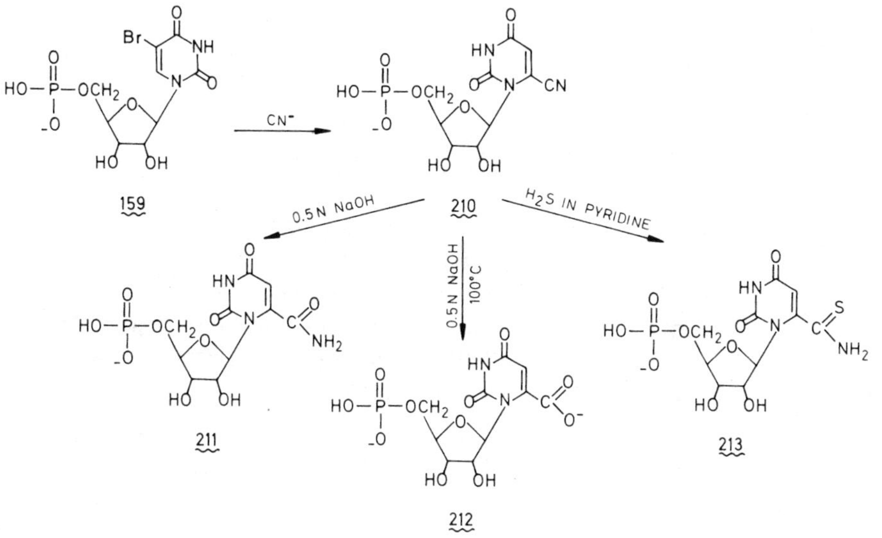

Scheme 2.10

(**159**) reacts with cyanide anions to form 6-cyanouridine 5′-phosphate (**210**). Alkaline hydrolysis of **210** at room temperature yielded 6-carboxamidouridine 5′-phosphate (**211**) and, under more rigorous conditions, orotidine 5′-phosphate (**212**). Reaction of **210** with hydrogen sulfide gave 6-thiocarboxamidouridine 5′-phosphate (**213**). A similar reaction of 5-bromouridine 2′,(3′)-phosphate with cyanide ions led to 6-cyanouridine 2′,(3′)-phosphate, which could be cyclized by means of ethyl chloroformate to the corresponding 2′,3′-cyclic phosphate.

2.3.5 Miscellaneous

Kochetkov et al. (113) synthesized a series of analogs of uridine 5′-(α-D-glucopyranosyl pyrophosphate) (**216**) by reaction of the corresponding pyrimidine nucleotide 5′-morpholidate with glucose-1-phosphate. For the sake of completion, analogs of the same kind synthesized by others have been included in Table 2.4. Catalytic hydrogenation of the 5,6-double bond in uridine 5′-phosphate (287), uridine 5′-diphosphate (288), and uridine 5′-triphosphate (288) occurs in the presence of rhodium on alumina. 5,6-Dihydrouridine 5′-phosphate (**214**) is a minor constituent of transfer RNA (27), whose function is still unclear. The chemical property of the 5,6-dihydrouracil ring to open in base and to reclose at neutral or acidic pH is noteworthy (289). Further reduction of 5,6-dihydrouracil derivatives with sodium borohydride leads to tetrahydrouracil compounds (290). The preparation of 2′-deoxy-tetrahydrouridine 5′-phosphate (**215**) was described by Maley et al. (290). Kochetkov et al. (295) examined the rates of catalytic hydrogenation and hydroxylaminolysis of uridine 5′-

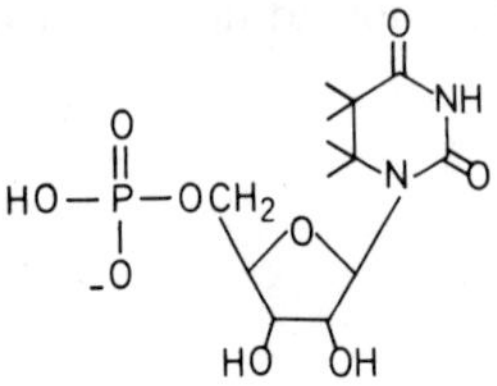

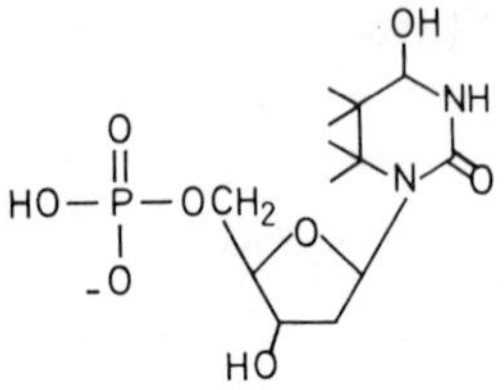

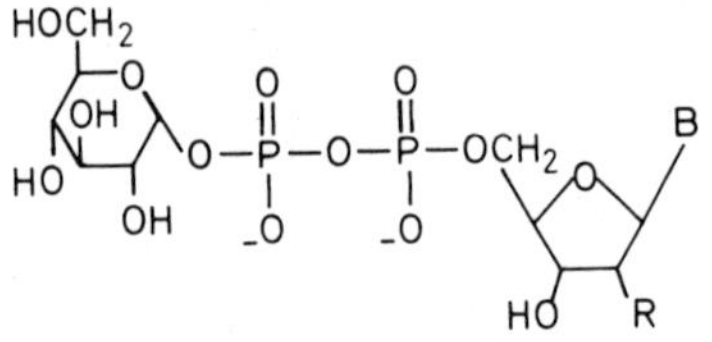

Table 2.4 Analogs of Uridine 5′-(α-D-Glucopyranosyl Pyrophosphate) (216)

B	R	Reference
N^3-Methyluracil	OH	113
Cytosine	OH	113
Isocytosine	OH	113
Uracil	H	113, 293
2-Thiouracil	OH	113
4-Thiouracil	OH	113
5-Hydroxyuracil	OH	291
5,6-Dihydrouracil	OH	291, 292
Thymine	OH	293
5-Fluorouracil	OH	294

phosphate or uridine 5′-(α-D-glucopyranosyl pyrophosphate) respectively. These authors observed that the latter always reacted more slowly and attributed this to a hydrogen bonded structure (**217**) in solution.

217

The biological properties of the above analogs of uridine 5′-(α-D-glucopyranosyl pyrophosphate) have been investigated (291, 294, 295).

2.4 RING ANALOGS OF PURINE NUCLEOTIDES

The azapurines and deazapurines comprise a class of nucleic-acid base analogs, in which the ring skeleton consists of a six-membered nitrogen heterocycle fused to a five-membered nitrogen heterocycle. Only those heterocyclic systems are considered to fall into this category, in which one

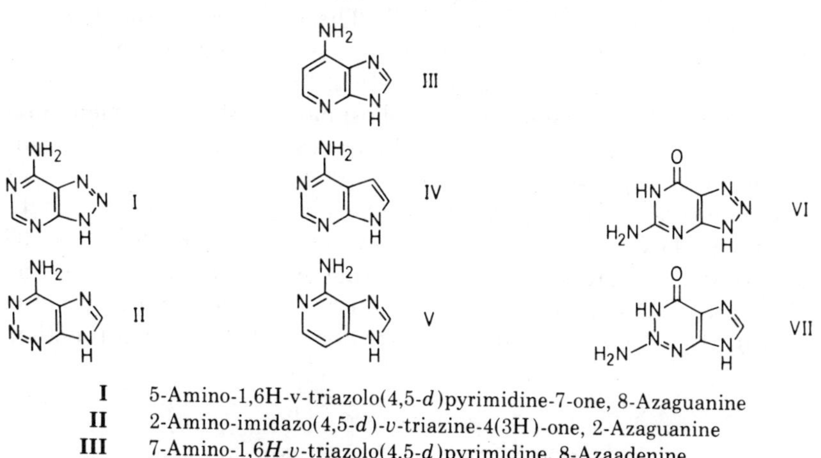

I	5-Amino-1,6H-v-triazolo(4,5-*d*)pyrimidine-7-one, 8-Azaguanine
II	2-Amino-imidazo(4,5-*d*)-*υ*-triazine-4(3H)-one, 2-Azaguanine
III	7-Amino-1,6*H*-*υ*-triazolo(4,5-*d*)pyrimidine, 8-Azaadenine
IV	4-Aminoimidazo(4,5-*d*)-*υ*-triazine, 2-Azaadenine
V	7-Amino-7*H*-pyrrolo(2,3-*d*)pyrimidine, 1-Deazaadenine
VI	4-Amino-7*H*-pyrrolo(2,3-*d*)pyrimidine, 7-Deazaadenine
VII	4 Aminoimidazo(4,5-*c*)pyridine, 3-Deazaadenine

nitrogen atom is either added to or omitted from the parent purine heterocycle. The biochemical literature prefers the descriptive names aza- or deazapurine respectively because they are much less complex and moreover easily reveal the structural relationship between analog and parent purine base. The structural formulas together with the correct chemical names of biochemically important aza- and deazapurine systems are given below (I–VII).

2.4.1 Azapurine Nucleotides

8-Azaguanine was first synthesized by Roblin et al. (296). These authors also observed inhibition of growth of *E. coli* by this analog which was prevented by guanine (296). 8-Azaguanine was the first purine analog shown to exhibit marked carcinostatic properties (297), although it did not prove useful in treatment of human cancer (298).

8-Azaguanine serves as a substrate for hypoxanthine-guanine phosphoribosyltransferase. This enzymic reaction was exploited by Way and Parks for the synthesis of 8-azaguanosine 5′-phosphate (299). 8-Azainosine 5′-phosphate was prepared in a similar fashion (300). The chemical synthesis of 8-azaguanosine 5′-phosphate (5-amino-3-β-D-ribofuranosyl-3H-*υ*-triazolo(4,5-d)pyrimidine-7(6H)-on 5′-phosphate) (**218**) was achieved by phosphorylation of 2′,3′-*O*-isopropylidene-8-azaguanosine with 2-cyanoethylphosphate in the presence of carbodiimide (301). Phosphorylation of 5′-*O*-trityl-8-azaguanosine by the same procedure led to 8-azaguanosine 2′-,(3′)-phosphate after removal of the blocking groups. Cyclization of the latter by dicyclohexylcarbodiimide yielded the corresponding 2′,3′-cyclic phosphate (302). The latter was also prepared by oxidation of the respective 2′,(3′)-phosphites as described by Holy and Smrt (303). Reaction of the morpholidate of **218** with inorganic pyrophosphate afforded 8-azaguanosine 5′-triphosphate (304). Treatment of the latter compound with myosin ATPase resulted in formation of 8-azaguanosine 5′-diphosphate (305).

Azaguanosine 5′-triphosphate can replace guanosine 5′-triphosphate as substrate for DNA-dependent RNA polymerase from *E. coli* (304, 306). This observation agrees with findings that 8-azaguanylic acid is incorporated into mRNA and transfer RNA in bacterial and mammalian cells (307–310). 2-Deoxy-8-azaguanosine 5′-triphosphate does not function as

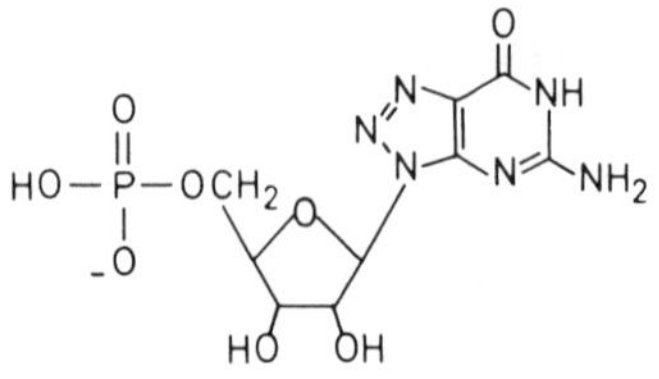

218

substrate for DNA polymerase I from *E. coli* (311). It is presently assumed that incorporation of 8-azaguanylic acid residues into mRNA either interferes with completion of mRNA synthesis or impairs the function of mRNA in protein synthesis (304, 312). Experimental attempts to demonstrate the inability of 8-azaguanine bases in synthetic mRNA's to participate in codon-anticodon interaction have failed (313, 314). The question arises as to what the structural features of 8-guanylic acid residues in RNA might be. Strong evidence for the syn conformation of 8-azaguanosine 5′-phosphate in RNA came from the observation that poly 8-azaguanylic acid was cleaved by pancreatic ribonuclease A (315, 316). The conformation in solution of 8-azaguanosine 5′-phosphate and 8-azaadenosine 5′-phosphate was investigated by ^{1}H- and ^{31}P-nmr (317). Lee et al.

SYN ⇌ ANTI (1)

2′-ENDO ⇌ 3′-ENDO (2)

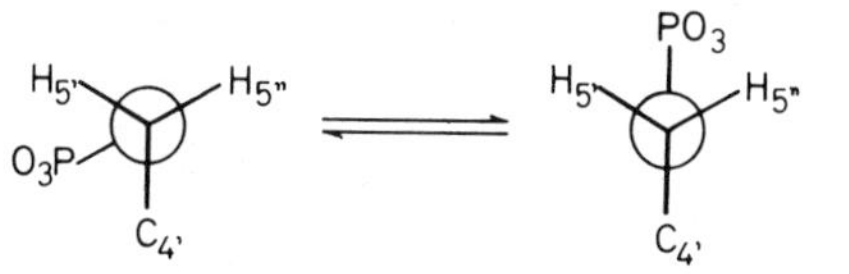

(3)

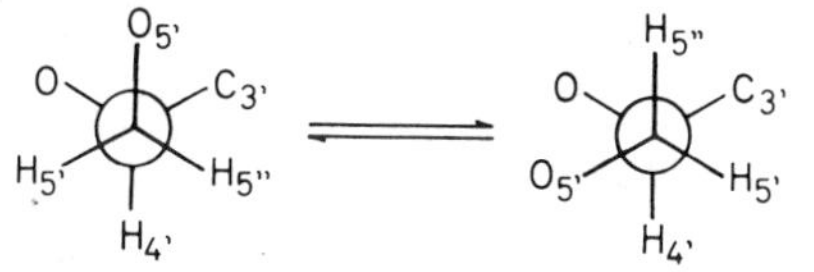

(4)

(317) observed that 8-azapurine nucleotides possess flexible structures in solution with four conformational equilibria involved (Eqs. 1–4). The equilibrium (3) represents the conformation about the C(5′)—O(5′) bond and equilibrium (4) the conformation about the C(4′)—C(5′) bond. The structures in solution are characterized by clear preferences for syn, 3′-endo and gauche′, gauche′ conformers. These differences in conformations when compared to the parent nucleotides guanosine 5′-phosphate and adenosine 5′-phosphate are thought to originate from electrostatic repulsions between N(8) and the 5′-phosphate group. Based on these observations, the above authors propose that RNA containing 8-azapurine nucleotides will assume an abnormal backbone conformation, which in turn might affect the biological function of this RNA. In this context an additional abnormal physical property of 8-azaguanine may also be involved. The pK_a of the acidic NH proton of the analog is 6.5, that is 3.3 pK-units below the pK_a value for the acidic proton of guanine (298).

When 5-amino-4-carbamyl-1 (β-D-ribofuranosyl)imidazole 5′-phosphate (**219**) is treated with sodium nitrite, cylization of **219** to 2-azainosine 5′-phosphate (**220**) occurs (318). Conversion of **220** into the respective thiono derivative and further transformation of the latter by known procedures to 2-azaadenosine 5′-phosphate should be feasible.

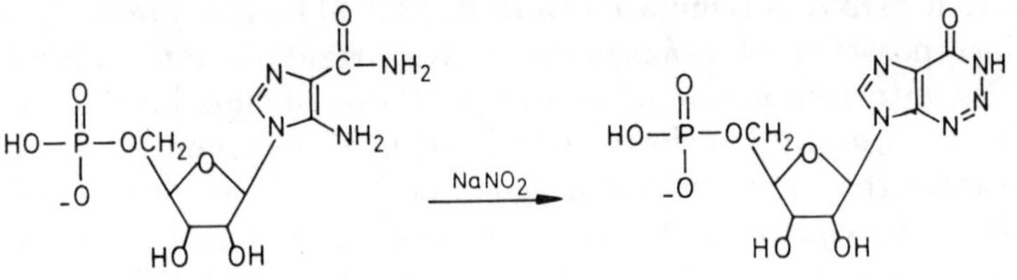

A promising synthetic approach has made 2-azaadenosine 3′,5′-cyclic phosphate (**224**) readily available (319). Alkaline hydrolysis of $1,N^6$-ethenoadenosine 3′,5′-cyclic phosphate leads to the ring-opened intermediate **222** which cyclizes to $1,N^6$-etheno-2-azaadenosine 3′,5′-cyclic phosphate (**223**) on treatment with sodium nitrite. Removal of the $1,N^6$-etheno group by means of *N*-bromosuccinimide furnishes 2-azaadenosine 3′,5′-cyclic phosphate (**224**). If this reaction sequence cannot be applied to adenosine-5′-phosphate, the possibility remains of enzymatically hydrolyzing **224** to 2-azaadenosine-5′-phosphate by phosphodiesterase. The synthesis of $1,N^6$-etheno-2-azaadenosine 5′-mono- and 5′-diphosphate was reported by Tsou and Yip (320).

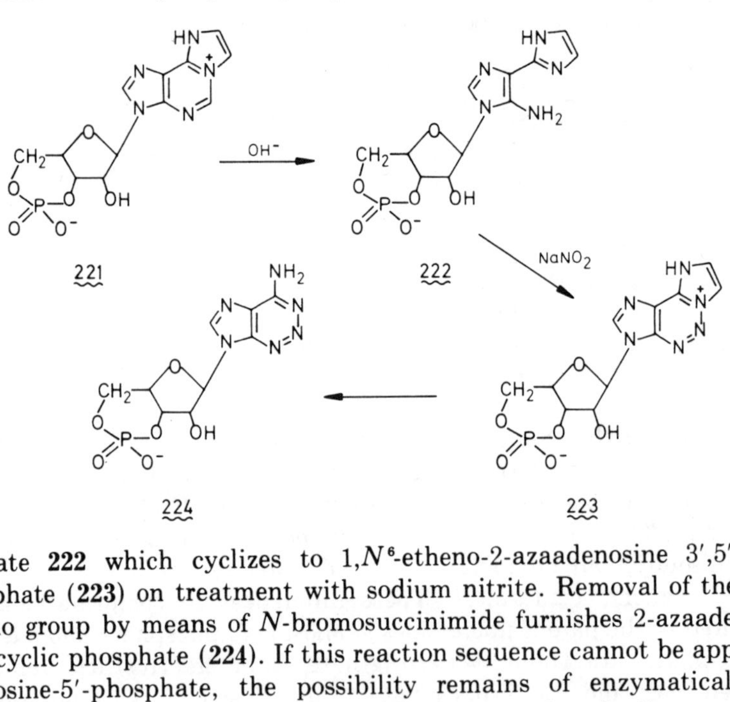

2.4.2 Deazapurine and Pyrazolo(4,5-d)pyrimidine Nucleotides

7-Deazaadenosine (tubercidine) (**225**) is an antibiotic produced by *Streptomyces tubercidus* (321). This antibiotic inhibits the growth of various microorganisms and is cytotoxic to mammalian cells. Phosphorylation of

7-deazaadenosine by P^1-diphenyl,P^2-morpholinopyrophosphorylchloride and subsequent reaction with inorganic pyrophosphate led to 7-deaza-adenosine 5′-triphosphate (**226**) (322). The nucleotide morpholidate can be anticipated as an intermediate in this reaction. The glycosidic bond of 7-deazaadenine nucleotides is resistant to both acid and enzymatic hydrolysis. Because of these properties, the conversion of 7-deazaadenosine to the corresponding nucleotides within cells is quite efficient (321). 7-Deazaadenosine was converted to 7-deazainosine by chemical deamination and the latter phosphorylated to 5′-mono- and 5′-diphosphate employing conventional procedures (323). Ikehara et al. (324) synthesized the isomeric mixture of 7-deazaadenosine 2′,(3′)-phosphates and isolated the 3′-phosphate in low yield by means of anion exchange chromatography.

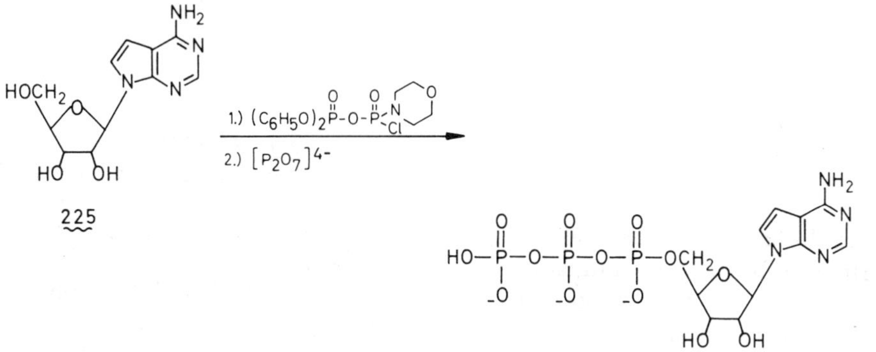

Two derivatives of 7-deazaadenosine, carrying either a cyano group (4-amino-5-cyanopyrrolo(2,3-d)pyrimidine)-7-β-D-ribofuranose (toyocomycin) or a carboxamido moiety (4-amino-5-carboxamidopyrrolo(2,3-d)pyrimidine)-7-β-D-ribofuranose (sangivamycin), are also produced as antibiotics by *Streptomyces*. The 5′-phosphates of toyocomycin (**227**) and sangivamycin (**228**) were obtained by phosphorylation of the unprotected nucleosides in triethyl phosphate with phosphorus oxychloride (325). The synthesis of 5′-mono, -di, and triphosphate of sangivamycin by routine procedures has been reported by Suhadolnik et al. (326). The 5′-mono-

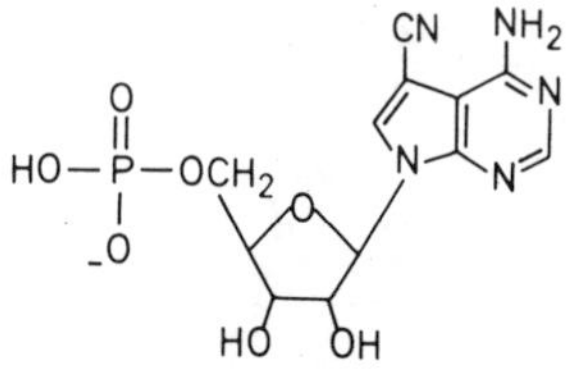

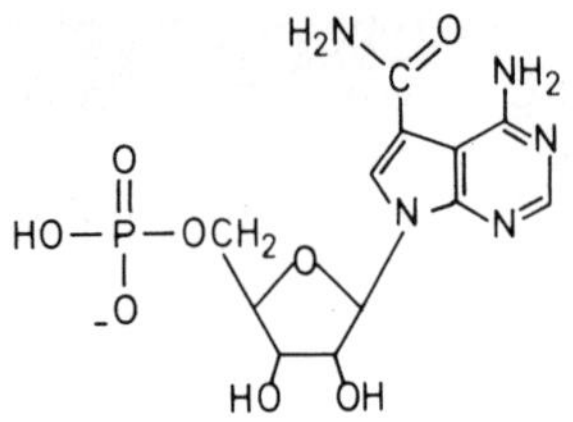

228

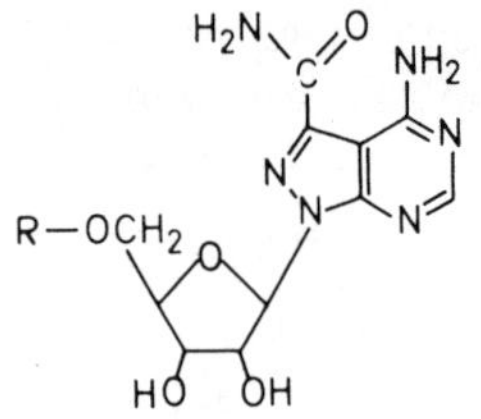

229, 230

229, R = P(O)(OH)O$^-$
230, R = P(O)O–O–P(O)(OH)O^{2-}

and diphosphate (**229, 230**) of (4-amino-1-β-D-ribofuranosyl) pyrazolo(3,4-d)pyrimidine-3-carboxamide, which might be regarded as a 6-aza analog of sangivamycin, were prepared by Hecht et al. (327). The 5′-phosphates of 1-deazaadenosine and 3-deazaadenosine were obtained by phosphorylation of the respective protected nucleosides with pyrophosphorylchloride (328). A combination of enzymic and chemical procedures led to the synthesis of 5′-diphosphates of 1-deazaadenosine, 3-deazaadenosine and 7-deazaadenosine (329).

Cook et al., discovered that 5-cyanomethyl-1-β-D-ribofuranosylimidazole-4-carboxamide (**231**) cyclizes to 3-deazaguanosine in the presence of aqueous sodium carbonate (330). This ring-closure reaction works equally

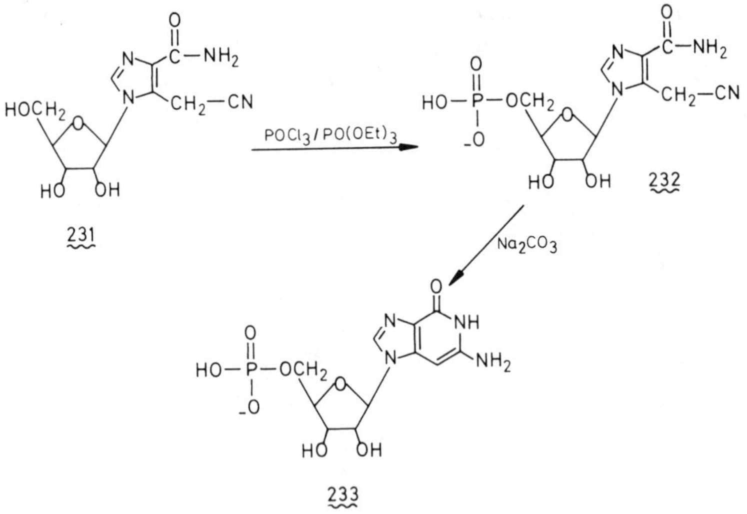

$POCl_3/PO(OEt)_3$

231 232

Na_2CO_3

233

well with the 5′-phosphate of **231** thus yielding 3-deazaguanosine 5′-phosphate (**233**).

Imidazolo(4,5-b)pyrimidine-3-β-D-ribofuranosyl-5′-phosphate (1-deazapurine-3-β-D-ribofuranosyl 5′-phosphate) was prepared and reacted with nicotinamide mononucleotide in the presence of dicyclohexyl carbodiimide to give nicotinamide-1-deazapurinedinucleotide (331).

The antibiotic 7-amino-3-(β-D-ribofuranosyl)pyrazolo(4,3-d)pyrimidine (formycin), an analog of adenosine, is produced by *Actinomyces*. Formycin inhibits the growth of microorganisms, the propagation of viruses and is cytotoxic to mammalian cells (332). The 5′-mono, -di, and triphosphate of formycin (**234–236**) have been obtained by a variety of phosphorylation

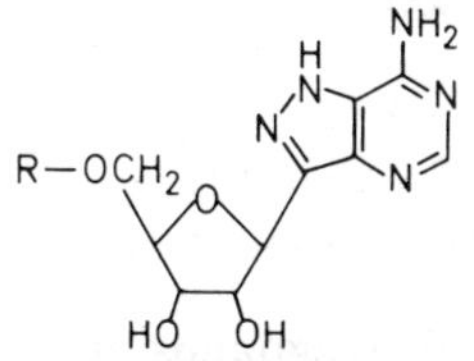

234, 235, 236

234, R = $P(O)O(OH)^-$

235, R = $P(O)O{-}O{-}P(O)(OH)O^{2-}$

236, R = $P(O)O{-}O{-}P(O)O{-}O{-}P(O)(OH)O^{3-}$

procedures (333–336). X-Ray structural analysis of formycin revealed a conformation about the glycosidic bond intermediate between syn and anti (337). Although there is no detailed information concerning its structure in solution, one might assume that this molecule would prefer the syn conformation. There is strong evidence from both biochemical and spectral studies that formycin residues occur in the syn conformation in polynucleotides (316, 334). From biochemical studies it appears that phosphate derivatives of formycin resemble adenine nucleotides in many aspects (304, 334, 336, 338, 339). Formycin, together with pseudouridine and showdomycin, belongs to the rare class of *C*-glycosides. Furthermore, it exhibits fluorescence emission of moderate quantum yield at ambient temperature and pH (334, 336).

Pyrrolo(4,3-d)pyrimidine-3-β-D-ribofuranoside (7-deazanebularine) is an analog of the nucleoside antibiotic purine-9-(β-D-ribofuranoside) (nebularine), produced by *Streptomyces*. 7-Deazanebularine is cytotoxic to mammalian cells and is incorporated into both cellular DNA and RNA (340). The 5′-mono, -di-, and -triphosphate (**237**, **238**, **239**) of 7-deazanebularine were synthesized by known procedures (341).

7-Deazanebularine 5′-triphosphate (**239**) proved to be an ambiguitive substrate for DNA-dependent RNA polymerase from *E. coli*, since it replaces both ATP and GTP with equal efficiency (341). In contrast, 7-deazanebularine when incorporated into synthetic mRNA substitutes only for adenine in all codon positions and thus shows no ambiguity in translation (342).

The synthesis of 2-amino-8-(β-D-ribofuranosyl)-imidazo(1,2-a)-s-triazin-4-one (5-aza-7-deazaguanosine), which belongs to a new class of purine analogs containing a bridgehead nitrogen atom, was first achieved

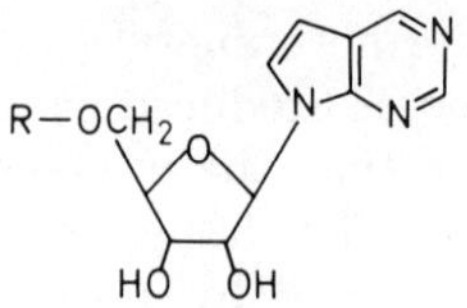

237, 238, 239

237, R = P(O)(OH)O^-

238, R = P(O)O–O–P(O)(OH)O^{2-}

239, R = P(O)O–O–P(O)O–O–P(O)(OH)O^{3-}

by Kim et al. (343). 5-Aza-7-deazaguanosine 5′-phosphate (**240**) was obtained by phosphorylation of the unprotected nucleoside in triethyl phosphate with phosphorus oxychloride.

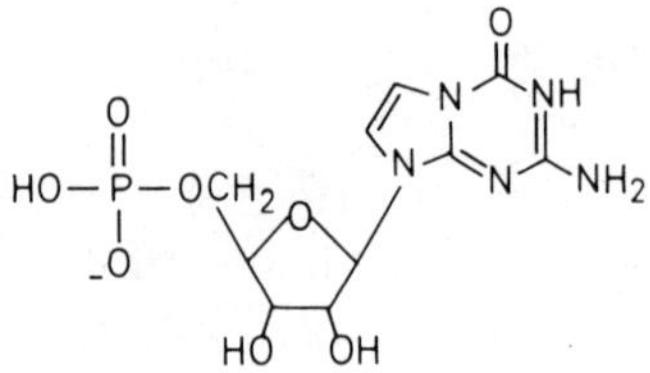

240

2.4.3 Miscellaneous

Adenine nucleotide analogs of a very original kind were designed by Leonard and coworkers. They accomplished the syntheses of 8-aminoimidazo(4,5-g)quinazoline (**241**), 9-aminoimidazo(4,5-f)quinazoline (**242**), and 6-aminoimidazo(4,5-f)quinazoline (**243**), in which the six-membered pyrimidine and the imidazole ring of adenine are separated by the formal insertion of a benzene ring (344, 345). The authors suggested the following descriptive names for the "stretched out" adenine analogs: *lin*-benzoadenine (**241**), *prox*-benzoadenine (**242**), and *dist*-benzoadenine

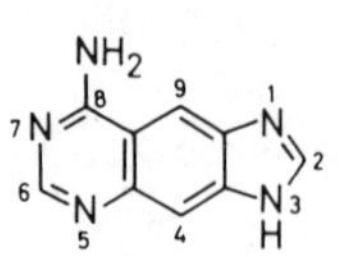

241 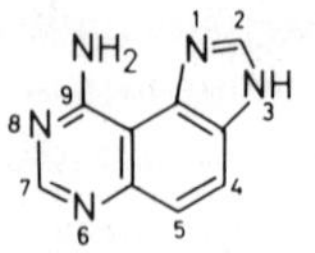 **242** 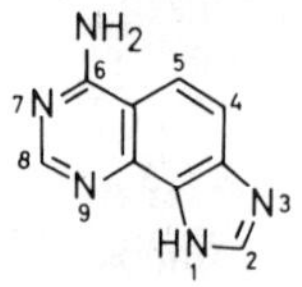 **243**

(**243**). *Lin* stands for linear, indicating the linear relationship of the three-ring systems in **241**; *prox* from proximal, and *dist* from distal, refer to the spatial relationship between the NH_2-group in **242** or **243** respectively and the imidazole ring.

In *lin*-benzoadenine, the pyrimidine and imidazole ring are laterally displaced by 2.4 Å. Phosphorylation of unprotected *lin*-benzoadenosine (**244**) to the corresponding 5′-phosphate (**245**) was carried out with pyrophosphoryl chloride (346). The phosphomorpholidate of **245** served as an intermediate in the preparation of the 5′-di and triphosphate of *lin*-benzoadenosine (346). **245** was hydrolyzed to the nucleoside by 5′-nucleotidase. The 5′-di and triphosphate of *lin*-benzoadenosine proved to be substrates for various kinases, although with less favorable kinetic parameters compared to the corresponding adenine nucleotides (346). *Lin*-benzoadenosine 3′,5′-cyclic phosphate (**247**) was synthesized by phosphorylation of **244** with trichloromethyl phosphonic acid dichloride and subsequent cyclization of *lin*-benzoadenosine 5′-trichloromethylphosphonate (**246**) by anionic displacement (346). **247** was slowly hydrolyzed to *lin*-benzoadenosine 5′-phosphate by phosphodiesterase from beef heart (346).

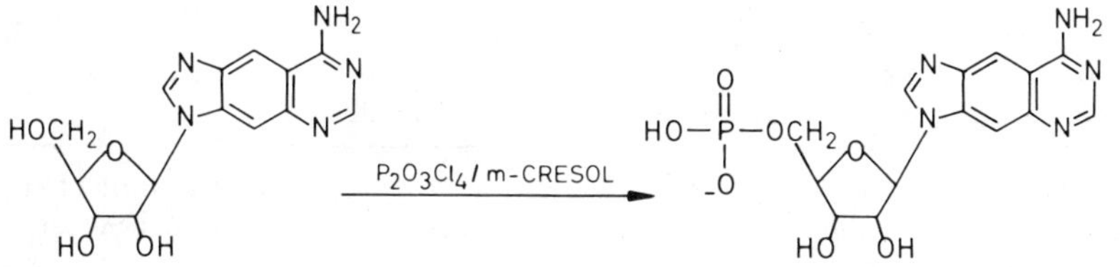

The pK_a for *lin*-benzoadenosine was determined to be 5.6, considerably higher than that of adenosine, the first protonation occurring at the quinazoline ring. The intense fluorescence emission of this compound makes the respective derivatives potentially useful adenine nucleotide

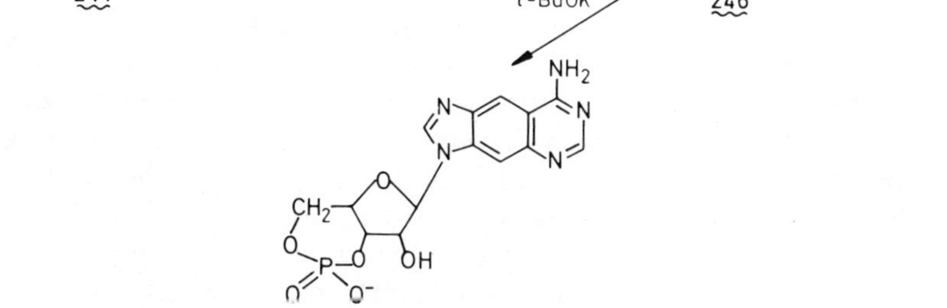

Table 2.5 Fluorescence Data of *lin*-Benzoadenosine

Solvent	Excitation (nm, uncorrected)	Emission	τ (ns)[a]	Q[b]
Water	320(sh), 332, 348	358, 372, 385	3.7	0.44
Ethanol	322(sh), 334, 350	358, 374, 390	3.7	0.44
Dioxan	322(sh), 336, 352	360, 376, 392	3.7	0.44

Source. Reference 347. Reprinted with permission from N. J. Leonard et al., *J. Am. Chem. Soc.* **98,** 3990 (1976). Copyright by the American Chemical Society.

[a] Life time of the excited state.

[b] Quantum yield.

analogs for spectroscopic studies (Table 2.5). Stacking interactions of *lin*-benzoadenine moieties with aromatic π-electron systems lead to a drastic reduction in fluorescence quantum yield (346).

5-Amino-1-(β-D-ribofuranosyl)imidazole-4-carboxylic acid 5′-phosphate (**249**) occupies a central position in the biosynthesis of purine nucleotides.

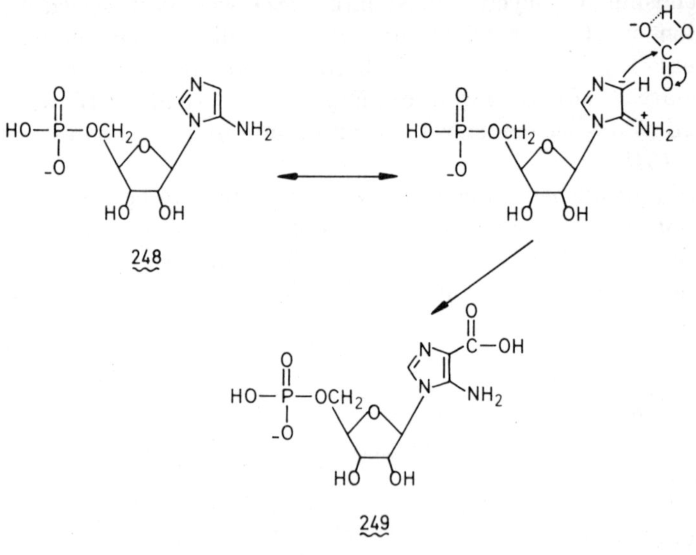

248

249

In the biosynthetic pathway, this compound is formed by carbonylation of 5-amino-1-(β-D-ribofuranosyl)imidazole 5′-phosphate (**248**). Cusack et al., observed that this reaction readily occurs even in the absence of an enzyme when **248** is heated in an aqueous solution of potassium hydrogen carbonate (348). The amide 1-β-D-ribofuranosyl-1,2,4-triazole-3-carboxamide (ribavirin) (**250**) has been synthesized by Robins et al. (349), who reported that it displays remarkable antiviral activity. Compound **250** undergoes facilitated transport into cells and is phosphorylated to the 5′-phosphate by deoxyadenosine or adenosine kinase respectively (350). The corresponding 5′-di and triphosphate have been identified in tissues of animals treated with **250** (351). 1-(β-D-Ribofuranosyl)-1,2,4-triazole-3-carboxamide 5′-phosphate (**251**) can be regarded as a structural analog of **248**. The synthesis of phosphate derivatives of **250** by known procedures has been described (352). The morpholidate of **251** served as precursor in the preparation of **253** and **254**. Cyclization of **251** in the presence of

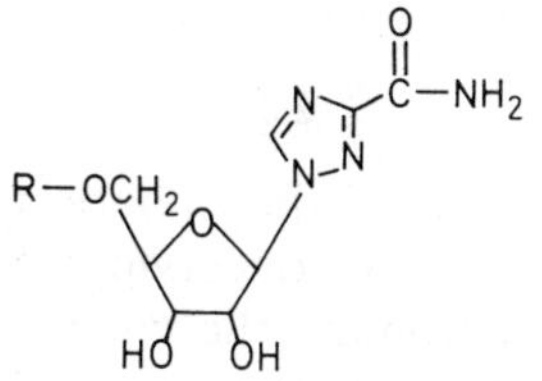

250, **251**, **252**, **253**, **254**

250, R = H
251, R = P(O)(OH)O$^-$
252, R = P(O)(NH$_2$)O$^-$
253, R = P(O)O–O–P(O)O(OH)$^{2-}$
254, R = P(O)O–O–P(O)O–O–P(O)(OH)O^{3-}

dicyclohexyl carbodiimide furnished 1-(β-D-ribofuranosyl)-1,2,4,-triazole 3′,5′-cyclic phosphate (352). Reaction of **250** with tri-*n*-butylammonium pyrophosphate at elevated temperatures gave 1-(β-D-ribofuranosyl)-1,2,4-triazole-carboxamide 2′,3′-cyclic phosphate, which was chemically hydrolyzed to the isomeric mixture of 2′,(3′)-monophosphates (352).

The structure in solution of **251** was determined by nmr spectroscopy (353). The preferred conformations are 2′-endo for the ribose puckering, anti for the conformation about the glycosidic bond and gauche, gauche for the conformation about the bonds C(4′)—C(5′), C(5′)—O(5′) (353). A set of abnormal conformational parameters is obviously not the explanation for the antiviral action of **250**. Biochemical studies provide evidence for the specific inhibition of viral RNA-polymerase by 1-(β-D-ribofuranosyl)-1,2,4-triazole-3-carboxamide 5′-triphosphate (354).

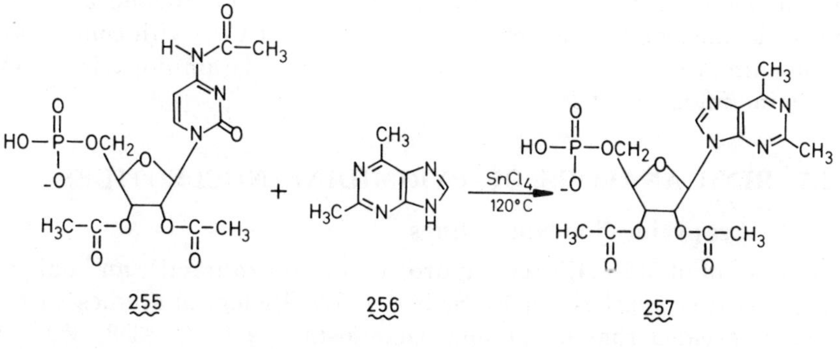

255 + **256** —SnCl$_4$, 120°C→ **257**

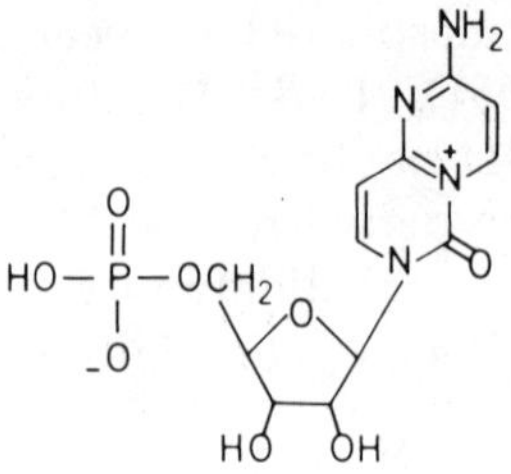

258

7-β-D-Ribofuranosyltheophyllin 5′-phosphate (**257**) was prepared by transglycosidation between *N*,2′,3′-triacetylcytidine 5′-phosphate (**255**) and theophyllin (**256**) (355). Although the yield of **257** was only 14% and additional complication results from the concomitant formation of small amounts of the α-anomer, this method should be useful with respect to its general applicability in the synthesis of purine nucleotide analogs.

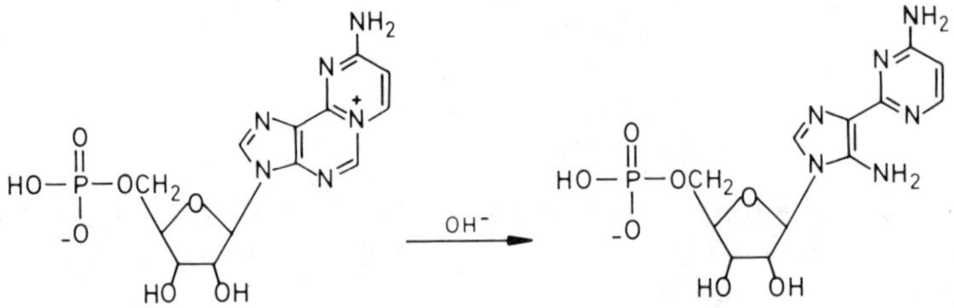

259 $\xrightarrow{OH^-}$ 260

Furukawa et al., discovered that cyanoethylene reacts with cytosine and adenine nucleotides in the presence of mercuric chloride to form a new heterocyclic ring system, which comprises N(3) and N(4) of the cytosine or N(1) and N(6) of the adenine ring respectively (356). Reaction of cytidine 5′-phosphate with cyanoethylene afforded 2-amino-7-(5-phospho-β-D-ribofuranosyl)-pyrimidine(1,2-c)pyrimidin 5-ium-6(7H)-one (**258**). **258** carries a formal positive charge at neutral pH as judged from electrophoresis. Adenosine 5′-phosphate reacts analogously with cyanoethylene to give 9-amino-3(5-phospho-β-D-ribofuranosyl)pyrimido(2,1-i)purin-5-ium (**259**). The compounds **258** and **259** are highly fluorescent. Compound **259** is susceptible to nucleophilic attack by hydroxyl ions at C(5) with concomitant ring opening to 1-(5-phospho-β-D-ribofuranosyl)-4-(aminopyrimidine-2-yl)5-aminoimidazole (**260**).

2.5 RING ANALOGS OF PYRIMIDINE NUCLEOTIDES

2.5.1 Azapyrimidine Nucleotides

1,2,4-Triazine-3,5(2,3,4,5-tetrahydro)-dione (6-azauracil), an analog of uracil, was first synthesized by Seibert (357). Biological studies with 6-azauracil revealed that it exhibits bacteriostatic activity (358, 359) and inhibits the growth of tumors (360–362). The conversion of 6-azauracil to 6-azauridine by microorganisms, most efficiently by *E. coli*, was employed for large scale preparation (363, 364). 6-Azauridine was shown to be phosphorylated to the respective 5′-mono, -di, and -triphosphate in *E. coli* (365, 366). In mammalian cells, 6-azauridine appears to be phosphorylated only as far as the 5′-phosphate (367).

Phosphorylation of 6-azauridine by a mixture of phosphoric acid and phosphorus pentoxide was reported by Handschuhmacher (368). 6-Azauridine 5′-phosphate (**261**) could be isolated from the reaction mixture. A more satisfactory synthesis of **261** as well as of the 5′-triphosphate was described by Holy et al. (369). Reaction of 6-azauridine (**262**) with triethyl phosphite gave 6-azauridine 2′,(3′)-phosphite (**263**)-oxidation, followed by

1.) $P(OEt)_3$
2.) H_2O

$Cl_3C-\overset{O}{\overset{\|}{C}}-CCl_3$

RNase A

261 262 263 264 265

cyclization of the latter with hexachloroacetone yielded 6-azauridine 2′,3′-cyclic phosphate (370). **264** was hydrolyzed to 6-azauridine 3′-phosphate (**265**) by ribonuclease A. 5-Methyl-6-azauridine 3′-phosphate was prepared in a similar way (277). Because of side reactions, this method of phosphorylation could not be applied to 4-thio-6-azauridine and 2-thio-6-azauridine respectively. Instead, the respective 2′,3′-cyclic phosphates **266** and **267** were obtained by phosphorylation with triethyl ammonium phosphate and trichloroacetonitrile (277).

6-Azauridine 5′-triphosphate does not serve as a substrate for DNA-dependent RNA-polymerase from *E. coli* (367). Experiments with synthetic trinucleoside diphosphates, which contain 6-azauridylic acid, showed that 6-azauracil cannot replace uracil in codon-anticodon interac-

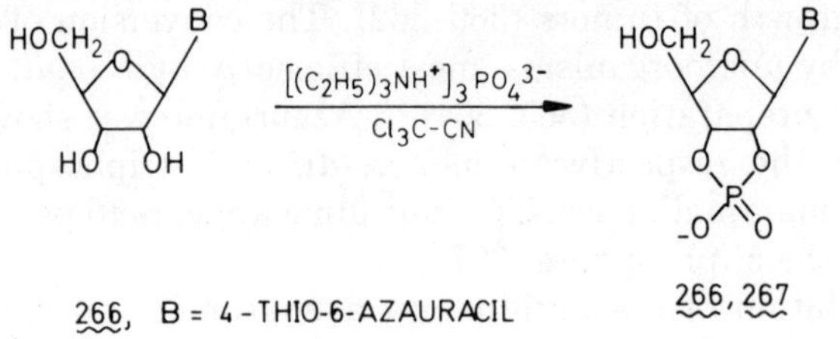

$[(C_2H_5)_3NH^+]_3 PO_4^{3-}$, $Cl_3C\text{-}CN$

266, 267

266, B = 4-THIO-6-AZAURACIL

267, B = 2-THIO-6-AZAURACIL

tions (367). The biological activity of 6-azauridine seems to result exclusively from the strong and reversible inhibition of orotidine 5′-phosphate decarboxylase by 6-azauridine 5′-phosphate (367). X-Ray diffraction (371) and nmr studies (372) disclosed that 6-azauridine 5′-phosphate and orotidine 5′-phosphate possess similar, abnormal gauche, trans conforma-

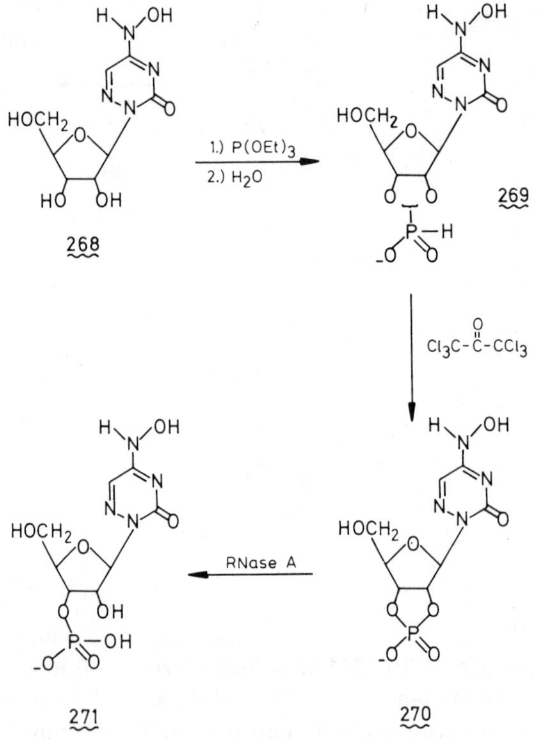

1.) $P(OEt)_3$ 2.) H_2O

268 269

$Cl_3C\text{-}CO\text{-}CCl_3$

RNase A

271 270

tions about the C(4′)—C(5′) bonds. This enables the former to bind reversibly as a substrate analog inhibitor to orotidine-5′-phosphate decarboxylase.

6-Azacytidine is phosphorylated to the 5′-phosphate within mammalian cells (373). The latter inhibits orotidine 5′-phosphate decarboxylase (373); mRNA containing 6-azacytidylic acid residues does not function in

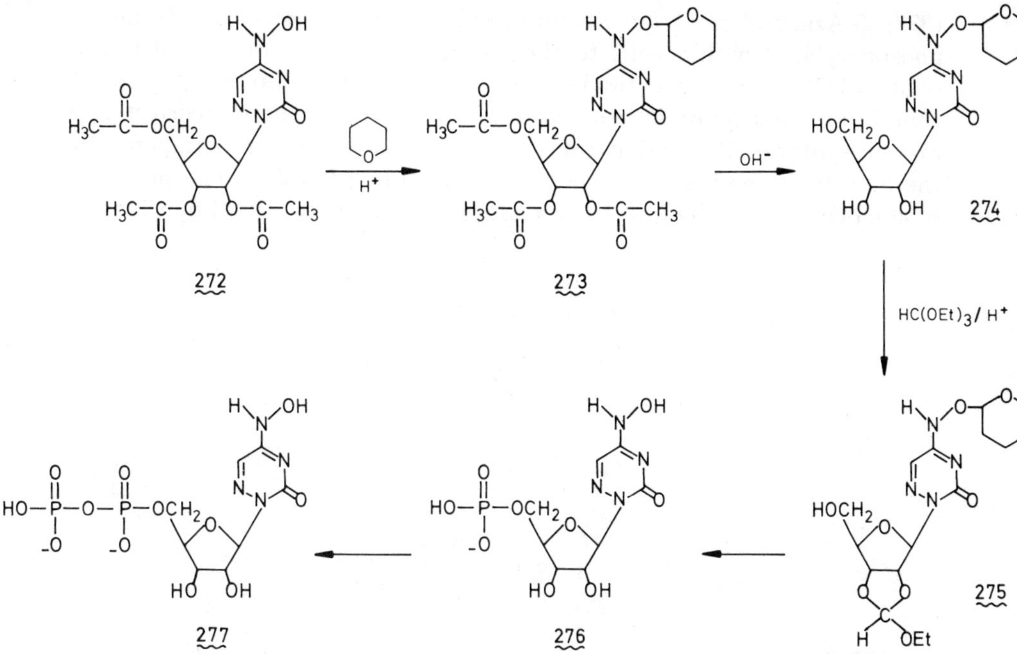

272 273 274 275 276 277

translation (367). The chemical synthesis of 6-azacytidine 5′-phosphate and 5′-diphosphate has been reported (374).

N[4]-Hydroxy-6-azacytidine 3′-phosphate (**271**) was prepared from the nucleoside **268** via the corresponding phosphite **269** and 2′,3′-cyclic phosphate **270**. The latter was cleaved by pancreatic ribonuclease A to the 3′-phosphate **271** (375). The synthesis of *N*[4]-hydroxy-6-azacytidine 5′-phosphate (**276**) proved to be cumbersome, because of the reactivity of the *N*[4]-hydroxy group (376). Phosphorylation of **275** and **276** involved standard procedures.

Chemical reactions with 6-azacytidine are complicated because of its limited stability towards alkali. In contrast *N*[4]-hydroxy-6-azacytidine proved to be much more stable against either alkali or acid (376).

The synthesis and the potent antimicrobial as well as cancerostatic activity of 5-azacytidine (**278**) were reported by Sorm et al. (377). Shortly thereafter, 5-azacytidine was discovered to be a nucleoside antibiotic

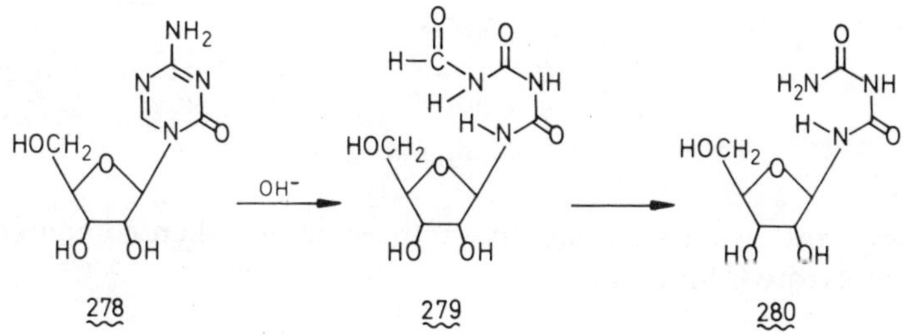

278 279 280

(378). 5-Azacytidine (1-β-D-ribofuranosyl-4-amino-s-triazine-2(1H)-one) is phosphorylated within cells to the respective 5′-mono,-di, and -triphosphate (367), and 5-azacytidylic acid is incorporated into RNA. mRNA containing 5-azacytidylic acid residues is unable to function correctly in protein synthesis (379). Experiments with 5-azacytidine are hampered by the lability of the 5-azacytosine ring at alkaline pH (380). The principal products of alkaline hydrolysis of 5-azacytidine were identified by Cihak et

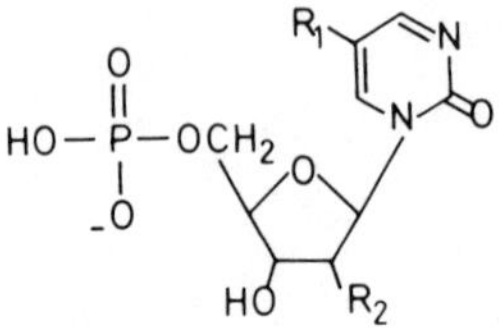

281, 282, 283, 284

281, R_1 = H R_2 = OH
282, R_1 = H R_2 = H
283, R_1 = CH_3 R_2 = OH
284, R_1 = CH_3 R_2 = H

al. (**279–280**) (380). The enzymic synthesis of 5-azacytidine 5′-phosphate was achieved by means of uridine-cytidine kinase. Conversion of the latter to the 5′-triphosphate required the action of cytidine monophosphate kinase and nucleoside diphosphate kinase (381). 5-Azacytidine 5′-triphosphate proved to be a substrate for DNA-dependent RNA polymerase from

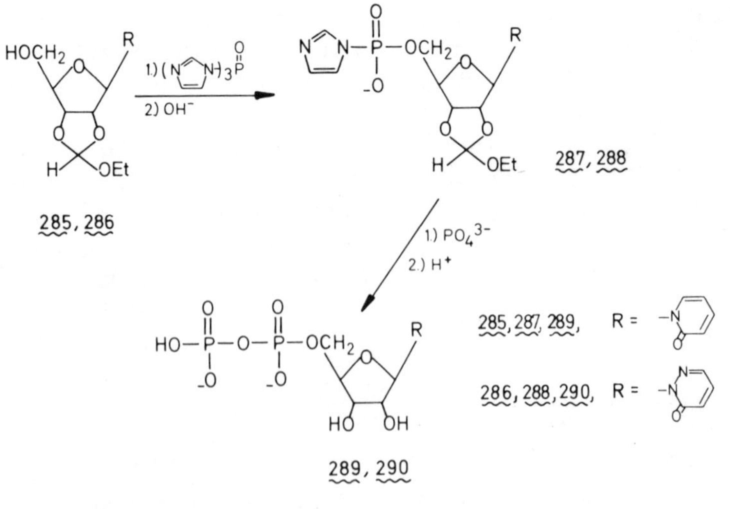

285, 286 → 287, 288 → 289, 290

285, 287, 289, R =

286, 288, 290, R =

E. coli with rather unfavorable kinetic parameters when compared to the natural substrate CTP (382).

2.5.2 Deazapyrimidine, Pyridon, and Pyridazon Nucleotides

1-(β-D-Ribofuranosyl)-2-pyrimidine selectively inhibits DNA synthesis in bacteria (383, 384). This observation stimulated interest in the chemical syntheses of related nucleosides as well as the syntheses of a series of 1-(β-D-ribofuranosyl)-2-pyrimidine 5′-phosphates (**281–284**) (385).

Pischl and Holy phosphorylated 1-(β-D-ribofuranosyl)-pyridon-(2) (**285**) and 2-(β-D-ribofuranosyl)-pyridazon-(3) (**286**), which were protected at their cis diol group, by means of tris-imidazolyl phosphinoxide (386). The resulting imidazolidates **287** and **288** were reacted with inorganic phosphate to give the 5′-diphosphates (**289**, **290**).

REFERENCES

1. B. Singer, *Prog. Nucl. Acids Res. Mol. Biol.* **15,** 219–284 (1975).
2. R. H. Hall, *The Modified Nucleosides in Nucleic Acids*, Columbia University Press, New York, 1971.
3. Y. Groner and J. Jurwitz, *Proc. Natl. Acad. Sci.* (*US*) **72,** 2930–2934 (1975).
4. R. C. Desrosiers, R. C. Friderici, and F. M. Rottman, *Biochemistry* **14,** 4367–4374 (1975).
5. Y. Furuichi, M. Morgan, A. J. Shatkin, W. Jelinek, M. Salditt-Georgieff, and J. E. Darnell, *Proc. Natl. Acad. Sci.* (*US*) **72,** 1904–1908 (1975).
6. C. M. Wei, A. Gershowitz, and B. Moss, *Biochemistry* **15,** 397–401 (1976).
7. O. K. Sharma, D. E. Hruby, and D. N. Beezley, *Biochem. Biophys. Res. Commun.* **72,** 1392–1398 (1976).
8. S. N. Neal, A. Schmidt, M. Tomaszewski, and A. Marcus, *Biochem. Biophys. Res. Commun.* **82,** 553–559 (1978).
9. J. M. Wu, C. P. Cheung, and R. J. Suhadolnik, *Biochem. Biophys. Res. Commun.* **78,** 1079–1086 (1977).
10. C. Auerbach and J. M. Robson, *Nature* **157,** 302 (1946).
11. R. M. Herriott, *J. Gen. Physiol.* **32,** 221–239 (1948).
12. P. Brookes and P. D. Lawley, *J. Chem. Soc.* **1961,** 3923–3928.
13. J. A. Haines, C. B. Reese, and Lord Todd, *J. Chem. Soc.* **1962,** 5281–5288.
14. B. E. Griffin and C. B. Reese, *Biochim. Biophys. Acta* **68,** 185–192 (1963).
15. J. A. Haines, C. B. Reese, and Lord Todd, *J. Chem. Soc.* **1964,** 1406–1412.
16. K. H. Scheit and A. Holy, *Biochim. Biophys. Acta* **149,** 344–354 (1967).
17. J. W. Jones and R. K. Robins, *J. Am. Chem. Soc.* **85,** 193–201 (1963).
18. K. H. Scheit and A. Holy, *Tetrahedron Lett.* **36,** 4303–4308 (1966).
19. P. Brookes and P. D. Lawley, *J. Chem. Soc.* **1960,** 539–545.
20. V. N. Shibaev and S. M. Spiridinova, in *Synthetic Procedures in Nucleic Acid Chemistry*, Vol. 1, pp. 461–462, W. W. Zorbach and R. S. Tipson, Eds., Wiley, New York, 1968.
21. P. D. Lawley and P. Brookes, *Biochem. J.* **92,** 190–200 (1964).
22. O. Dimroth, *Ann.* **373,** 336 (1910).

23. E. I. Budowski, E. D. Swerdlov, R. P. Shibaeva, G. S. Monastirskaja, and N. K. Kochetkov, *Mol. Biol.* (*USSR*) **2,** 329 (1968).
24. P. Brookes and P. D. Lawley, *J. Chem. Soc.* **1962,** 1348–1351.
25. R. L. C. Brimacombe and C. B. Reese, *J. Chem. Soc.* **1966,** 588–592.
26. J. T. Kusmierek and D. Shugar, *Acta Biochim. Pol.* **20,** 365–381 (1973).
27. F. Pochon and A. M. Michelson, *Biochim. Biophys. Acta* **149,** 99–106 (1967).
28. N. K. Kochetkov, E. I. Budowski, and V. N. Shibaev in *Synthetic Procedures in Nucleic Acid Chemistry* Vol. 1, pp. 497–499, W. Zorbach and R. S. Tipson, Eds., Wiley, New York, 1968.
29. N. K. Kochetkov, E. I. Budowsky, and V. N. Shibaev, *Izv. Akad. Nauk. SSSR, Otd. Khim. Nauk.* **1962,** 1035.
30. W. Szer and D. Shugar, *Acta Biochim. Pol.* **8,** 235 (1961).
31. K. H. Scheit, *Biochim. Biophys. Acta* **119,** 425–426 (1966).
32. R. Thedford, M. H. Fleysher, and R. H. Hall, *J. Med. Chem.* **8,** 486–491 (1965).
33. F. Seela and F. Cramer, *Chem. Ber.* **109,** 82–89 (1976).
34. F. Cramer, K. Randerath, and E. A. Schäfer, *Biochim. Biophys. Acta* **72,** 150–156 (1963).
35. F. Cramer and H. Seidel, *Biochim. Biophys. Acta* **72,** 157–161 (1963).
36. A. Yamazaki, I. Kumashiro, and T. Takenishi, *Chem. Pharm. Bull.* (Tokyo) **17,** 1128–1133 (1969).
37. H. Seidel, *Biochim. Biophys. Acta* **138,** 98–106 (1967).
38. H. J. Rhaese, *Biochim. Biophys. Acta* **166,** 311–326 (1968).
39. M. A. Stevens and G. B. Brown, *J. Am. Chem. Soc.* **80,** 2729–2762 (1958).
40. M. A. Stevens, H. W. Smith, and G. B. Brown, *J. Am. Chem. Soc.* **81,** 1734–1738 (1959).
41. O. M. Friedman, G. N. Mahapatra, B. Dash, and R. Stevenson, *Biochim. Biophys. Acta* **103,** 286–297 (1965).
42. L. L. Gerchman, J. Dombrowski, and D. B. Ludlum, *Biochim. Biophys. Acta* **272,** 672–675 (1972).
43. W. A. H. Grimm and N. J. Leonard, *Biochemistry* **6,** 3625–3631 (1967).
44. R. Y. Schmitz, F. Skoog, A. Vienze, G. C. Walker, L. H. Kiekegaard, and N. J. Leonard, *Phytochemistry* **14,** 1479–1484 (1975).
45. C. I. Hong, G. L. Tritsch, A. Mittelman, P. Hebborn, and G. B. Chedda, *J. Med. Chem.* **18,** 465–473 (1975).
46. A. D. Broom, M. E. Uchic, and J. T. Uchic, *Biochim. Biophys. Acta* **425,** 278–286 (1976).
47. V. Armanath and A. D. Broom, *Biochemistry* **15,** 4386–4389 (1976).
48. C. I. Hong and G. B. Chedda, *J. Med. Chem.* **16,** 956–959 (1973).
49. H. Fraenkel-Conrat, *Biochim. Biophys. Acta* **15,** 307–309 (1954).
50. A. M. Michelson and M. Grunberg-Manago, *Biochim. Biophys. Acta* **91,** 92–104 (1964).
51. D. E. Hoard, *Biochim. Biophys. Acta* **40,** 62–70 (1960).
52. S. Lewin and D. A. Humphreys, *J. Chem. Soc.* (*B*) **1967,** 562–565.
53. R. Haselkorn and P. Doty, *J. Biol. Chem.* **236,** 2738–2745 (1961).
54. K. H. Scheit, *Tetrahedron Lett.* **15,** 1031–1039 (1965).

55. N. E. Broude, E. I. Budowski, and N. K. Kochetkov, *Mol. Biol. (USSR)* **1,** 214–219 (1967).
56. R. Shapiro, B. I. Cohen, S. J. Shiuey, and H. Maurer, *Biochemistry* **8,** 238–245 (1969).
57. R. Shapiro and J. Hachmann, *Biochemistry* **5,** 2799–2807 (1966).
58. J. A. Secrist, III, J. R. Barrio, N. J. Leonard, and G. Weber, *Biochemistry* **11,** 3499–3506 (1972).
59. N. K. Kochetkov, V. N. Shibaev, ànd A. A. Kost, *Tetrahedron Lett.* **22,** 1993–1996 (1971).
60. G. R. Penzer, *Eur. J. Biochem.* **34,** 297–305 (1973).
61. A. H.-J. Wang, L. G. Dammann, J. R. Barrio, and I. C. Paul, *J. Am. Chem. Soc.* **96,** 1205–1213 (1974).
62. R. D. Spencer, G. Weber, G. L. Tolman, J. R. Barrio, and N. J. Leonard, *Eur. J. Biochem.* **45,** 425–429 (1974).
63. J. R. Barrio, G. L. Tolman, N. J. Leonard, R. D. Spencer, and G. Weber, *Proc. Natl. Acad. Sci. (US)* **70,** 941–943 (1973).
64. V. G. Neef and F. M. Huennekens, *Biochemistry* **15,** 4042–4047 (1976).
65. J. R. Barrio, J. A. Secrist, III, and N. J. Leonard, *Proc. Natl. Acad. Sci. (US)* **69,** 2039–2042 (1972).
66. B. A. Gruber and N. J. Leonard, *Proc. Natl. Acad. Sci. (US)* **72,** 3966–3969 (1975).
67. N. J. Leonard and G. L. Tolman, *Ann. N.Y. Acad. Sci.* **255,** 43–58 (1975).
68. P. D. Sattsangi, N. J. Leonard, and C. R. Frihart, *J. Org. Chem.* **42,** 3292–3296 (1977).
69. J. R. Barrio, J. A. Secrist, III, and N. J. Leonard, *Biochem. Biophys. Res. Commun.* **46,** 597–604 (1972).
70. J. R. Barrio, L. G. Dammann, L. H. Kirkegaard, R. L. Switzer, and N. J. Leonard, *J. Am. Chem. Soc.* **95,** 961–962 (1973).
71. A. H.-J. Wang, J. R. Barrio, and I. C. Paul, *J. Am. Chem. Soc.* **98,** 7401–7408 (1976).
72. J. R. Barrio, P. D. Sattsangi, B. A. Gruber, L. G. Dammann, and N. J. Leonard, *J. Am. Chem. Soc.* **98,** 7408–7414 (1976).
73. J. C. Greenfield, N. J. Leonard, and R. I. Gumport, *Biochemistry* **14,** 698–706 (1975).
74. S. D. Rose, *Biochim. Biophys. Acta* **361,** 231–235 (1974).
75. G. B. Elion, E. Burgi, and G. H. Hitchings, *J. Am. Chem. Soc.* **74,** 411–414 (1952).
76. G. B. Elion and G. H. Hitchings, *J. Am. Chem. Soc.* **77,** 1676 (1955).
77. J. H. Burchenal, M. L. Murphy, R. R. Ellison, D. A. Karnofsky, M. P. Sykes, T. C. Tan, A. L. Mermann, M. Yugeoglu, W. P. L. Myers, and I. Krakoff, *Ann. N.Y. Acad. Sci.* **60,** 359–368 (1954).
78. P. Roy-Burman, in *Recent Results in Cancer Research*, Vol. 25, Springer, Berlin, 1970.
79. J. A. Montgomery, Rational Design of Purine Nucleoside Analogs in *Handbook of Experimental Pharmacology XXXVIII/1*, pps. 76–111, O. Eichler, A. Farah, H. Hertzen, and A. D. Welch, Eds., Springer, Berlin, 1974.
80. J. J. Fox, J. Wempen, A. Hampton, and I. L. Doerr, *J. Am. Chem. Soc.* **80,** 1669–1675 (1958).

81. J. A. Montgomery and H. J. Thomas, *J. Org. Chem.* **26,** 1926–1929 (1961).
82. A. Hampton and M. H. Maguire, *J. Am. Chem. Soc.* **83,** 150–157 (1961).
83. M. Saneyoshi, *Chem. Pharm. Bull. (Tokyo)* **19,** 493–498 (1971).
84. F. Eckstein, *Tetrahedron Lett.* **13,** 1157–1160 (1967).
85. F. Perini and A. Hampton, *J. Heterocycl. Chem.* **7,** 969–971 (1970).
86. H. R. Rackwitz and K. H. Scheit, *Chem. Ber.* **107,** 2284–2294 (1974).
87. J. A. Carbon, *Biochem. Biophys. Res. Commun.* **7,** 366–369 (1962).
88. A. J. Murphy, J. A. Duke, and L. Stowring, *Arch. Biochem. Biophys.* **137,** 297–298 (1970).
89. J. L. Darlix, P. Fromageot, and E. Reich, *Biochemistry* **12,** 914–919 (1973).
90. K. H. Scheit, unpublished.
91. T. Ueda, K. Miura, and T. Kasai, *Chem. Pharm. Bull. (Tokyo)* **26,** 2122–2127 (1978).
92. U. Thewalt and C. E. Bugg, *J. Am. Chem. Soc.* **94,** 8892–8898 (1972).
93. I. L. Doerr, I. Wempen, D. A. Clarke, and J. J. Fox, *J. Org. Chem.* **26,** 3401–3409 (1961).
94. J. J. Fox, D. van Praag, I. Wempen, I. L. Doerr, L. Cheong, J. E. Knoll, M. L. Eidinoff, A. Bendich, and G. B. Brown, *J. Am. Chem. Soc.* **81,** 178–187 (1959).
95. V. N. Shibaev, M. A. Grachev, and S. M. Spirodonova, *Proc. Nucl. Acid Chem.*, **1,** 503–505 (1968).
96. M. Saneyoshi and F. Sawada, *Chem. Pharm. Bull. (Tokyo)* **17,** 181–190 (1969).
97. K. H. Scheit, *Chem. Ber.* **101,** 1141–1147 (1968).
98. M. Saneyoshi, *Chem. Pharm. Bull. (Tokyo)* **16,** 1400–1402 (1968).
99. T. Ueda, K. Miura, K. Imazawa, and K. Odajima, *Chem. Pharm. Bull. (Tokyo)* **22,** 2377–2382 (1974).
100. T. Ueda, Y. Iida, K. Ikeda, and Y. Mizuno, *Chem. Pharm. Bull. (Tokyo)* **16,** 1788–1794 (1968).
101. P. Faerber and K. H. Scheit, *Chem. Ber.* **104,** 456–460 (1971).
102. N. K. Kochetkov, E. I. Budowsky, and V. N. Shibaev, *Proc. Nucl. Acid Chem.*, **1,** 500–502 (1968).
103. H. Vorbrüggen, P. Strehlke, and G. Schulz, *Angew. Chem.* **81,** 997 (1969).
104. P. Lengyel and R. W. Chambers, *J. Am. Chem. Soc.* **82,** 752–753 (1960).
105. K. H. Scheit and P. Faerber, *Eur. J. Biochem.* **50,** 549–555 (1975).
106. W. Bähr, P. Faerber, and K. H. Scheit, *Eur. J. Biochem.* **33,** 535–544 (1973).
107. H. Hayatsu, *J. Am. Chem. Soc.* **91,** 5693–5694 (1969).
108. K. H. Scheit and P. Faerber, *J. Carbohydr. Nucleosides Nucleotides* **1,** 375–379 (1974).
109. K. H. Scheit and P. Faerber, *Eur. J. Biochem.* **24,** 385–392 (1971).
110. H. R. Rackwitz and K. H. Scheit, *Eur. J. Biochem.* **72,** 191–200 (1977).
111. M. Sprinzl, K. H. Scheit, and F. Cramer, *Eur. J. Biochem.* **34,** 306–310 (1973).
112. P. Faerber and K. H. Scheit, *Chem. Ber.* **103,** 1307–1311 (1970).
113. N. K. Kochetkov, E. I. Budowsky, V. N. Shibaev, G. I. Yelisseeva, M. A. Grachev, and V. P. Demushkin, *Tetrahedron* **19,** 1207–1218 (1963).
114. F. Sawada and F. Ishii, *J. Biochem.* **64,** 161–165 (1968).

115. T. Samejima, M. Kita, M. Saneyoshi, and F. Sawada, *Biochim. Biophys. Acta* **179,** 1-9 (1969).
116. F. Pochon, C. Balny, K. H. Scheit, and A. M. Michelson, *Biochim. Biophys. Acta* **228,** 49-56 (1971).
117. E. Shefter and H. G. Mautner, *J. Am. Chem. Soc.* **89,** 1249-1253 (1967).
118. P. Faerber, W. Saenger, K. H. Scheit, and D. Suck, *FEBS-Lett.* **10,** 41-45 (1970).
119. W. Saenger and K. H. Scheit, *J. Mol. Biol.* **50,** 153-169 (1970).
120. W. Saenger, D. Suck, and K. H. Scheit, *FEBS-Lett.* **5,** 262-264 (1969).
121. S. W. Hawkinson and C. L. Coulter, *Acta Crystallogr. sect. B*, **27,** 34-42 (1970).
122. G. H. Y. Lin, M. Sundaralingam, and S. K. Arora, *J. Am. Chem. Soc.* **93,** 1235-1241 (1971).
123. M. P. Schweizer, J. T. Witkowski, and R. K. Robins, *J. Am. Chem. Soc.* **93,** 277-279 (1971).
124. K. H. Scheit and W. Saenger, *FEBS-Lett.* **2,** 305-308 (1969).
125. K. H. Scheit, *Biochim. Biophys. Acta* **209,** 445-454 (1970).
126. W. Bähr, H. Sommer, and K. H. Scheit, *Biochim. Biophys. Acta* **287,** 427-437 (1972).
127. K. H. Scheit, *Tetrahedron Lett.* **2,** 113-118 (1967).
128. E. Sato and Y. Kanaoka, *Biochim. Biophys. Acta* **232,** 213-216 (1971).
129. J. A. Secrist, III, J. R. Barrio, and N. J. Leonard, *Biochem. Biophys. Res. Commun.* **45,** 1262-1269 (1971).
130. K. H. Scheit, *Biochim. Biophys. Acta* **195,** 294-298 (1966).
131. E. B. Ziff and J. R. Fresco, *J. Am. Chem. Soc.* **90,** 7338-7342 (1968).
132. H. Hayatsu and M. Yano, *Tetrahedron Lett.* **9,** 755-758 (1969).
133. M. Yano and H. Hayatsu, *Biochim. Biophys. Acta* **199,** 303-315 (1970).
134. M. Sono and H. Hayatsu, *Chem. Pharm. Bull. (Tokyo)* **21,** 995-1000 (1973).
135. M. Pleiss, H. Ochiai, and P. A. Cerutti, *Biochem. Biophys. Res. Commun.* **34,** 70-76 (1969).
136. M. G. Pleiss and P. A. Cerutti, *Biochemistry* **10,** 3093-3099 (1971).
137. K. H. Scheit, *Biochim. Biophys. Acta* **166,** 285-293 (1968).
138. K. H. Scheit and P. Faerber, *Eur. J. Biochem.* **33,** 545-550 (1973).
139. L. B. Townsend and G. H. Milne, *J. Heterocycl. Chem.* **7,** 753-754 (1970).
140. G. H. Milne and L. B. Townsend, *J. Med. Chem.* **17,** 263-268 (1974).
141. C. Y. Shiue and S. H. Chu, *J. Chem. Soc. Chem. Commun.* **9,** 319-320 (1975).
142. C. Y. Shiue and S. H. Chu, *J. Heterocycl. Chem.* **12,** 493-500 (1975).
143. C. Y. Shiue and S. H. Chu, *J. Org. Chem.* **40,** 2971-2974 (1975).
144. C. P. Pal, *J. Am. Chem. Soc.* **99,** 1973-1974 (1977).
145. H. G. Mautner, *J. Am. Chem. Soc.* **78,** 5292-5294 (1956).
146. R. A. Lang, R. K. Robins, and L. B. Townsend in *Synthetic Procedures in Nucleic Acid Chemistry*, Vol. 1, p. 228, W. W. Zorbach, R. S. Tipson, Eds., Wiley, New York, 1968.
147. M. Ikehara, I. Tazawa, and T. Fukui, *Chem. Pharm. Bull. (Tokyo)* **17,** 1019-1024 (1969).
148. R. E. Holmes and R. K. Robins, *J. Am. Chem. Soc.* **86,** 1242-1245 (1964).

149. R. Shapiro and S. C. Agarwal, *Biochem. Biophys. Res. Commun.* **24,** 401–405 (1966).
150. D. B. McCormick and G. E. Opar, *J. Med. Chem.* **12,** 333–334 (1969).
151. M. Ikehara, I. Tazawa, and T. Fukui, *Biochemistry* **8,** 736–743 (1969).
152. A. M. Kapuler and E. Reich, *Biochemistry* **10,** 4050–4061 (1971).
153. M. Ikehara and S. Uesugi, *Chem. Pharm. Bull. (Tokyo)* **17,** 348–354 (1969).
154. E. I. Budowsky, V. N. Shibaev, S. M. Spridinova, and N. K. Kochetkov, *Izv. Akad. Nauk. SSSR, Ser. Khim.* **1971,** 1280–1281.
155. S. S. Tavale and H. M. Sobell, *J. Mol. Biol.* **48,** 109–123 (1970).
156. J. O. Folayan and D. W. Hutchinson, *Biochim. Biophys. Acta* **474,** 329–333 (1977).
157. B. E. Haley and J. F. Hoffman, *Proc. Natl. Acad. Sci. (US)* **71,** 3367–3371 (1974).
158. I. L. Cartwright, D. W. Hutchinson, and V. W. Armstrong, *Nucleic Acids Res.* **3,** 2331–2339 (1976).
159. F. E. Evans and N. O. Kaplan, *J. Biol. Chem.* **251,** 6791–6797 (1976).
160. J. Lavayre, M. Ptak, and M. Leng, *Biochem. Biophys. Res. Commun.* **65,** 1355–1362 (1975).
161. C. R. Lowe, *Eur. J. Biochem.* **73,** 265–274 (1977).
162. T. Naka and M. Honjo, *Chem. Pharm. Bull. (Tokyo)* **24,** 2052–2056 (1976).
163. M. Ikehara, E. Ohtsuka, and S. Uesugi, *Chem. Pharm. Bull. (Tokyo)* **21,** 444–445 (1973).
164. S.-H. Chu, C.-Y. Shiue, and M.-Y. Chu, *J. Med. Chem.* **18,** 559–564 (1975).
165. R. Koberstein, L. Cobianchi, and H. Sund, *FEBS-Lett.* **64,** 176–180 (1976).
166. G. Schäfer, E. Schrader, G. Rowohl-Quisthoudt, S. Penades, and M. Rimpler, *FEBS-Lett.* **64,** 185–189 (1976).
167. A. M. Kapuler, C. Monny, and A. M. Michelson, *Biochim. Biophys. Acta* **217,** 18–29 (1970).
168. M. Hattori, J. Frazier, and H. Todd Miles, *Biochemistry* **14,** 5033–5045 (1975).
169. M. Ikehara and K. Murao, *Chem. Pharm. Bull. (Tokyo)* **16,** 1330–1336 (1968).
170. W. Freist, F. v. d. Haar, M. Sprinzl, and F. Cramer, *Eur. J. Biochem.* **64,** 389–393 (1976).
171. H.-J. Schäfer, P. Scheurich, and K. Dose, *Hoppe Seyler's Z. Physiol. Chem.* **357,** 278 (1976).
172. R. J. Wagenvoord, J. van der Kraan, and A. Kemp, *Biochim. Biophys. Acta* **460,** 17–24 (1977).
173. P. Scheurich, H.-J. Schäfer, and K. Dose, *Hoppe Seyler's Z. Physiol. Chem.* **358,** 298 (1977).
174. S.-H. Chu, C.-Y. Shiue, and M.-Y. Chu, *J. Med. Chem.* **17,** 406–409 (1974).
175. R. Koberstein, *Eur. J. Biochem.* **67,** 223–229 (1976).
176. H. J. Brentnall and D. W. Hutchinson, *Tetrahedron Lett.* **25,** 2595–2596 (1972).
177. M. Ikehara, S. Uesugi, and K. Yoshida, *Biochemistry* **11,** 830–836 (1972).
178. Y. Kawazoa, M. Maeda, and K. Nushi, *Chem. Pharm. Bull. (Tokyo)* **20,** 1341–1342 (1972).
179. L. F. Christensen, R. B. Meyer, Jr., J. P. Miller, L. N. Simon, and R. K. Robins, *Biochemistry* **14,** 1490–1496 (1976).

180. J. W. Littlefield and D. B. Dunn, *Biochem. J.* **70,** 642–651 (1958).
181. J. Davoll and B. A. Lowy, *J. Am. Chem. Soc.* **74,** 1563–1566 (1952).
182. A. Yamazaki, J. Kumashiro, and T. Kakenishi, *J. Org. Chem.* **33,** 2583–2586 (1968).
183. M. Ikehara, M. Hattori, and T. Fukui, *Eur. J. Biochem.* **31,** 329–334 (1972).
184. A. Yamazaki, I. Kumashiro, and T. Takenishi, *J. Org. Chem.* **32,** 3258–3260 (1967).
185. M. Ikehara and M. Hattori, *Biochim. Biophys. Acta* **272,** 27–32 (1972).
186. Kin-ichi Imai, R. Marumoto, K. Kobayashi, Y. Yoshioka, J. Toda, and M. Honjo, *Chem. Pharm. Bull. (Tokyo)* **19,** 576–586 (1971).
187. A. Yamazaki, J. Kumashiro, and T. Takenishi, *Chem. Pharm. Bull. (Tokyo)* **16,** 338–344 (1968).
188. T. Sato, in *Synthetic Procedures in Nucleic Acid Chemistry*, Vol. 1, p. 264, W. W. Zorbach, R. S. Tipson, Eds., Wiley, New York, 1968.
189. J. Davoll, B. Lythgoe, and A. R. Todd, *J. Chem. Soc.* **1948,** 1685–1687.
190. J. A. Montgomery and K. Hewson, *J. Heterocycl. Chem.* **1,** 213–214 (1964).
191. G. Gough, M. H. Maguire, and F. Michal, *J. Med. Chem.* **12,** 494–498 (1969).
192. B. Jastorff and W. Freist, *Angew. Chem.* **84,** 711–712 (1972).
193. A. Yamazaki, T. Saito, Y. Yamada, and I. Kumashiro, *Chem. Pharm. Bull. (Tokyo)* **17,** 2581–2585 (1969).
194. E. Freese, *Proc. Natl. Acad. Sci. (US)* **45,** 622–633 (1959).
195. R. E. Koch, *Proc. Natl. Acad. Sci. (US)* **68,** 773–776 (1971).
196. D. C. Ward, E. Reich, and L. Stryer, *J. Biol. Chem.* **244,** 1228–1237 (1969).
197. K. H. Scheit, unpublished.
198. C. Janion and D. Shugar, *Acta Biochim. Pol.* **20,** 271–284 (1973).
199. H. R. Rackwitz and K. H. Scheit, *Eur. J. Biochem.* **72,** 191–200 (1977).
200. K. H. Scheit, *J. Carbohydr. Nucleosides Nucleotides* **1,** 385–399 (1974).
201. W. R. McClure and K. H. Scheit, *FEBS-Lett.* **32,** 207–269 (1973).
202. J. Davoll and B. A. Lowy, *J. Am. Chem. Soc.* **73,** 1650–1655 (1951).
203. F. B. Howard, J. Frazier, and H. T. Miles, *J. Biol. Chem.* **241,** 4293–4295 (1966).
204. K. H. Scheit, unpublished.
205. F. Ishikawa, J. Frazier, and H. Todd Miles, *Biochemistry* **12,** 4790–4798 (1973).
206. F. Cramer and G. Schlingloff, *Tetrahedron Lett.* **43,** 3201–3206 (1964).
207. Z. Kazimierczuk and D. Shugar, *Acta Biochim. Pol.* **20,** 395–402 (1973).
208. H. H. Mantsch, I. Goia, M. Kezdi, O. Bârzu, M. Dânsoreanu, and G. Jebeleanu, *Biochemistry* **14,** 5593–5601 (1975).
209. T. Golas, M. Fikus, Z. Kazimierczuk, and D. Shugar, *Eur. J. Biochem.* **65,** 183–192 (1976).
210. G. Wiegand and R. Kaleja, *Eur. J. Biochem.* **65,** 473–479 (1976).
211. N. Yamaji, Y. Ynasa, and M. Kato, *Chem. Pharm. Bull. (Tokyo)* **24,** 1561–1567 (1976).
212. R. B. Meyer, Jr., D. A. Shuman, and R. K. Robins, *J. Am. Chem. Soc.* **96,** 4962–4966 (1974).
213. N. Yamaji, K. Suda, Y. Onoue, and M. Kato, *Chem. Pharm. Bull. (Tokyo)* **25,** 3239–3246 (1977).
214. N. Yamaji, K. Tahara, and M. Kato, *Chem. Pharm. Bull. (Tokyo)* **26,** 2391–2395 (1978).

215. T. K. Bradshaw and D. W. Hutchinson, *Chem. Soc. Rev.* **6,** 43–62 (1977).
216. C. Heidelberger, *Progr. Nucl. Acid Res. Mol. Biol.* **4,** 1–50 (1965).
217. C. Heidelberger, *Ann. Rev. Pharmacol.* **7,** 101–124 (1967).
218. W. H. Prusoff, *Pharmacol. Rev.* **19,** 209–250 (1967).
219. D. C. Remy, A. V. Sunthankar, and C. Heidelberger, *J. Org. Chem.* **27,** 2491–2500 (1962).
220. S. Slapikoff and P. Berg, *Biochemistry* **6,** 3654–3658 (1967).
221. I. Wempen and J. J. Fox, in *Synthetic Procedures in Nucleic Acid Chemistry,* Vol. 1, p. 425, W. W. Zorbach, R. S. Tipson, Eds., Wiley, New York, 1968.
222. N. C. Yung, J. H. Burchenal, R. Fecher, R. Duschinsky, and J. J. Fox, *J. Am. Chem. Soc.* **83,** 4060–4065 (1961).
223. J. O. Folayan and D. W. Hutchinson, *Biochim. Biophys. Acta* **340,** 194–198 (1974).
224. M. J. Robins and S. R. Naik, *J. Am. Chem. Soc.* **93,** 5277–5278 (1971).
225. M. J. Robins, G. Ramani, and M. McCoss, *Can. J. Chem.* **53,** 1302–1306 (1975).
226. W. H. Prusoff, W. L. Holmes, and A. D. Welch, *Cancer Res.* **13,** 221–225 (1953).
227. W. H. Prusoff, *Biochim. Biophys. Acta* **32,** 295–296 (1959).
228. R. Letters and A. M. Michelson, *J. Chem. Soc.* **1962,** 71–76.
229. A. M. Michelson, in *Synthetic Procedures in Nucleic Acid Chemistry,* Vol. 1, pp. 491–492, W. W. Zorbach, R. S. Tipson, Eds., Wiley, New York, 1968.
230. A. M. Michelson, J. Dondon, and M. Grunberg-Manago, *Biochim. Biophys. Acta* **55,** 529–540 (1962).
231. M. J. Bessman, I. R. Lehman, J. Adler, S. B. Zimmermann, E. S. Sims, and A. Kornberg, *Proc. Natl. Acad. Sci.* (*US*) **44,** 633-640 (1958).
232. A. M. Michelson, *J. Chem. Soc.* **1958,** 1957–1963.
233. J. Smrt and F. Sorm, *Coll. Czech. Chem. Commun.* **25,** 553–558 (1960).
234. M. A. W. Eaton and D. W. Hutchinson, *Biochemistry* **11,** 3162–3167 (1972).
235. M. Grunberg-Manago and A. M. Michelson, *Biochim. Biophys. Acta* **80,** 431–440 (1964).
236. F. B. Howard, J. Frazier, and H. T. Miles, *J. Biol. Chem.* **244,** 1291-1302 (1969).
237. K. Kikugawa, I. Kawada, and M. Ichino, *Chem. Pharm. Bull.* (*Tokyo*) **23,** 35–41 (1975).
238. P. K. Chang and A. D. Welch, *J. Med. Chem.* **6,** 428–430 (1963).
239. A. M. Michelson and C. Monny, *Biochim. Biophys. Acta* **149,** 88–98 (1967).
240. D. W. Visser and P. Roy-Burman, in *Synthetic Procedures in Nucleic Acid Chemistry,* Vol. 1, pp. 493–496, W. W. Zorbach, R. S. Tipson, Eds., Wiley, New York, 1968.
241. M. A. W. Eaton and D. W. Hutchinson, *Biochim. Biophys. Acta* **319,** 281–287 (1973).
242. F. Maley, *Arch. Biochem. Biophys.* **96,** 550–556 (1962).
243. A. H. Alegria, *Biochim. Biophys. Acta* **149,** 317–324 (1967).
244. K. H. Scheit, *Chem. Ber.* **99,** 3884–3891 (1966).
245. H. R. Rackwitz and K. H. Scheit, *J. Carbohydr. Nucleosides Nucleotides* **2,** 407–412 (1975).

246. K. H. Scheit and W. Kampe, *Chem. Ber.* **98,** 1045–1048 (1965).
247. D. W. Visser, in *Synthetic Procedures in Nucleic Acid Chemistry*, Vol 1, pp. 407–408, W. W. Zorbach, R. S. Tipson, Eds., Wiley, New York, 1968.
248. M. Roberts and D. W. Visser, *J. Am. Chem. Soc.* **74,** 668–669 (1952).
249. R. Lührmann, U. Schwarz, and H. G. Gassen, *FEBS-Lett.* **32,** 55–58 (1973).
250. W. Hillen, and H. G. Gassen, *Biochim. Biophys. Acta* **407,** 347–356 (1975).
251. L. Thiry, *Virology* **28,** 543–554 (1966).
252. S. Roy-Burman, P. Roy-Burman, and D. W. Visser, *Biochem. Pharmacol.* **19,** 2745–2756 (1970).
253. J. O. Folayan and D. W. Hutchinson, *Tetrahedron Lett.* **51,** 5077–5080 (1973).
254. L. Szabo, T. I. Kalman, and T. J. Bardos, *J. Org. Chem.* **35,** 1434–1437 (1970).
255. T. J. Bardos, J. Aradi, Y. K. Ho, and T. I. Kalman, *Ann. N.Y. Acad. Sci.* **255,** 522–531 (1975).
256. R. M. K. Dale, D. C. Livingston, and D. C. Ward, *Proc. Natl. Acad. Sci. (US)* **70,** 2238–2242 (1973).
257. R. M. K. Dale, E. Martin, D. C. Livingston, and D. C. Ward, *Biochemistry* **14,** 2447–2457 (1975).
258. R. M. K. Dale, D. C. Ward, D. C. Livingston, and E. Martin, *Nucleic Acids Res.* **2,** 915–930 (1975).
259. C. van Broeckhoven and R. de Wachter, *Arch. Int. Physiol. Biochem.* **85,** 200–201 (1977).
260. C. van Broeckhoven and R. de Wachter, *Nucleic Acids Res.* **5,** 2133–2151 (1978).
261. V. K. Shibaev, G. I. Eliseeva, and N. K. Kochetkov, *Doklady Akad. Nauk SSSR* **203,** 860 (1972).
262. G. F. Huang and P. F. Torrence, *J. Org. Chem.* **42,** 3821–3824 (1977).
263. V. W. Armstrong, H. Sternbach, and F. Eckstein, *Biochemistry* **15,** 2086–2091 (1976).
264. V. W. Armstrong, J. K. Dattagupta, F. Eckstein, and W. Saenger, *Nucleic Acids Res.* **3,** 1791–1810 (1976).
265. B. E. Griffin, A. R. Todd, and A. Rich, *Proc. Natl. Acad. Sci. (US)* **44,** 1123–1128 (1958).
266. D. Shugar and W. Szer, *J. Mol. Biol.* **5,** 580–582 (1962).
267. R. Thedford, M. H. Fleysher, and R. H. Hall, *J. Med. Chem.* **8,** 486–491 (1965).
268. M. Swierkowski and D. Shugar, *J. Mol. Biol.* **47,** 57–67 (1970).
269. W. Szer, *Biochem. Biophys. Res. Commun.* **20,** 182–186 (1965).
270. T. Kulikowski and D. Shugar, *Biochim. Biophys. Acta* **374,** 164–175 (1974).
271. T. Kulikowski and D. Shugar, *Acta Biochim. Pol.* **18,** 209–236 (1971).
272. M. Swierkowski and D. Shugar, *Acta Biochim. Pol.* **16,** 263–277 (1969).
273. J. Giziewicz and D. Shugar, *Acta Biochim. Pol.* **22,** 87–98 (1975).
274. A. Holy, *Coll. Czech. Chem. Commun.* **37,** 1555–1576 (1972).
275. J. M. Carpenter and G. Shaw, *J. Chem. Soc.* **1970,** 2016–2022.
276. A. Kampf, R. L. Barfknecht, P. J. Shaffer, S. Osaki, and M. P. Mertes, *J. Med. Chem.* **19,** 903–908 (1976).
277. A Holy and R. Bald, *Coll. Czech. Chem. Commun.* **36,** 2809–2823 (1971).

278. C. Heidelberger, *Cancer Res.* **30,** 1549–1569 (1970).
279. P. Reyes and C. Heidelberger, *Mol. Pharmacol.* **1,** 14–30 (1965).
280. D. V. Santi and T. T. Sakai, *Biochemistry* **10,** 3598–3607 (1971).
281. P. V. Danenberg and C. Heidelberger, *J. Med. Chem.* **16,** 712–714 (1973).
282. D. Suck and W. Saenger, *J. Am. Chem. Soc.* **94,** 6520–6526 (1972).
283. A. Holy, R. Bald, and F. Sorm, *Coll. Czech. Chem. Commun.* **37,** 592–602 (1972).
284. J. G. Moffatt, *J. Am. Chem. Soc.* **85,** 1118–1123 (1963).
285. Z. Kućerova, A. Holy, and R. Bald, *Coll. Czech. Chem. Commun.* **37,** 2052–2058 (1972).
286. T. Ueda, M. Yamamoto, A. Yamane, M. Imazawa, and H. Inoue, *J. Carbohydr. Nucleosides Nucleotides* **5,** 261–271 (1978).
287. W. E. Cohn and D. G. Doherty, *J. Am. Chem. Soc.* **78,** 2863–2866 (1956).
288. P. Roy-Burman in *Synthetic Procedures in Nucleic Acid Chemistry*, Vol. 1, pp. 461–462, W. W. Zorbach and R. S. Tipson, Eds., Wiley, New York, 1968.
289. R. D. Batt, K. J. Martin, J. McT. Ploeser, and J. Murray, *J. Am. Chem. Soc.* **76,** 3663–3665 (1954).
290. F. Maley and G. F. Maley, *Arch. Biochem. Biophys.* **144,** 723–799 (1971).
291. P. Roy-Burman, S. Roy-Burman, and D. W. Visser, *J. Biol. Chem.* **243,** 1692–1697 (1968).
292. V. N. Shibaev and G. I. Eliseeva, in *Synthetic Procedures in Nucleic Acid Chemistry*, Vol. 1, p. 470, W. W. Zorbach and R. S. Tipson, Eds., Wiley, New York, 1968.
293. N. K. Kochetkov, E. I. Budowski, V. N. Shibaev, and G. I. Eliseeva, *Izv. Akad. Nauk. SSSR, Otdel. Khim. Nauk.* **1966,** 1779–1785.
294. N. D. Goldberg, J. L. Dahl, and R. E. Parks, Jr., *J. Biol. Chem.* **238,** 3109–3114 (1963).
295. E. I. Budowski, T. N. Drushinina, G. I. Eliseeva, N. D. Gabrielyan, N. K. Kochetkov, V. N. Shibaev, and G. L. Shdanov, *Biochim. Biophys. Acta* **122,** 213–224 (1966).
296. R. O. Roblin, Jr., J. O. Lampen, J. P. English, Q. P. Cole, and J. R. Vaughan, *J. Am. Chem. Soc.* **67,** 290–294 (1945).
297. G. W. Kidder, V. C. Dewey, R. E. Parks, Jr., and G. L. Woodside, *Cancer Res.* **11,** 204–211 (1951).
298. K. C. Agarwal and R. E. Parks, Jr., in *Antineoplastic and Immunosuppressive Agents*, Vol. 2, p. 464, A. C. Sartorelli and D. G. Johns, Eds., Springer, Berlin 1975.
299. J. L. Way and R. E. Parks, Jr., *J. Biol. Chem.* **231,** 467–480 (1958).
300. R. L. Miller and D. L. Adamczyk, *Biochem. Pharmacol.* **25,** 883–888 (1976).
301. J. K. Roy, D. C. Kram, J. L. Dahl, and R. E. Parks, Jr., *J. Biol. Chem.* **236,** 1158–1162 (1961).
302. H. J. Thomas, K. Hewson, and J. A. Montgomery, *J. Org. Chem.* **27,** 192–194 (1961).
303. A. Holy and J. Smrt, *Coll. Czech. Chem. Commun.* **31,** 1528–1543 (1966).
304. J. L. Darlix, P. Fromageot, and E. Reich, *Biochemistry* **10,** 1525–1531 (1971).
305. D. H. Levin, *Biochim. Biophys. Acta* **61,** 75–81 (1962).
306. F. M. Kahan and J. Hurwitz, *J. Biol. Chem.* **237,** 3778–3785 (1962).
307. H. G. Mandel and R. Markham, *Biochem. J.* **69,** 297–306 (1958).

308. D. H. Levin, *J. Biol. Chem.* **238,** 1098–1104 (1963).

309. M. Karon, S. Weissman, C. Meyer, and P. Henry, *Cancer Res.* **25,** 185–192 (1965).

310. D. H. Levin, *Biochemistry* **5,** 1618–1624 (1965).

311. R. E. Parks, Jr. and K. C. Agarwal, in *Antineoplastic and Immunosuppressive Agents,* Vol. 2, p. 462, A. C. Sartorelli and D. G. Johns, Eds., Springer, Berlin, 1975.

312. S. W. Kwan and T. E. Webb, *J. Biol. Chem.* **242,** 5542–5548 (1967).

313. D. Grunberger, L. Meissner, A. Holy, and F. Sorm, *Biochim. Biophys. Acta* **119,** 432–433 (1966).

314. D. Grunberger, C. O'Neal, and M. Nirenberg, *Biochim. Biophys. Acta* **119,** 581–585 (1966).

315. D. H. Levin, *Biochim. Biophys. Acta* **61,** 75–81 (1962).

316. D. C. Ward, W. Fuller, and E. Reich, *Proc. Natl. Acad. Sci. (US)* **62,** 581–588 (1969).

317. C. H. Lee, F. E. Evans, and R. H. Sarma, *J. Biol. Chem.* **250,** 1290–1296 (1975).

318. M. Kawana, G. A. Ivanovicz, R. J. Rousseau, and R. K. Robins, *J. Med. Chem.* **15,** 841–843 (1972).

319. N. Yamaji and M. Kato, *Chem. Lett.* **4,** 311–314 (1975).

320. K. C. Tsou and K. F. Yip, *Biopolymers* **13,** 987–993 (1974).

321. P. Roy-Burman, Analogs of Nucleic Acid Components, in *Recent Results in Cancer Research,* p. 70, P. Rentchnick, Ed., Springer, Berlin, 1970.

322. S. Nishimura, F. Harada, and M. Ikehara, *Biochim. Biophys. Acta* **129,** 301–309 (1966).

323. P. F. Torrence, E. DeClercq, J. A. Waters, and B. Witkop, *Biochemistry* **13,** 4400–4408 (1974).

324. M. Ikehara, F. Harada, and E. Ohtsuka, *Chem. Pharm. Bull. (Tokyo)* **14,** 1338–1346 (1966).

325. J. P. Miller, K. H. Boswell, K. Muneyama, R. L. Tolman, M. B. Scholten, R. K. Robins, L. N. Simon, and D. A. Shuman, *Biochem. Biophys. Res. Commun.* **55,** 843–849 (1973).

326. R. J. Suhadolnik, T. Uematsu, H. Uematsu, and R. G. Wilson, *J. Biol. Chem.* **243,** 2761–2766 (1968).

327. S. M. Hecht, R. Frye, D. Werner, T. Fukui, and S. D. Hawrelak, *Biochemistry* **15,** 1005–1015 (1976).

328. Y. Mizuno, S. Kitano, and A. Nomura, *Chem. Pharm. Bull. (Tokyo)* **23,** 1664–1670 (1975).

329. L. Hagenberg, H. G. Gassen, and H. Matthaei, *Biochem. Biophys. Res. Commun.* **50,** 1104–1112 (1973).

330. P. D. Cook, R. J. Rousseau, A. M. Mian, R. B. Meyer, Jr., P. Dea, G. Ivanovicz, D. G. Streeter, J. T. Witkowski, M. G. Stout, L. N. Simon, R. W. Sidwell, and R. K. Robins, *J. Am. Chem. Soc.* **97,** 2916–2917 (1975).

331. C. Woenckhaus and G. Pfleiderer, *Biochem. Z.* **341,** 495–501 (1965).

332. P. Roy-Burman, Analogs of Nucleic Acid Components, in *Recent Advances in Cancer Research,* p. 75, P. Rentchnick, Ed., Springer, Berlin, 1970.

333. M. Ikehara, K. Murao, F. Harada, and S. Nishimura, *Biochim. Biophys. Acta* **155,** 82–90 (1968).

334. D. C. Ward, A. Cerami, E. Reich, G. Acs, and L. Altwerger, *J. Biol. Chem.* **244,** 3243–3250 (1969).

335. M. Ikehara, and T. Tezuka, *Nucleic Acid Res.* **1,** 907–917 (1974).

336. A. Maelicke, M. Sprinzl, F.v.d. Haar, T. A. Khwaja, and F. Cramer, *Eur. J. Biochem.* **43,** 617–625 (1974).

337. P. Prusiner, T. Brennan, and M. Sundaralingam, *Biochemistry* **12,** 1196–1203 (1973).

338. S. Uesugi, T. Tezuka, and M. Ikehara, *Biochemistry* **14,** 2903–2906 (1975).

339. M. Ikehara, K. Murao, F. Harada, and S. Nishimura, *Biochim. Biophys. Acta* **174,** 696–703 (1969).

340. B. Brdar and E. Reich, *J. Biol. Chem.* **247,** 725–730 (1972).

341. D. C. Ward and E. Reich, *J. Biol. Chem.* **247,** 705–719 (1972).

342. D. Grunberger, D. C. Ward, and E. Reich, *J. Biol. Chem.* **247,** 720–724 (1972).

343. S. H. Kim, D. G. Bartholomew, L. B. Allen, R. K. Robins, G. R. Revankar, and P. Dea, *J. Med. Chem.* **21,** 883–889 (1978).

344. N. J. Leonard, A. G. Morrice, and M. A. Sprecker, *J. Org. Chem.* **40,** 356–363 (1975).

345. A. G. Morrice, M. A. Sprecker, and N. J. Leonard, *J. Org. Chem.* **40,** 363–366 (1975).

346. N. J. Leonard, D. I. C. Scopes, P. van der Lijn, and J. R. Barrio, *Biochemistry* **17,** 3677–3685 (1978).

347. N. J. Leonard, M. A. Sprecker, and A. G. Morrice, *J. Am. Chem. Soc.* **98,** 3987–3994 (1976).

348. N. J. Cusack, G. Shaw, and G. J. Lichtfield, *J. Chem. Soc.* **1971** 1501–1507.

349. R. W. Sidwell, J. H. Huffman, G. P. Khare, L. B. Allen, J. T. Witkowski, and R. K. Robins, *Science* **177,** 705–706 (1972).

350. D. G. Streeter, L. N. Simon, R. K. Robins, and J. P. Miller, *Biochemistry* **13,** 4543–4549 (1974).

351. J. P. Miller, L. J. Kigwana, D. G. Streeter, L. N. Simon, and J. Roboz, *Ann. N.Y. Acad. Sci.* **284,** 211–229 (1977).

352. L. B. Allen, K. H. Boswell, T. A. Khwaja, R. B. Meyer, Jr., R. W. Sidwell, J. T. Witkowski, L. F. Christensen, and R. K. Robins, *J. Med. Chem.* **21,** 742–746 (1978).

353. P. Dea, M. P. Schweizer, and G. P. Kreishman, *Biochemistry* **13,** 1862–1867 (1974).

354. B. Eriksson, E. Helgstrand, N. G. Johansson, A. Larsson, A. Misiorny, J. O. Noren, L. Philipson, K. Stenberg, G. Stening, S. Stridh, and B. Öberg, *Antimicrob. Agents Chemother.* **11,** 946–951 (1977).

355. M. Miyaki, A. Saito, and B. Shimizu, *Chem. Pharm. Bull.* (*Tokyo*) **18,** 2459–2468 (1970).

356. Y. Furukawa, O. Miyashita, and M. Honjo, *Chem. Pharm. Bull.* (*Tokyo*) **22,** 2552–2556 (1974).

357. W. Seibert, *Ber,* **80,** 494–502 (1947).

358. F. Sorm and J. Skoda, *Coll. Czech. Chem. Commun.* **21,** 487–488 (1956).

359. R. E. Handschuhmacher and A. D. Welch, *Cancer Res.* **16,** 965–969 (1956).

360. F. Sorm and H. Keilova, *Experientia* **14,** 215 (1958).

361. J. J. Jaffe, R. E. Handschuhmacher, and A. D. Welch, *Yale J. Biol. Med.* **30,** 168–176 (1957).

362. G. B. Elion, N. Nathan, and G. H. Hitchings, *Cancer Res.* **18,** 802–817 (1958).
363. J. Skoda, V. F. Hess, and F. Sorm, *Coll. Czech. Chem. Commun.* **22,** 1330–1333 (1957).
364. R. Schindler and A. D. Welch, *Science* **125,** 548–549 (1957).
365. J. Skoda and F. Sorm, *Coll. Czech. Chem. Commun.* **24,** 1331–1337 (1959).
366. J. Skoda and F. Sorm, *Biochim. Biophys. Acta* **28,** 659–660 (1958).
367. J. Skoda, in *Antineoplastic and Immunosuppressive Agents*, Vol. 2, pp. 348–372, A. C. Sartorelli and D. G. Johns, Eds., Springer, Berlin, 1975.
368. R. E. Handschuhmacher, *J. Biol. Chem.* **235,** 764–768 (1960).
369. A. Holy and F. Sorm, *Coll. Czech. Chem. Commun.* **31,** 1562–1568 (1966).
370. A. Holy, in *Synthetic Procedures in Nucleic Acid Chemistry*, Vol. 1, pp. 482–486, W. W. Zorbach and R. S. Tipson, Eds., Wiley, New York, 1968.
371. W. Saenger and D. Suck, *Nature* **242,** 610–612 (1973).
372. F. E. Hruska, D. J. Wood, R. J. Mynott, and R. H. Sarma, *FEBS-Lett.* **31,** 153–155 (1973).
373. R. E. Handschuhmacher, J. Skoda, and F. Sorm, *Coll. Czech. Chem. Commun.* **28,** 2983–2990 (1963).
374. J. Beranek and F. Sorm, *Coll. Czech. Chem. Commun.* **28,** 469–480 (1963).
375. M. P. Mertes, A. Holy, and J. Smrt, *Coll. Czech. Chem. Commun.* **33,** 3313–3323 (1968).
376. M. P. Mertes and J. Smrt, *Coll. Czech. Chem. Commun.* **33,** 3304–3312 (1968).
377. F. Sorm, A. Piskala, A. Cihak and J. Vesely, *Experientia* **20,** 202–203 (1964).
378. L. J. Hanka, J. S. Evans, D. J. Mason, and A. Dietz, *Antimicrob. Agents Chemother.* 619–624 (1966).
379. V. Paces, J. Doskocil, and F. Sorm, *Biochim. Biophys. Acta* **161,** 352–360 (1968).
380. A. Cihak, J. Skoda, and F. Sorm, *Coll. Czech. Chem. Commun.* **29,** 300–308 (1964).
381. T. T. Lee and R. L. Momparler, *Anal. Biochem.* **71,** 60–67 (1976).
382. T. T. Lee and R. L. Momparler, *Biochem. Pharmacol.* **26,** 403–406 (1977).
383. T. B. Øyen, *Biochim. Biophys. Acta* **186,** 237–243 (1969).
384. I. Vortruba, A. Holy, and H. Pischl, *Coll. Czech. Chem. Commun.* **37,** 2213–2220 (1972).
385. R. Wightman and A. Holy, *Coll. Czech. Chem. Commun.* **38,** 1381–1397 (1973).
386. H. Pischl and A. Holy, *Coll. Czech. Chem. Commun.* **33,** 2066–2073 (1968).

THREE

Nucleotides with Uncommon Glycosidic Bonds

This chapter deals with nucleotide analogs that possess glycosidic bonds other than those normally found in natural nucleotides. The nucleotides discussed are comprised of analogs with *N*-glycosidic bonds to nitrogen atoms normally not involved in glycosidic bond formation, as well as pseudouridylic acid where the aglycon is linked to the sugar via a C—C bond. Nucleotides derived from the naturally occurring C-glycoside formycin were discussed in Section 2.5.2.

3.1 NUCLEOTIDES WITH UNCOMMON *N*-GLYCOSIDIC BONDS

Hatfield et al. (1, 2) isolated 3-β-D-ribosyluric acid as well as the corresponding 5′-phosphate from beef erythrocytes. They also demonstrated the presence of an enzyme that catalyzes the formation of 3-β-D-ribosyluric acid 5′-phosphate from uric acid and phosphoribosyl pyrophosphate (3). Thus 3-β-D-ribosylpurine derivatives are obviously natural compounds.

Leonard and Laursen (4) synthesized 3-β-D-ribofuranosyladenine (**1**) by alkylation of adenine with 1-bromo-2,3,5-tri-*O*-benzoyl-D-ribose. Although the preferred site of alkylation of adenine is usually N(3) (5–7), the N(3)- and N(9)-substituted adenines (adenosine) were isolated in comparable yields. The structure and configuration of **1** were established by ultraviolet absorption and ^{1}H-nmr spectroscopy. Leonard and coworkers chose the descriptive name 3-isoadenosine for this compound. The relative lability toward acid and base in comparison to adenosine dictated the choice among the available phosphorylation procedures. Phosphorylation of unprotected 3-isoadenosine by 2-cyanoethyl phosphate and dicyclohexylcarbodiimide furnished a mixture of isomeric mono- and diphosphates. The monophosphates could be separated from the diphosphates and

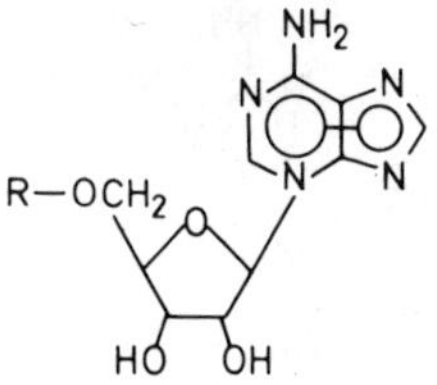

1, 2, 3, 4

1, R = H
2, R = P(O)(OH)O$^-$
3, R = P(O)O-O-P(O)(OH)O^{2-}
4, R = P(O)O-O-P(O)O-O-P(O)(OH)O^{3-}

unphosphorylated starting material by ion-exchange chromatography and the 5′-monophosphate was then obtained by fractional crystallization from the monophosphate mixture (4). In a more direct approach 2′,3′-*O*-ethoxy-ethylene-3-isoadenosine was phosphorylated by the same procedure, affording exclusively 3-isoadenosine 5′-phosphate (**2**) after removal of the blocking group. Treatment of the isomeric 3-isoadenosine monophosphates by crude snake venom 5′-nucleotidase selectively hydrolyzed the 5′-monophosphate and the remaining mixture of 3-isoadenosine 2′(3′)-phosphates could be cyclized to afford the corresponding 2′,3′-cyclic phosphate (**5**). 3-Isoadenosine 5′-phosphate was reacted with diphenylphosphor-

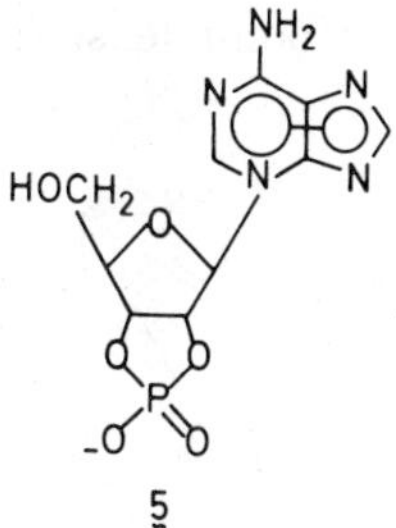

5

ylchloride to give the intermediate P^1-diphenyl P^2-(3-β-D-ribofuran-osyladenine-5′-)-pyrophosphate, from which 3-isoadenosine 5′-diphosphate (**3**), 3-isoadenosine 5′-triphosphate (**4**), and P^1-ribosylnicotineamide-5′ P^2-(3-β-D-ribofuranosyladenine)-5′-pyrophosphate (**6**) were obtained by the anion-displacement procedure (8). 3-Isoadenosine supported the growth of an adenine requiring strain of *E. coli*, but failed to support growth of a mammalian cell line rendered dependent on an exogenous purine source by a folic acid antagonist (9). However, it displayed cytoxicity to mammalian cells comparable to that of 6-azauridine (9). 3-Isoadenosine 5′-triphosphate substituted for ATP in the hexokinase-catalyzed phosphorylation of glucose (8) and the 5′-diphosphate (**3**) proved to be a substrate for polynucleotide phosphorylase (8, 10). With a variety of dehy-

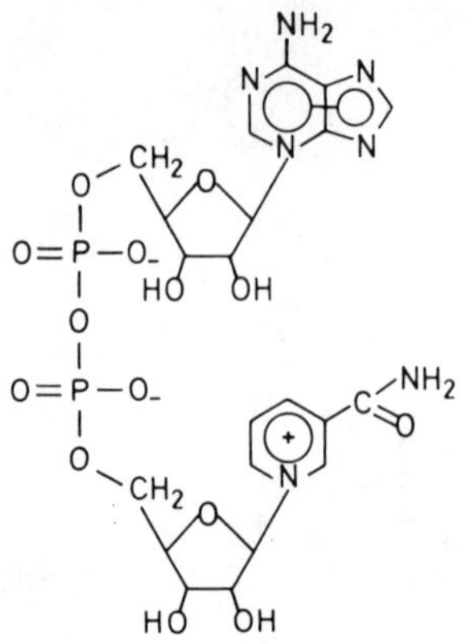

6

drogenases, compound **6** serves as an efficient substitute of nicotinamide-adenine-dinucleotide (8). Although the formation of adenine-uracil base pairs in interactions between poly 3-isoadenylic acid and poly uridylic acid could be established, polynucleotides containing 3-isoadenylic residues do not function as templates in ribosomal protein synthesis (10).

Bald and Holy (11) synthesized various derivatives of 3-(β-D-ribofuranosyl)uracil 2′,3′-cyclic phosphate (**7–9**) employing the triethyl phosphite procedure. The compounds **7**, **8**, and **9** turned out to be extremely poor substrates for pancreatic ribonuclease A as well as ribonuclease T2.

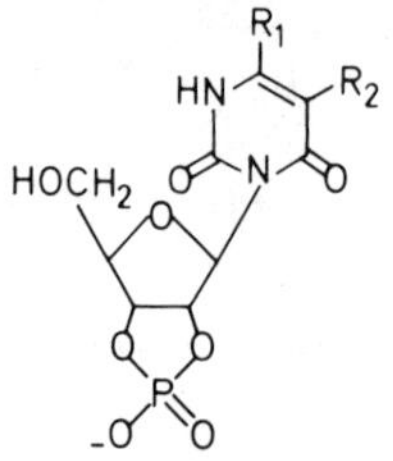

7, 8, 9

3.2 PSEUDOURIDYLIC ACID

5-(β-D-Ribofuranosyl)uracil (pseudouridine) (**10**) is an essential minor constituent of transfer RNA (12). Protonation of the ring oxygen atom of the furanose moiety or abstraction of a proton from N(1) of the uracil residue in **10** leads to a ring-opened intermediate, which then undergoes isomerization (13).

The ease with which **10** isomerizes on treatment with either acid or base makes the synthesis of pseudouridine phosphates a difficult task (Scheme 3.1). Chambers prepared pseudouridine 5′-mono- and diphosphate as

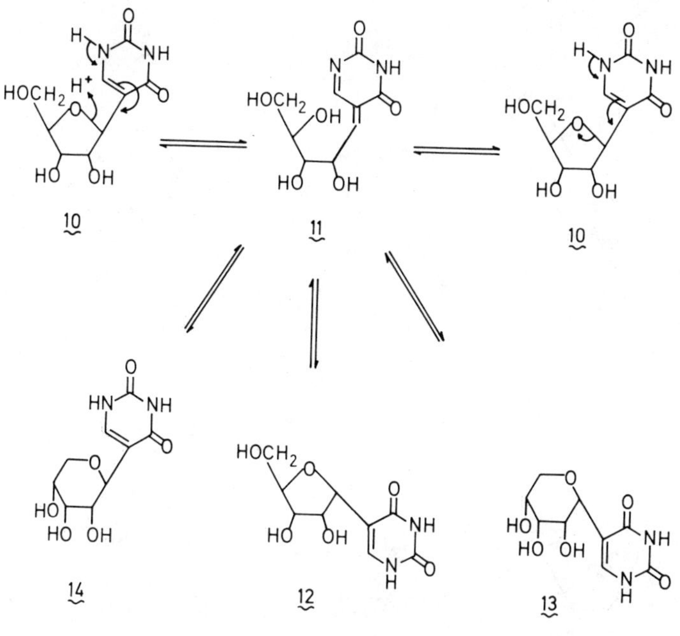

Scheme 3.1

follows (14). Protected **10** was phosphorylated by 2-cyanoethyl phosphate and dicyclohexylcarbodiimide and after removal of blocking groups, a mixture of 5-(β-D-ribofuranosyl)uracil 5′-phosphate (74%) (**15**) and 5-(α-D-ribofuranosyl)uracil 5′-phosphate (13%) (**16**) was obtained. This mixture was converted to the respective 5′-phosphoamidates, which in turn were

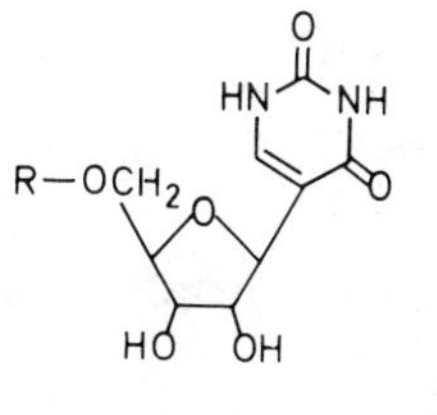

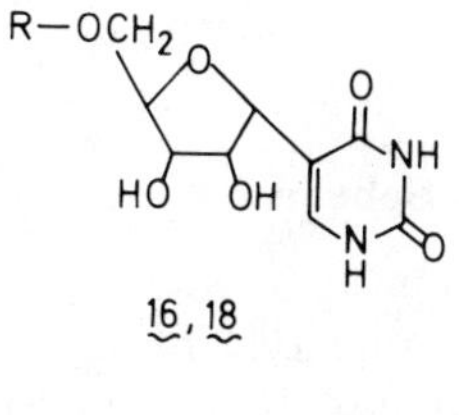

15, 16 R = P(O)(OH)O^{-}

17, 18 R = P(O)O-O-P(O)(OH)O^{2-}

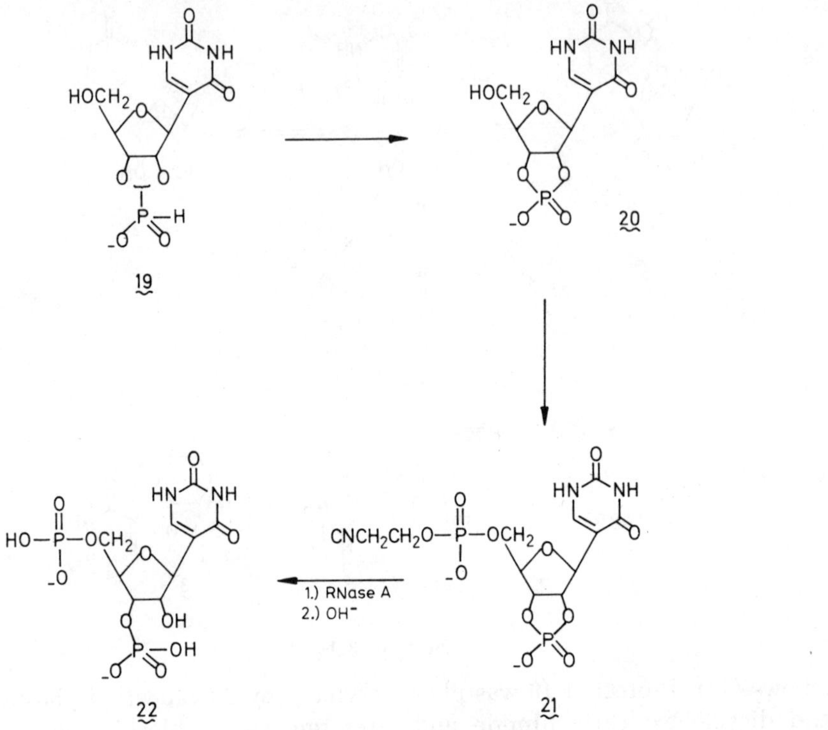

reacted with inorganic phosphate to give the 5′-diphosphates **17** and **18** which could be separated by ion-exchange chromatography. The chemical synthesis of **17** was likewise described by Pochon et al. (15). Cyclization of pseudouridine 2′,(3′)-phosphate (**19**), isolated from a hydrolysate of RNA (16), led to formation of pseudouridine 3′,2′-cyclic phosphate (**20**). Phos-

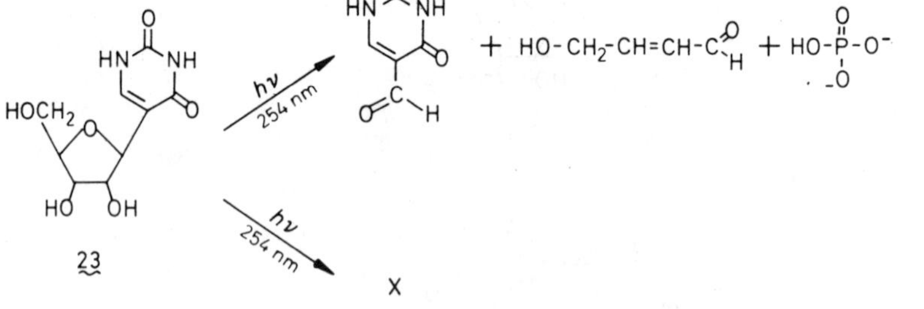

Scheme 3.2

phorylation of **20** by 2-cyanoethyl phosphate and dicyclohexylcarbodiimide gave **21.** Subsequent hydrolysis by ribonuclease A and removal of the blocking groups yielded pseudouridine 5′,3′-diphosphate

(**22**) (17). The desired material (**22**) was separated from the concomitantly formed isomers by anion-exchange chromatography. Pseudouridine 5′-triphosphate was synthesized both enzymatically by means of a crude extract from yeast (18, 19) as well as chemically (20).

According to the mechanism depicted in Scheme 3.2, pseudouridine 3′-phosphate (**23**) undergoes two photochemical reactions when irradiated at 254 nm (21). The first reaction involves photolysis of **23** to 5-formyluracil, 4-hydroxycrotonaldehyde, and inorganic phosphate. The second reaction is a rearrangement of **23** to a product of unknown structure with very similar properties when compared to the starting material.

REFERENCES

1. H. S. Forrest, D. Hatfield, and J. M. Lagowski, *J. Chem. Soc.* **1961** 963–968.
2. D. Hatfield and H. S. Forrest, *Biochim. Biophys. Acta* **62,** 185–187 (1962).
3. D. Hatfield, R. R. Rinehart, and H. S. Forrest, *J. Chem. Soc.* **1963** 899-902.
4. N. J. Leonard and R. A. Laursen, *Biochemistry* **4,** 354-365 (1965).
5. N. J. Leonard and J. A. Deyrup, *J. Am. Chem. Soc.* **84,** 2148–2160 (1962).
6. N. J. Leonard, and T. Fujii, *J. Am. Chem. Soc.* **85,** 3719 (1963).
7. J. W. Jones and R. K. Robins, *J. Am. Chem. Soc.* **84,** 1914–1919 (1962).
8. N. J. Leonard and R. A. Laursen, *Biochemistry* **4,** 365–376 (1965).
9. K. Gerzon, I. S. Johnson, G. B. Boder, J. C. Cline, P. J. Simpson, C. Speth, N. J. Leonard, and R. A. Laursen, *Biochim. Biophys. Acta* **119,** 445–461 (1966).
10. A. M. Michelson, C. Monny, R. A. Laursen, and N. J. Leonard, *Biochim. Biophys. Acta* **119,** 258–267 (1966).
11. R. W. Bald and A. Holy, *Coll. Czech. Chem. Commun.* **36,** 3657–3669 (1971).
12. R. H. Hall, *The Modified Nucleosides in Nucleic Acids*, Columbia University Press, New York, 1971.
13. *Organic Chemistry of Nucleic Acids*, N. K. Kochetkov and E. I. Budowski, Eds., Plenum Press, London, 1972.
14. R. W. Chambers, V. Kurkov, and R. Shapiro, *Biochemistry* **2,** 1192–1203 (1963).
15. F. Pochon, A. M. Michelson, M. Grunberg-Manago, W. E. Cohn, and L. Dondon, *Biochim. Biophys. Acta* **80,** 441–447 (1964).
16. W. E. Cohn, *Biochem. Prep.* **8,** 116 (1961).
17. M. Tomasz and R. W. Chambers, *Biochemistry* **4,** 1720–1724 (1965).
18. I. H. Goldberg and M. Rabinowitz, *J. Biol. Chem.* **238,** 1793–1800 (1963).
19. I. H. Goldberg and M. Rabinowitz, *Biochim. Biophys. Acta* **54,** 202–204 (1961).
20. F. M. Kahan and J. Hurwitz, *J. Biol. Chem.* **237,** 3778-3785 (1962).
21. L. D. H. Shulman, I. Kućan, B. Edelman, and R. W. Chambers, *Biochemistry* **12,** 201–208 (1973).

FOUR

Nucleotides with Modified Phosphate Groups

Replacement of the oxygen atom of a P—O—P bond or a P—O—C bond in nucleotides offers great potential for possible modifications. Many nucleotide analogs belonging to either category **I** or **II** have been prepared. Compounds of type **II** in which Y represents a methylene group are separately treated as nucleoside phosphonates. Analogs of nucleoside 5′-

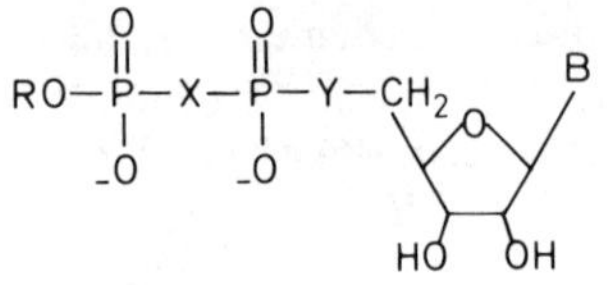

I, II

I, X = CH_2,NH Y = O R = H or $P(O)(OH)O^-$

II, X = O Y = CH_2, NH, S R = H or $P(O)(OH)O^-$

triphosphates belonging to type **I** comprise those with a substituent *X* inserted between the α and β, as well as the β and γ phosphate residues. Those with P—S—P instead of P—O—P bonds have not yet been reported. A further modification of the phosphate moiety involves replacement of a nonbridging oxygen atom by a sulfur atom to give a nucleoside phosphorothioate (**III**). This substitution is obviously not restricted to the α phosphate group in nucleoside 5′-polyphosphates. Because of the interesting chemical and biochemical properties of analogs of type **III**, this

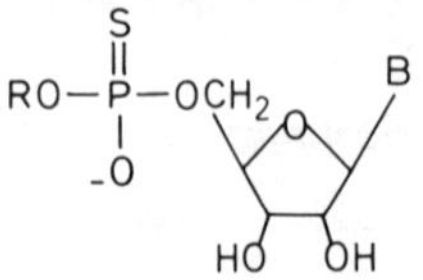

III

R = H , $P(O)(OH)O^-$, $P(O)O\text{-}O\text{-}P(O)(OH)O^{2-}$

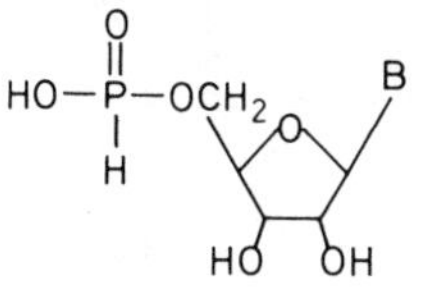

$$\mathrm{HO{-}P(=O)(H){-}OCH_2}\text{-ribose(2'-OH, 3'-OH)-B}$$

IV

modification turned out to be very useful. The nucleoside phosphites (**IV**) have not proven to be of particular importance as far as their biochemical properties are concerned. They merit consideration, however, because of their usefulness as intermediates in the chemical synthesis of certain nucleoside phosphates.

4.1 NUCLEOTIDE ANALOGS WITH ALTERED P—O—P BONDS

Myers et al., reported the first synthesis of an ATP analog with a modified P—O—P bond (1). Reaction of methylenediphosphonate with adenosine 5′-phosphormorpholidate (**1**) or condensation of adenosine 5′-phosphate (**2**) with methylenediphosphonate in the presence of dicyclohexylcarbodiimide led to 5′-adenylyl methylenediphosphonate (**3**). The descriptive abbreviation β,γ-methylene ATP indicates the position of the methylene grouping within the tripolyphosphate chain. Adenosine 5′-methylenediphosphonate (**4**) (α,β-methylene ADP) was obtained by phosphorylation of protected adenosine with methylenediphosphonate in the presence of an activating reagent (2). The synthesis of 5′-guanylyl methylenediphosphonate (**5**) (β,γ-methylene GTP) by reaction of guanosine 5′-phosphormorpholidate with methylenediphosphonate was described by Hershey et al. (3). The corresponding analogs of ATP, GTP, and GDP possessing α,β-methylene substituents instead of oxygen, namely adenosine 5′-methylenediphosphono-P^2-phosphate (**6**) (4), guanosine 5′-methylenediphosphono-P^2-phosphonate (**7**) (5), and guanosine 5′-methylenediphosphonate (**8**) (5) were prepared by methods similar to those described above. Trowbridge et al., synthesized the bismethylene analog of trimetaphosphate (**9**) and discovered that the latter readily reacted with alcohols under acid catalysis to form monoester triphosphates (51). Reaction of 2′,3′-*O*-isopropylideneadenosine (**10**) with **9** led to adenosine 5′-bis(dihydroxyphosphinylmethyl)phosphinate (**11**), an ATP analog in which both P—O—P linkages were replaced by P—CH_2—P bonds.

Yount et al., substituted the β,γ P—O—P bond in ATP by a P—NH—P bond, employing the reaction of a suitably activated adenosine 5′-phosphate with imidodiphosphate (6). The preparation of the respective GTP analog, 5′-guanylylimidodiphosphate by conventional procedures has been reported (7). As is to be expected, on account of the P—NH—P bond 5′-adenylylimidodiphosphate (**12**) is subject to more rapid hydrolysis at acidic pH than is ATP. It is also slowly degraded at neutral pH. The principal products of hydrolysis are inorganic phosphate and adenosine 5′-phosphoramidate (6). A comparison of bond angles and bond distances

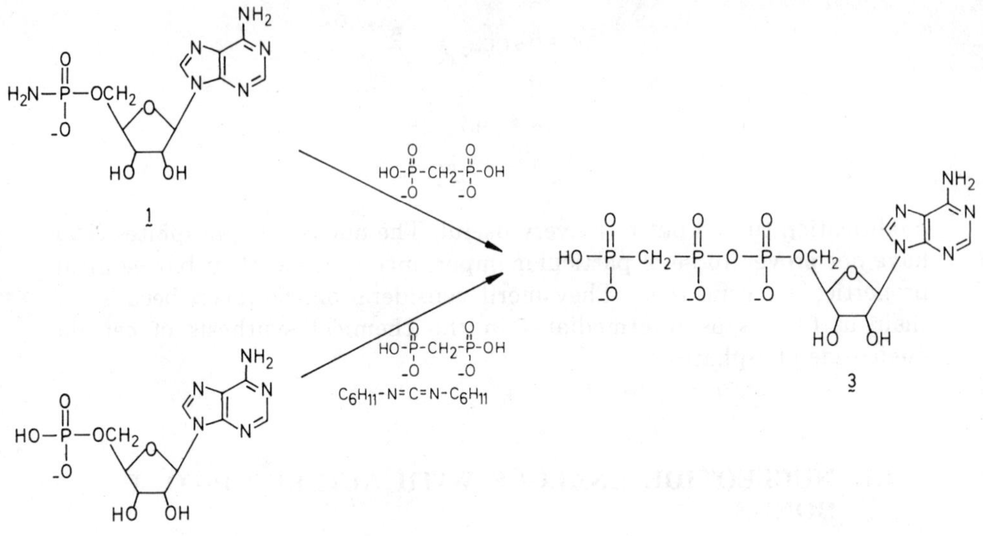

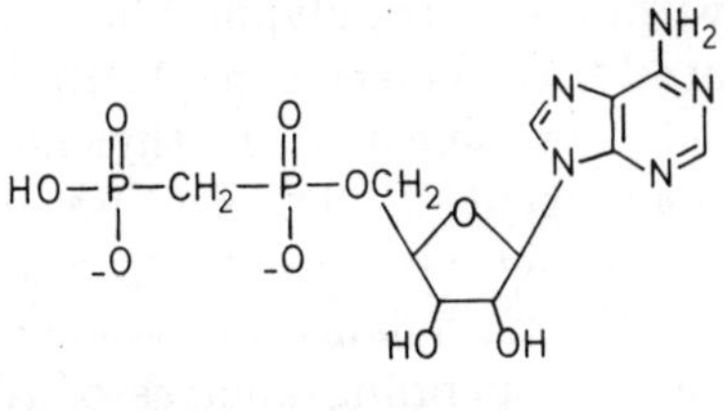

of inorganic pyrophosphate, imidodiphosphate (8) and methylenediphosphonate (6), as depicted in Fig. 4.1, reveals that the bond angles in P—NH—P and P—O—P are similar. They both differ from the bond angle in P—CH_2—P. However, because of the longer P—C bond, the P—P distances are essentially the same in all three compounds. In agreement

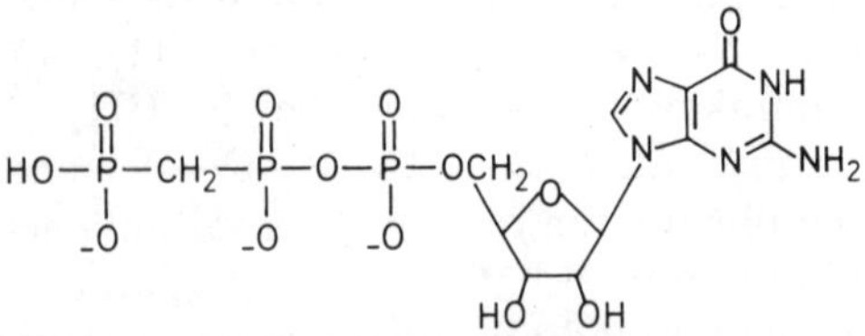

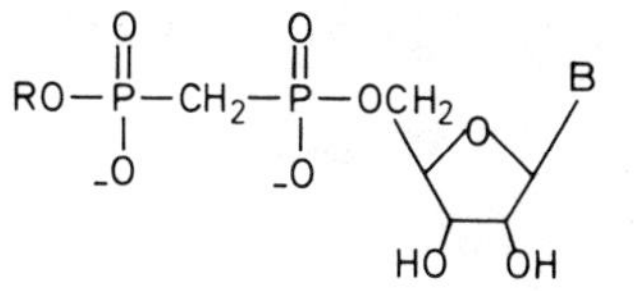

6, 7, 8

6, R= P(O)(OH)O^- B= ADENINE

7, R= P(O)(OH)O^- B= GUANINE

8, R= H B= GUANINE

with expectations, the last pK_a of β,γ-methylene ATP (**3**), β,γ-imido ATP (**12**), and ATP decreases from 8.4 to 7.7 to 7.1 respectively. As the electronegativity of the bridging atom increases, the attached phosphate group becomes a stronger acid (6). In this respect β,γ-imido ATP (**12**) resembles ATP more closely. It is interesting to note that both **3** and **12**

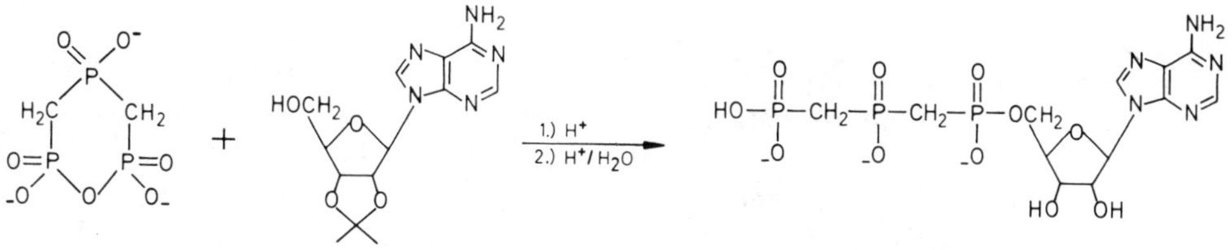

1.) H^+
2.) H^+/H_2O

9 10 11

bind divalent cations with higher affinities than ATP. The importance of analogs of nucleoside 5′-polyphosphates with either imido or methylene bridges result from the resistance of the respective bonds toward enzymatic attack. They can be used either as reversible inhibitors in enzymatic reactions that require a nucleoside 5′-polyphosphate as substrate or in

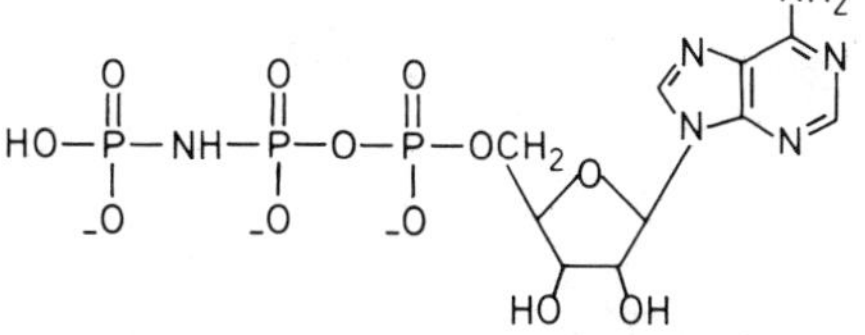

12

defining those interactions that depend only on the binding and not on the cleavage of a P—O—P linkage. This has been amply demonstrated by Yount et al., who studied the interaction of the ATP analogs β,γ-methylene ATP (**3**), and β,γ-imido ATP (**12**) with heavy meromyosin, myosin, and actomyosin (9). The GTP analog β,γ-methylene GTP (3) and β,γ-imido GTP (7) were found to be competitive inhibitors of the ribosome-dependent GTPase. α,β-Methylene ATP (**6**) (10) and α,β-methylene GTP (5) proved useful in investigations of adenylate cyclase

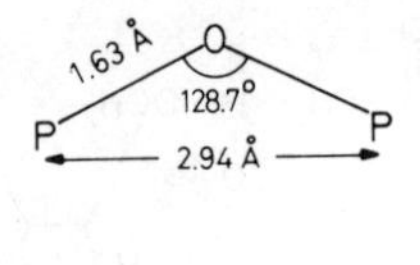

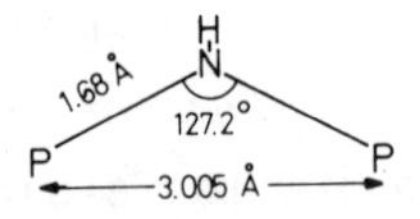

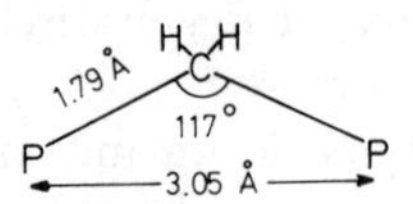

Figure 4.1 A comparison of bond angles and bond distances in inorganic phosphate, imidodiphosphate, and methylene diphosphonate.

mechanisms. No attempt is undertaken in this book to review the vast body of information regarding the biochemical properties of the above analogs. This subject was competently reviewed by Yount (11). The biochemical studies reported, however, seem to indicate that the imido analogs of nucleoside 5′-polyphosphates, in contrast to the methylene analogs, more closely mimic the parent oxygen compounds.

13

Bennett et al. (12) attempted the synthesis of guanosine 3′,5′-bis(methylenediphosphonate), an analog of guanosine 3′,5′-bis(pyrophosphate). Reaction of unprotected guanosine with methylenediphosphonate in the presence of dicyclohexylcarbodiimide led to the expected mixture of 3′,5′- and 2′,5′-isomers (**13**), which could be separated by ion-exchange chromatography. Phosphorylation of 2′-*O*-tetrahydropyranylguanosine with methylenediphosphonate occurred selectively at the 5′- and 3′-OH groups. Under the acidic conditions necessary to remove the 2′-*O*-tetrahydropyranyl group some equilibration occurred but the 3′,5′-compound could still be isolated as the major isomer (87%).

4.2. THIOPHOSPHATE ANALOGS OF NUCLEOTIDES

Nucleoside phosphorothioates were first described by Eckstein (13, 14). Thiophosphorylation of nucleosides was carried out in a convenient manner with trisimidazolyl-1-phosphinesulfide (**18**). Acid hydrolysis with aqueous acetic acid at elevated temperature removed both the blocking group as well as the imidazole moieties from the intermediates **19–22** and furnished the nucleoside phosphorothioates **23–26**. This procedure was also successfully applied to the thiophosphorylation of thymidine and 2′-deoxy-*N*⁴-dimethylaminomethylenecytidine to give the 5′-phosphorothioates of thymidine and 2′-deoxycytidine respectively. Selective thiophosphorylation of adenosine to adenosine 5′-phosphorothioate (**23**) by thiophosphorylchloride in triethylphosphate, introduced by Murray et al. (15),

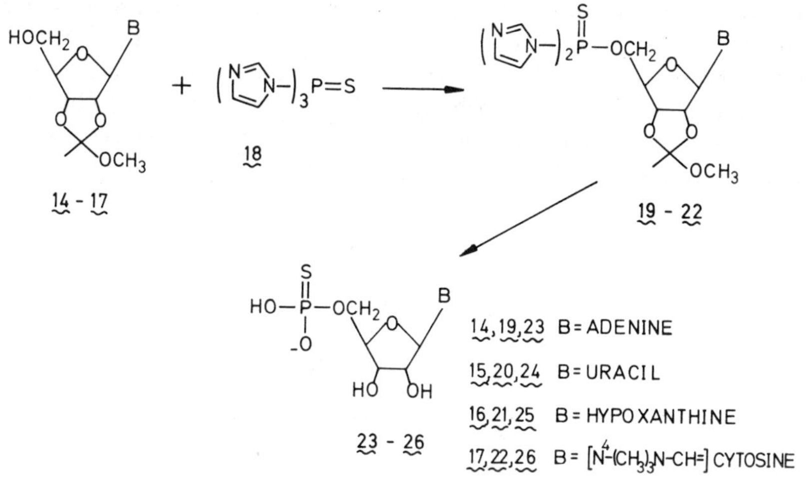
HOCH2
B
OCH3
14 - 17
+
P=S
18
S
P—OCH2
B
OCH3
19 - 22
S
HO—P—OCH2
-O
B
HO
OH
23 - 26
14,19,23 B = ADENINE
15,20,24 B = URACIL
16,21,25 B = HYPOXANTHINE
17,22,26 B = [N4-(CH3)2N-CH=] CYTOSINE

seems to be, however, the method of choice for the preparation of nucleoside 5′-phosphorothioates.

Monothiophosphate itself proved to be capable of thiophosphorylation of nucleosides at elevated temperatures (16). The yields of the formed nucleoside-*O*-phosphorothioates were in the range of 50 to 90%. The selectivity 5′-OH versus 3′-OH of thiophosphorylation was found to be 2:1 in the 2′-deoxynucleosides tested (16). Evidence was presented that the active thiophosphorylating species was monothiopyrophosphate, formed by intial dimerization of monothiophosphate. Indeed, independently prepared monothiopyrophosphate thiophosphorylated nucleosides at room temperatures in good yields, giving *O*-phosphorothioate esters exclusively (16).

Uridine 2′,3′-cyclic phosphorothioate (**27**) and guanosine 2′,3′-cyclic phosphorothioate (**28**) were obtained by thiophosphorylation of the respective 5′-*O*-acetylnucleosides with trisimidazolyl-1-phosphinesulfide (17).

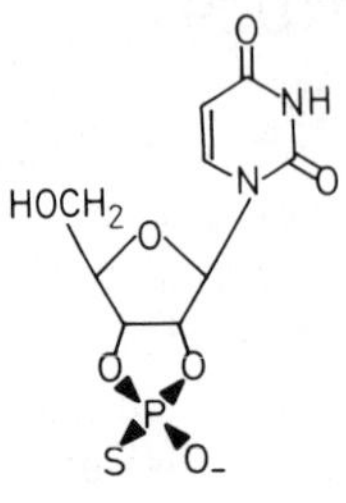

27, B = URACIL
28, B = GUANINE

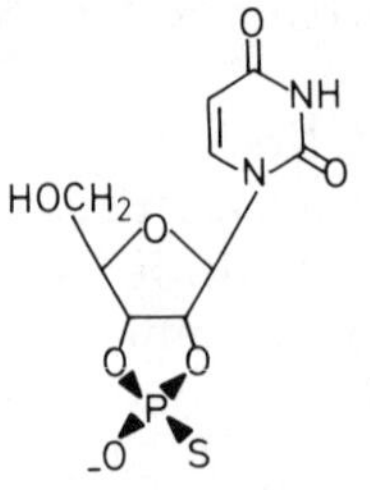

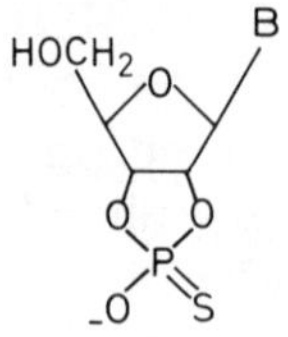

The cyclic phosphorothioates **27** and **28** exist as pairs of diastereomers. The diastereomers of **27** could be separated by crystallization and X-ray diffraction studies revealed the endo configuration for the crystalline isomer (18). Exo and endo refer to the spatial relationship between the sulfur atom and the ribose moiety. From the crystal structure of **27** it was found that the negative charge was located at the oxygen atom. Although both diasteromers of **27** were substrates for pancreatic ribonuclease A, the endo isomer displayed a much higher affinity for the enzyme (19). Compound **27** was employed in an elegant investigation of the stereochemistry of transesterification and hydrolysis catalyzed by ribonuclease (20–22).

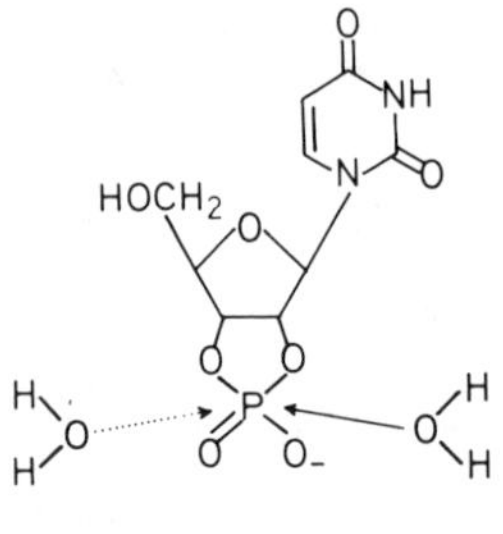

The 2′,3′-cyclic phosphate ring in **29** can be opened by water or an alcohol molecule either by attack "in line" (arrow, dotted line) or "adjacent" (arrow, solid line) to the leaving 2′-OH group. It is generally assumed that the hydrolysis of five-membered cyclic phosphates involves a pentacoordinate intermediate in which both the attacking nucleophile and the leaving group occupy apical positions in a trigonal bipyramid (20–22). The "adjacent" mechanism would require a pseudorotation to translocate the attacking nucleophile into an apical position. Hydrolysis of the endo isomer of **27** in ^{18}O-water by pancreatic ribonuclease allowed the solution of this problem. Chemical cyclization of the uridine 3′-phosphorothioate (**30**) afforded **27** as a mixture of diastereomers, which were separated by crystallization. If hydrolysis of **27** by ribonuclease had proceeded via the "in line" mechanism, all of the ^{18}O should have been present in the exo isomer of **27**. This was indeed the experimental observation.

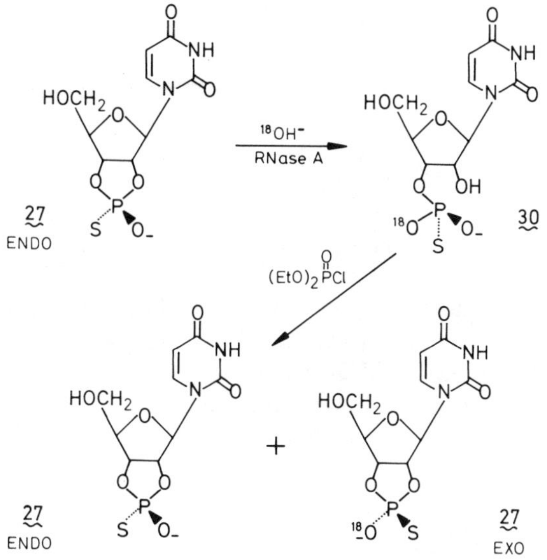

The transesterification step of ribonuclease action was studied in an analogous manner. Methanolysis of uridine 3′-phosphorothioate (**30**) endo isomer, catalyzed by ribonuclease, gave uridine 3′-phosphorothioate methyl ester. The latter was crystallized and shown by X-ray structural analysis to possess the R configuration. This result is in agreement with an "in line" mechanism of RNase A for the transesterification step (22).

The diastereomers of guanosine 2′,3′-cyclic phosphorothioate (**28**) could not be separated by crystallization (1). However, by comparison with the pure isomers of **27**, the respective resonance signals of the exo and endo isomers could be assigned in the ^{31}P-nmr spectrum of **28**. Only the endo isomer of **28** was shown to be a substrate for ribonuclease T_1, whereas the exo isomer acted as a competitive inhibitor. Methanolysis of **28** catalyzed by ribonuclease T_1 led to formation of guanosine 3′-phosphorothioate methyl ester with the R configuration. These results indicate that the mechanism of hydrolysis and transesterification of ribonuclease T_1 follows an "in line" mechanism (17). Cyclization of adenosine 5′-*O*,*O*-bis(*p*-nitrophenyl)-phosphorothioate to adenosine 3′,5′-cyclic phosphorothioate (**31**) was achieved by reaction with potassium *tert*-butoxide (23). The ^{31}P-nmr spectrum of **31** disclosed the existence of two diastereomers, which in analogy to **27** were endo (S configuration) and exo (R configuration). Compound **31** is very slowly hydrolyzed by phosphodiesterase from beef heart with no marked differences between the diastereomers (23). The slow hydrolysis of **31** is not caused by low affinity to the enzyme, since it is a potent competitive inhibitor of the enzymatic hydrolysis of adenosine 3′,5′-cyclic phosphate. It also proved to be effective in the stimulation of protein kinases (23).

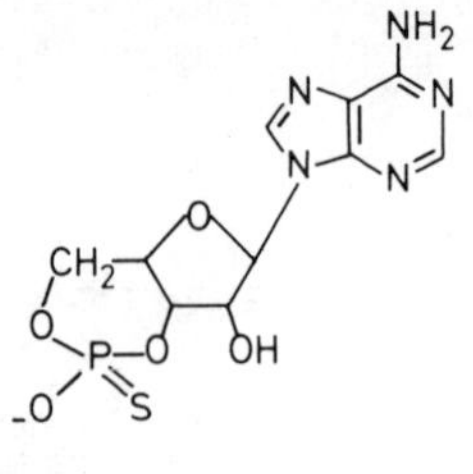

31

Nucleoside phosphorothioates are extremely poor substrates for phosphatases from a variety of sources (21, 24). They are, however, competitive inhibitors with reasonable affinities.

Thiophosphorylation of 3′-*O*-acetylthymidine with trisimidazolyl-1-phosphinesulfide gave thymidine 5′-phosphoroimidazolidate (**32**) after

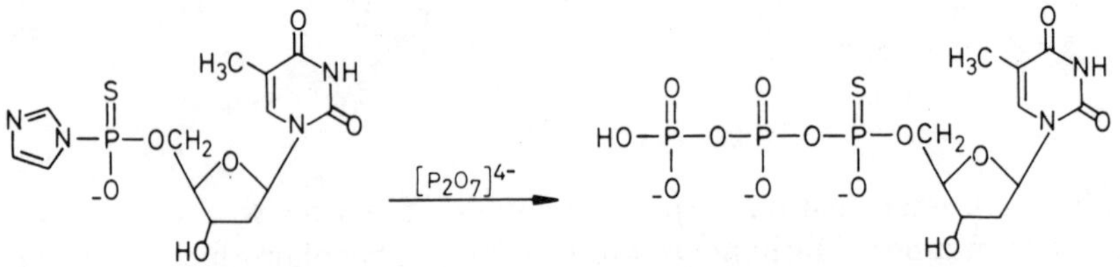

$[P_2O_7]^{4-}$

32 33

alkaline hydrolysis. Compound **32** served as an intermediate in the preparation of thymidine 5′-(*O*-1-thiotriphosphate) (**33**) (25). Protected uridine was thiophosphorylated in a similar fashion (25). It was later observed that activation of a nucleoside 5′-phosphorothioate by diphenylphosphorylchloride and subsequent reaction with either inorganic phos-

$H_2N-C(=O)-CH_2-CH_2-S-P(=O)(O^-)-OH$ +

OH^-

38 39

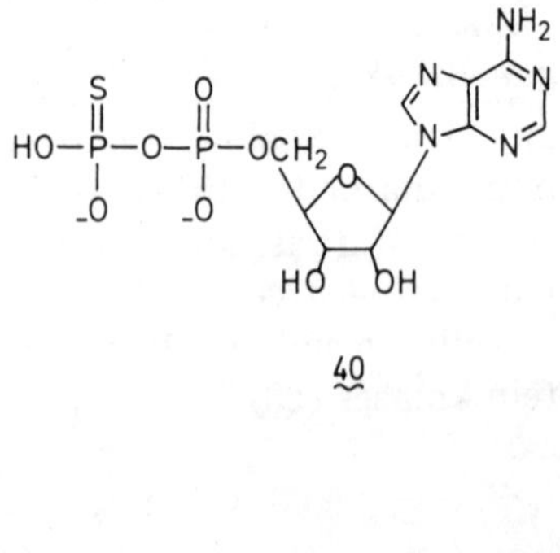

40

phate or pyrophosphate is a more suitable synthetic route to nucleoside 5′-(*O*-1-thiopolyphosphates). Reaction of *S*-2-carbamoylethylthiophosphate (**38**) with P^1-diphenyl P^2-(5′-*O*-adenosine)diphosphate (**39**) gave adenosine 5′-(*O*-2-thiodiphosphate) (**40**) (27) after removal of the carbamoylethyl blocking group. Activation of *S*-2-carbamoylethylthiophosphate by diphenylphosphorylchloride followed by reaction with adenosine 5′-diphosphate afforded adenosine 5′-(*O*-3-thiotriphosphate) (**41**) (27). The 5′-(*O*-3-

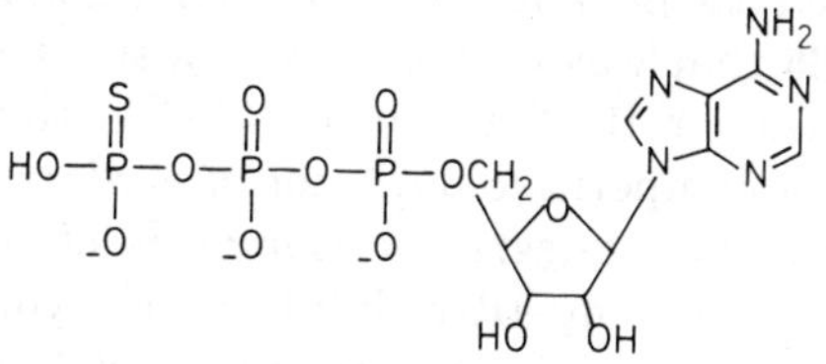

41

thiotriphosphates) of uridine and guanosine have been similarly prepared (27). The same strategy, namely the reaction of P^1-diphenyl P^2-(2-cyanoethyl) pyrophosphate with adenosine 5′-(*O*-2-thiodiphosphate) allowed the synthesis of adenosine 5′-(*O*-2-thiotriphosphate) (**42**) (26). The procedures employed in the preparations of compounds **40**, **41**, and **42** were as well applicable to the synthesis of the corresponding derivatives of other nucleosides.

The position of the phosphorothioate group in thiophosphate analogs of nucleoside 5′-polyphosphates can be identified by ^{31}P-nmr spectroscopy (26).

In compounds such as ATP or ADP the substitution by sulfur of an oxygen that is neither involved in an ester bond nor located at the terminal phosphate moiety introduces a further chiral center into the molecule. They therefore exist as pairs of diastereomers. Attempts to separate the diastereomers of **36**, **37**, and **42** by crystallization have failed

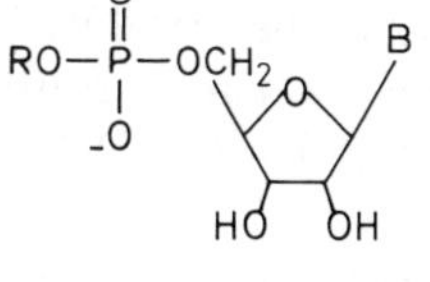

34 - 37

	R	B	Reference
34	$P(O)(OH)O^-$	Uracil	25, 26
35	$P(O)O{-}O{-}P(O)(OH)O^{2-}$	Uracil	25
36	$P(O)(OH)O^-$	Adenine	25
37	$P(O)O{-}O{-}P(O)(OH)O^{2-}$	Adenine	26

(26). Even ^{31}P-nmr spectroscopy did not provide evidence for their existence (26). However, enzymatic reactions have allowed their differentiation. Isomeric adenosine 5′-(*O*-1-thiotriphosphate) (**37**) was degraded by myokinase to approximately 50% in a fast reaction. Eckstein and Goody (26) assumed that the remaining **37** was a single isomer and arbitrarily designated it as isomer B. About 45% at most of chemically synthesized adenosine 5′-(*O*-1-thiodiphosphate) (**36**) could be phosphorylated by pyruvate kinase and phosphoenol pyruvate to **37**. The product was completely degraded by myokinase in a fast reaction and therefore assumed to be adenosine 5′-(*O*-1-thiotriphosphate) isomer A. That portion of **36** that was not utilized as a substrate by pyruvate kinase was designated isomer B of adenosine 5′-(*O*-1-thiodiphosphate) (**36**). It turned out to be a competitive inhibitor of moderate affinity for pyruvate kinase. Similarly, the undegraded isomer B of adenosine 5′-(*O*-1-thiotriphosphate) (**37**) is a competitive inhibitor of myokinase (26). Employing the same methods, degradation by myokinase of **42** and phosphorylation of **37** by pyruvate kinase, it was possible to stereospecifically synthesize isomers A and B of adenosine 5′-(*O*-2-thiotriphosphate) (**42**) (26). Eckstein

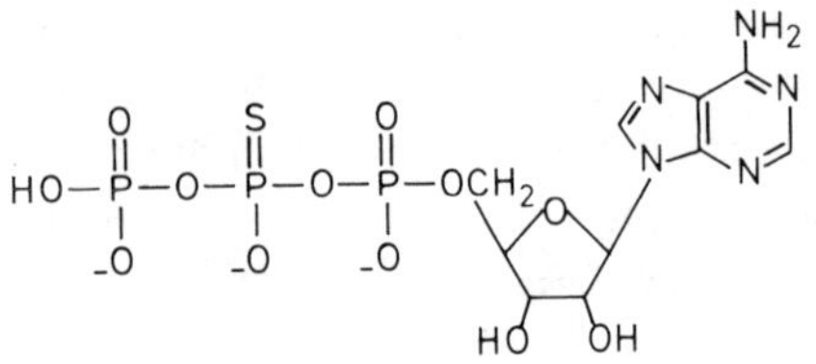

HO—P(=O)(O⁻)—O—P(=S)(O⁻)—O—P(=O)(O⁻)—OCH₂–[ribose, HO OH]–adenine (NH₂)

42

and Goody proposed a stereochemical reason for the difference in the behaviour of the diastereomers toward kinases. The ADP-Mg^{2+} complex that is the substrate for pyruvate, involves interactions with the α- and β-phosphate residues (28). If one accepts that Mg^{2+} cations do not generally interact with sulfur, but rather with oxygen atoms of phosphorothioates (29), the structures for the R and S configuration of adenosine 5′-(*O*-1-thiodiphosphate) differ remarkably. Although the affinities of the isomers of **36** to pyruvate kinase might be similar, the spatial position of the β-phosphate groups relative to the simultaneously bound phosphoenolpyruvate can be such that P—O—P bond formation is excluded for one isomer.

High-pressure liquid chromatography resolves chemically synthesized **36** into two fractions that behave differently toward ADP-utilizing enzymes (22). This technique could therefore be a general means of separating the diastereomers of thiophosphate analogs of nucleoside 5′-polyphosphates.

The properties of nucleotide thiophosphate analogs with various enzymes are summarized in Tables 4.1 and 4.2. Diastereomer A of adenosine 5′-(*O*-1-thiotriphosphate) (**37**) was employed by Eckstein et al. (31) in an original approach to elucidate the stereochemistry of polymerization

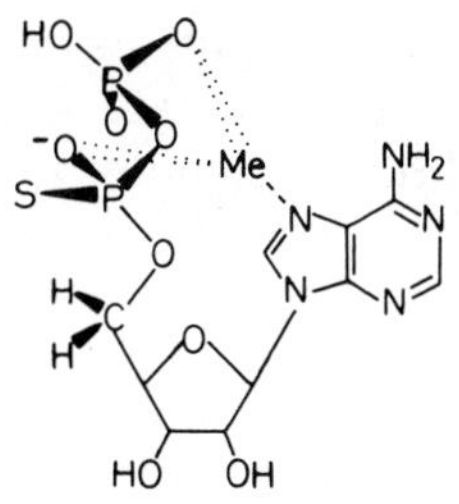

36 S-CONFIGURATION

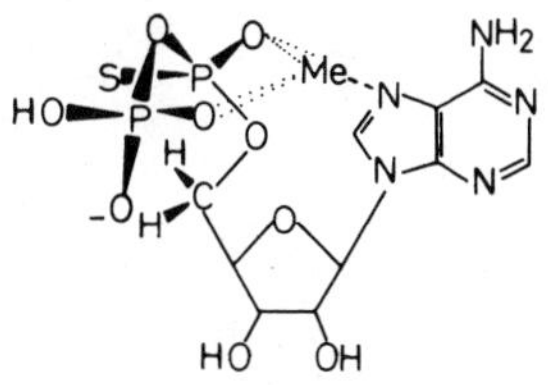

36 R-CONFIGURATION

by DNA-dependent RNA polymerase from *E. coli*. As a model reaction the polymerization of **37** and UTP on the alternating template poly[d(A–T)·d(A–T)] was chosen. Three classes of mechanisms of phosphodiester bond formation can be envisaged:

1. Those that lead to inversion of configuration.
2. Those that result in retention of configuration.
3. Those that lead to racemization of configuration at the α-phosphate of the nucleoside 5′-triphosphate.

Table 4.1 Properties of Thiophosphate Analogs with Various Enzymes

Enzyme	Compounds and Properties[a]
Alkaline phosphatase (*E. coli*)	ADPβS; competitive inhibitor
DNA-dependent RNA	UTPγS; substrate
polymerase (*E. coli*)	ATPγS; substrate
Myosin ATPase (Rabbit)	ATPγS; substrate
Methionyl-tRNA synthetase (Yeast)	ATPγS; substrate
Phenylalanine-tRNA synthetase (*E. coli*)	ATPγS; substrate
DNA polymerase II (*E. coli*)	dATPγS; substrate
Polynucleotide phosphorylase (*M. luteus*)	ADPβS; substrate

Source. Reference 30.

[a] The greek letters indicate the location of the phosphorothioate group within the polyphosphate chain.

Phosphodiester bond formation can be assumed to proceed via a pentacoordinated phosphorus intermediate as outlined in Schemes 4.1 and 4.2. In the case of an "in line" mechanism, where the incoming 3′-OH

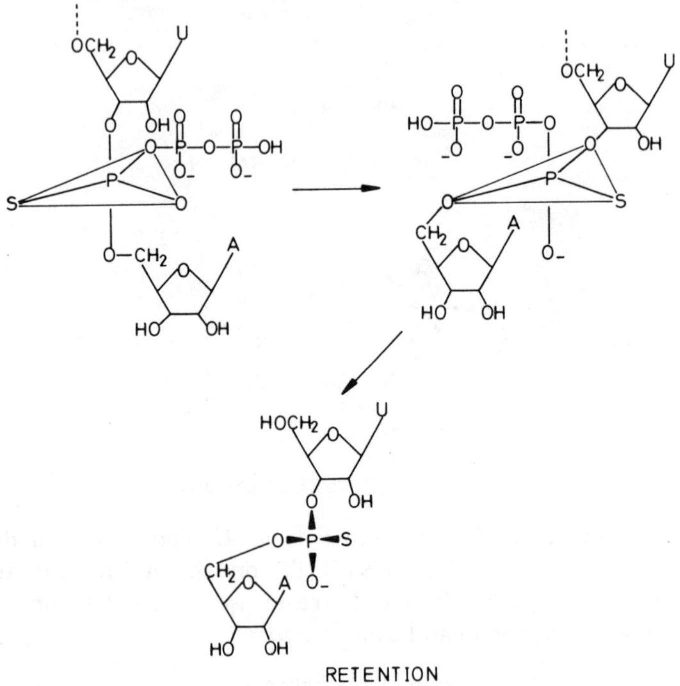

RETENTION

Scheme 4.1

group as well as the leaving pyrophosphosphate both occupy apical positions, inversion of configuration should occur. If an "adjacent" mechanism occurs, the trigonal bipyramid must undergo a pseudorotation to place the leaving group (pyrophosphate) in an apical position. This would lead to overall retention of configuration. Retention of configuration would also be observed, if an enzyme-nucleotide intermediate were formed by nucleophilic attack of a functional group at the enzyme active site with the α-phosphate moiety of the substrate and subsequent phosphorylation of the 3′-OH group of the growing RNA chain by this enzyme-nucleotide intermediate. The product of enzymatic polymerization, the alternating poly r(A_s-U) in which the internucleotidic linkage between the 3′-position of uridine and the 5′-position of adenosine is a phosphorothioate, was degraded by the combined action of phosphodiesterase and pancreatic ribonuclease A. The enzymatic hydrolysis finally yielded adenosine 5′-phosphate and uridine 2′,3′-cyclic phosphorothioate, the latter proving to be pure endo isomer. Since it is known that transesterification of endo uridine 2′,3′-cyclic phosphorothioate leads to a phosphorothioate ester

Table 4.2 Substrate Properties of Nucleotide Thiophosphate Analogs

Reactants	Enzyme	Products
AMP + ADPαS	Adenylate kinase	ADPαS, ATPαS, ATP, ADP
AMPS + ATP	Adenylate kinase	ADPαS, ATPαS, AMPS, ADP, AMP
ADP + ADPβS	Adenylate kinase	ATPγS, ATP, AMP
ADPβS	Adenylate kinase	No reaction
ADP + UTPγS	Nucleosidediphosphate kinase	AMP, ATP, ATPγS, UDP
GDP + UTPγS	Nucleosidediphosphate kinase	GTPγS, UDP
s^6IDP + UTPγS	Nucleosidediphosphate kinase	s^6ITPγS, UDP
cl^6IDP + GTPγS	Nucleosidediphosphate kinase	cl^6ITPγS, GDP
ADPβS + ATP	Nucleosidediphosphate kinase	No reaction

Source. Reference 30.

with the R configuration, the phosphorothioate ester bond formed by DNA-dependent RNA polymerase must also have the R configuration (see Scheme 4.3). It could later be shown that ATPαS isomer A has the S configuration (32), thus indicating that the mechanism of phosphodiester bond formation by RNA polymerase involves inversion of configuration.

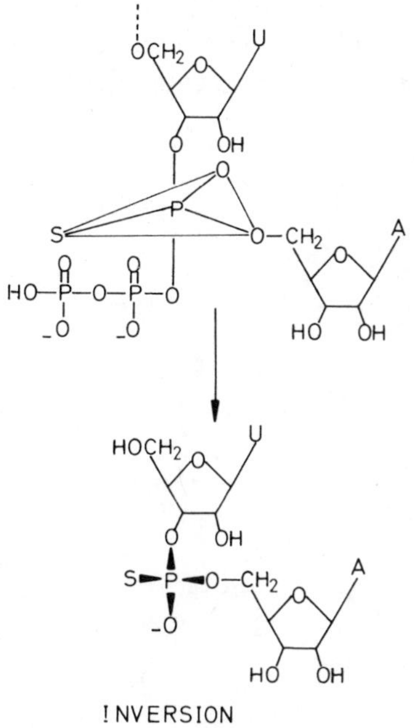

Scheme 4.2

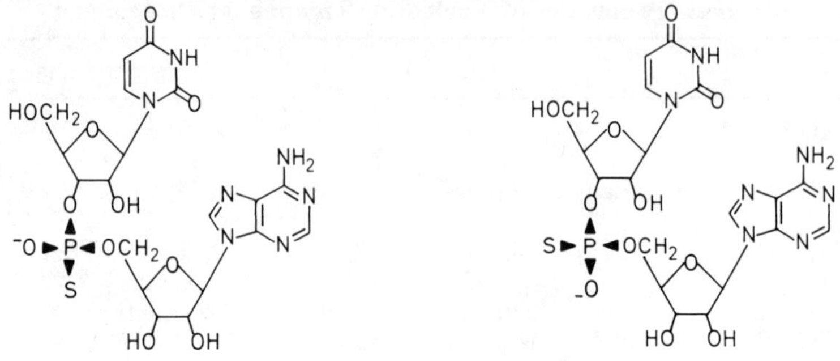

S CONFIGURATION

R CONFIGURATION

Scheme 4.3

An analogous attempt was undertaken by Eckstein et al. (33) to unravel the stereochemical course of internucleotide bond formation by tRNA nucleotidyltransferase from yeast.

The absolute stereochemistry of the diasteromers of adenosine 5′-(*O*-2-thiotriphosphate) (**42**) was established by Jaffe and Cohn (34). The Mg^{2+}-complexes of the A and B isomer of **42** could be distinguished by ^{31}P-nmr spectroscopy (35). Moreover, it was observed that Mg^{2+}-cations chelate **42** through the phosphate oxygen atoms, while Cd^{2+}-cations chelate **42** and adenosine 5′-(*O*-3-thiotriphosphate) through the sulfur atoms (34). Therefore, with respect to the geometry of the phosphate-metal ion binding the Mg^{2+}-complex of **42** isomer B is equivalent to the Cd^{2+}-complex of

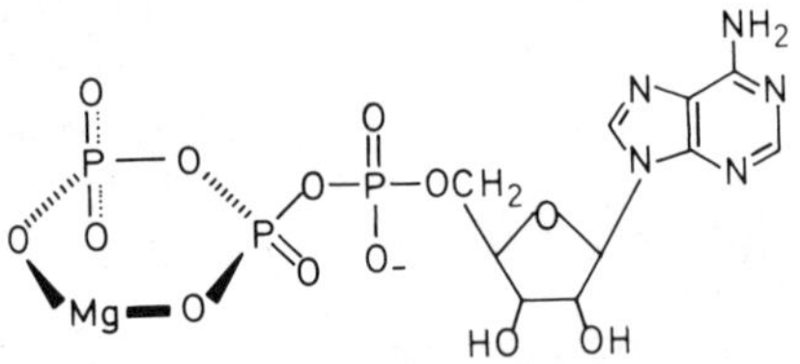

43

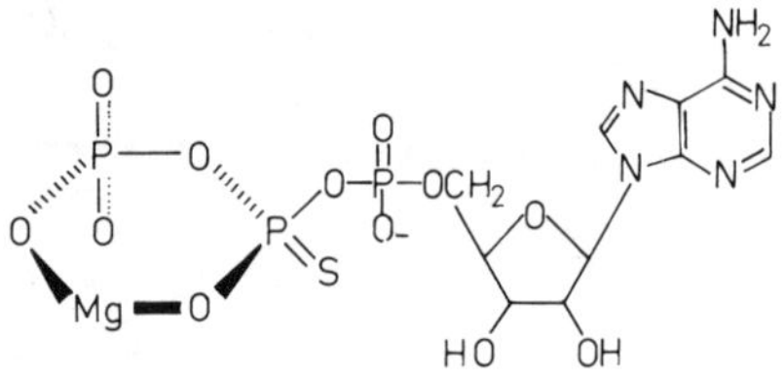

44 B-ISOMER

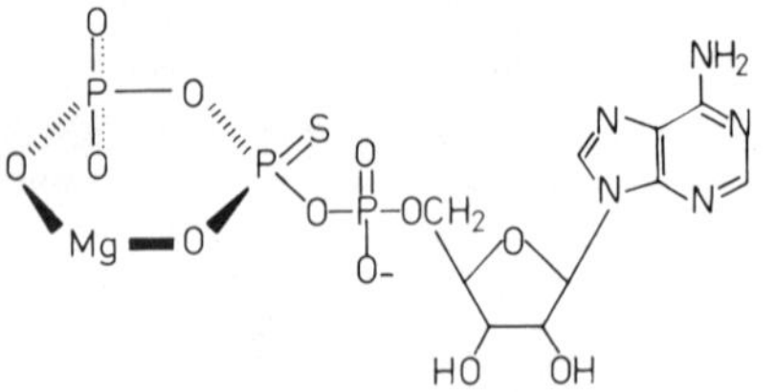

45 A-ISOMER

42 isomer A. As a consequence, the stereospecificity of hexokinase and pyruvate kinase for adenosine 5′-(*O*-2-thiotriphosphate) (**42**) should be inverted, if Cd^{2+} replace Mg^{2+} as chelating cations. This was indeed observed with hexokinase (34). It is interesting to note that pyruvate kinase in principle gave similar results, although the stereospecificity of this enzyme is less pronounced than that of hexokinase. The absolute configuration of $Co^{(III)}(NH_3)_4ATP$ has been determined by X-ray structural analysis (36), and since this compound is a substrate of hexokinase, Cornelius and Cleland concluded that the active configuration of the natural substrate Mg^{2+}-ATP should be identical (**43**) (36). Since the B isomer of adenosine 5′-(*O*-2-thiotriphosphate) (**42**) is the preferred substrate for hexokinase, the stereochemistry of the respective Mg^{2+}-complex should be similar to **43**. The stereochemistry of the A and B isomers of adenosine 5′-(*O*-2-thiotriphosphate) is depicted in formulas **44** and **45**.

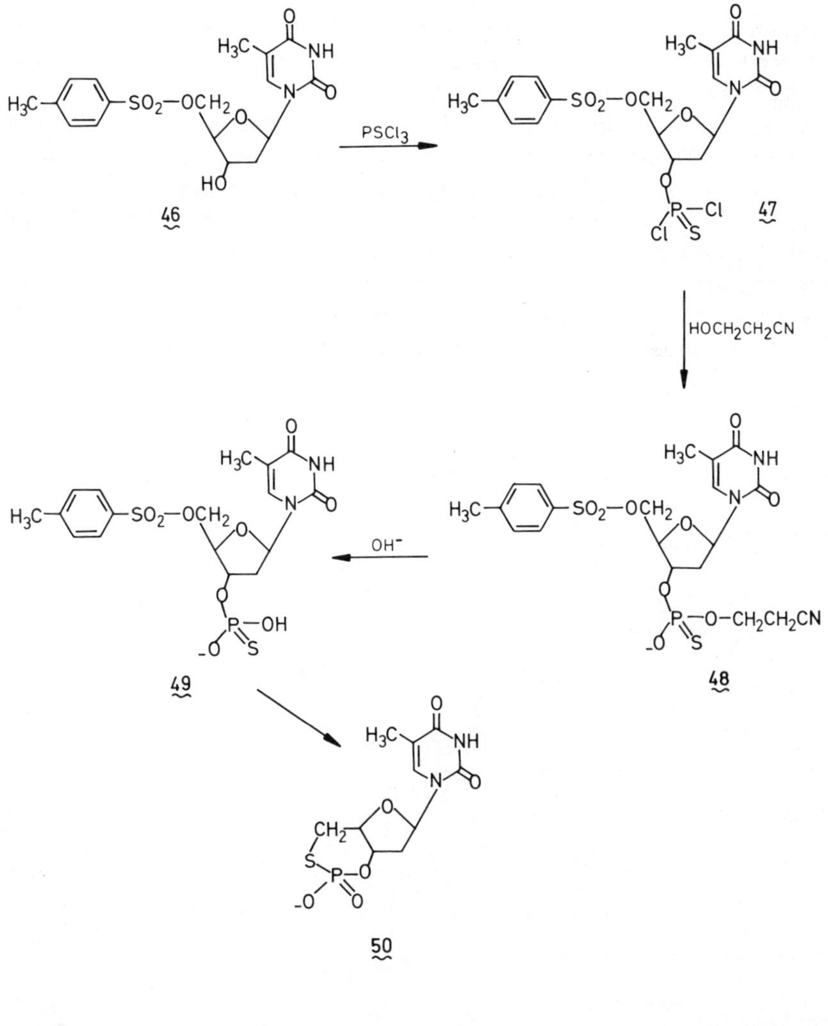

In addition to their interesting biological properties, nucleoside phosphorothioates offer the possibility of specific chemical reactions at the phosphorothioate moieties. Thus they react with mercurials (37) and thiol reagents (27). Alkylation of phosphorothioates leads to formation of *S*-alkyl phosphorothioates (27, 38). Nagyvary et al., observed that displacement of tosyl groups in 5′-tosylnucleoside 3′-phosphorothioate cyanoethyl esters occurred with formation of the respective nucleoside *S*-thiophosphate esters (39, 40). Treatment of **48** with dilute alkali at room temperature led to the respective 3′-phosphorothioate **49**, which then cyclized to 5′-deoxy-5′-thiothymidine 3′,5′-cyclic phosphorothioate (**50**). If **48** was kept in concentrated solution at elevated temperature, oligomerization occurred. The oligonucleotides formed contained the unnatural P—S—C(5′) linkage. The displacement of tosyl groups by phosphorothioates was also employed in a stepwise fashion for the synthesis of oligonucleotides (40). Nucleoside phosphorothioates were found to react with acrylonitrile preferentially to form *S*-2-(cyanoethyl) esters, although the corresponding *O*-ester was always produced as a byproduct in this reaction (40).

Inorganic dithiophosphate (**51**) was introduced by Bradbury and Nagyvary (41) as a mild thiophosphorylating agent. O^2,2′-Anhydro-1-β-D-arabinosylcytosine (**52**) reacted with **51** to give predominantly O^2,2′-*O*-anhydro-1-β-D-arabinosylcytosine 3′-phosphorothioate (**53**). The latter gives 2′-deoxy-2′-thiocytidine 2′,3′-cyclic phosphorothioate (**54**) in a spontaneous intramolecular rearrangment. ^{1}H-Nmr spectroscopy disclosed 2′-endo and anti conformation for **54**, a surprising observation, since cytidine 2′,3′-cyclic phosphate was shown to adopt the syn conformation (42).

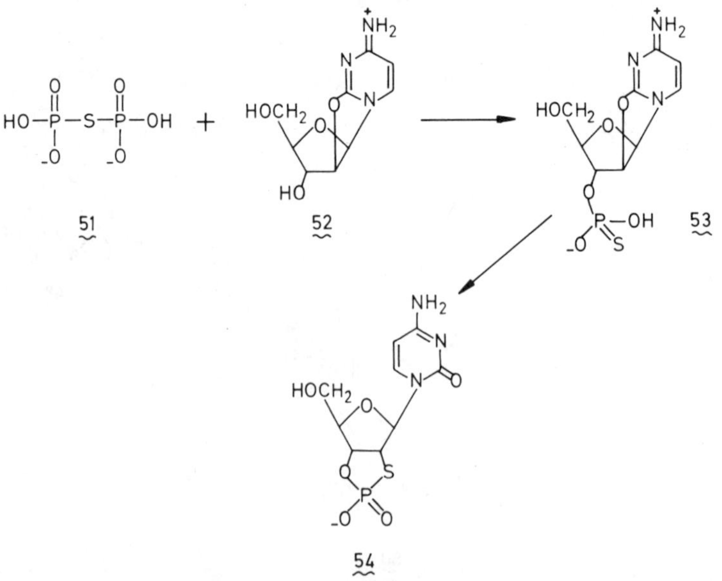

51 52 53

54

Although chemical hydrolysis of **54** was considerably faster than that of cytidine 2′,3′-cyclic phosphate, the former was not hydrolyzed by pancreatic ribonuclease A. It is not known if it is an inhibitor of ribonuclease A.

4.3. NUCLEOTIDES WITH ALTERED P—O—C BONDS

Michelson reported the synthesis of oligouridylic acids containing P—S—C(5′) bonds by polymerization of 5′-deoxy-5′-thiouridine 2′,3′-cyclic phosphate (43). The first nucleotide with a P—S—C(5′) bond, 5′-deoxy-5′-thioinosine 5′-phosphate (**55**), was prepared by alkylation of inorganic thiophosphate with 5′-deoxy-5′-iodoinosine (44). The analog **55**

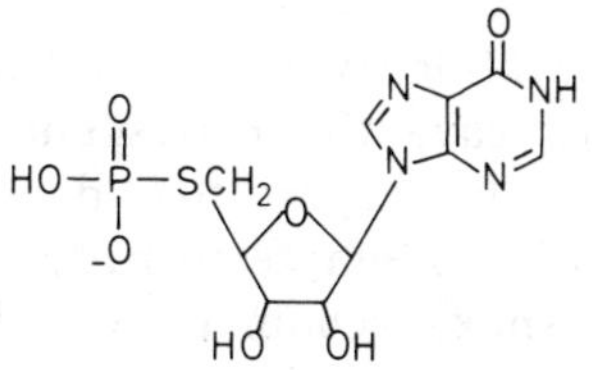

55

proved to be as good a substrate for inosine 5′-phosphate dehydrogenase as inosine 5′-phosphate itself (44). The same procedure was applied to the synthesis of the 5′-deoxy-5′-thiophosphates of uridine (**56**), adenosine (**57**), and thymidine (**58**) (45, 46). They were converted to the corresponding 5′-deoxy-5′-thiotriphosphates **59**, **60**, and **61** by conventional methods (45, 46). 5′-Deoxy-5′-thiouridine 5′-diphosphate (**62**) was formed on treatment of **59** with myokinase (44). Unexpectedly, the UTP and ATP analogs **59** and **60** respectively, were neither substrate nor inhibitor of DNA-dependent RNA polymerase from *E. coli* (45). Evidence was derived from ^{1}H-nmr spectroscopy that nucleotides with P—S—C(5′) bonds do not exist in the gauche-gauche conformation normally adopted by the naturally occuring nucleotides. This led Scheit et al., to assume that gauche-gauche conformation about the C(4′)—C(5′) bond is an essential prerequisite for substrates of RNA polymerase from *E. coli* (45). Similar observations were made in the case of the UDP analog **62** and the dTTP analog **61** that were neither substrates nor inhibitors of polynucleotide phosphorylase from *E. coli* (47) or DNA polymerase I (46) from *E. coli* respectively. In contrast, 5′-deoxy-5′-thioadenosine 5′-triphosphate (**60**) substituted rather effectively for ATP as substrate for hexokinase, thus indicating that a distinct stereochemical orientation of the C(4′)—C(5′) bond is not a stringent requirement for this enzyme (45).

Jastorff and Hettler reported the first attempt to synthesize nucleotide analogs with P—N—C bonds (48). Phosphorylation of 5′-deoxy-5′-

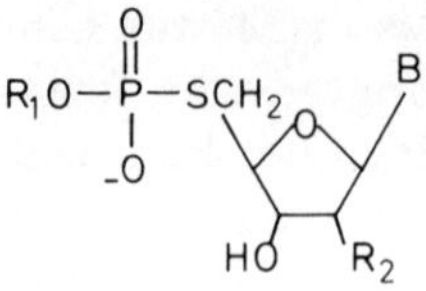

56-62

	R_1	B	R_2
56	H	Uracil	OH
57	H	Adenine	OH
58	H	Thymine	H
59	P(O)O—O—OP(O)(OH)O^{2-}	Uracil	OH
60	P(O)O—O—P(O)(OH)O^{2-}	Adenine	OH
61	P(O)O—O—P(O)(OH)O^{2-}	Thymine	H
62	P(O)(OH)O^-	Uracil	OH

aminothymidine (**64**) and 5′-deoxy-5′-aminoadenosine (**63**) by means of diesterphosphorylchlorides gave the corresponding diesterphosphoramidates (**65–70**) that could be hydrolyzed to 5′-deoxy-5′-aminonucleoside 5′-*N*-phosphate esters (**71–76**). Attempts to hydrolyze the latter to the 5′-*N*-monophosphates with snake venom phosphodiesterase proved unsuc-

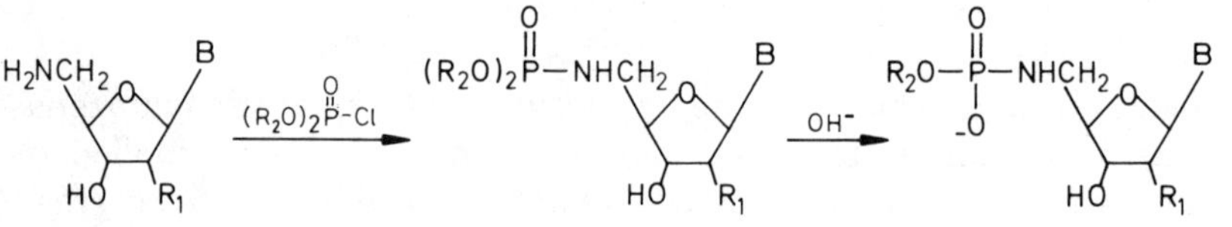

63, 64 → 65–70 → 71–76

	R_1	R_2	B		R_1	R_2	B
63	OH		Adenine	**67, 73**	OH	$CH_2—C_6H_4NO_2$	Adenine
64	H		Thymine	**68, 74**	H	$CH_2—C_6H_4NO_2$	Thymine
65, 71	OH	Cl_3CCH_2		**69, 75**	OH	$C_6H_4NO_2$	Adenine
66, 72	H	Cl_3CCH_2		**70, 76**	H	$C_6H_4NO_2$	Thymine

cessful since the products turned out to be extremely labile in the pH range 6.5–10, degrading to phosphate and 5′-deoxy-5′-aminonucleosides. Only hydrolysis of **76** with 1 N NaOH at 100° and subsequent electrophoresis and chromatography furnished evidence for the existence of 5′-deoxy-5′-aminothymidine 5′-*N*-phosphate (**77**). Hettler and Jastorff (48)

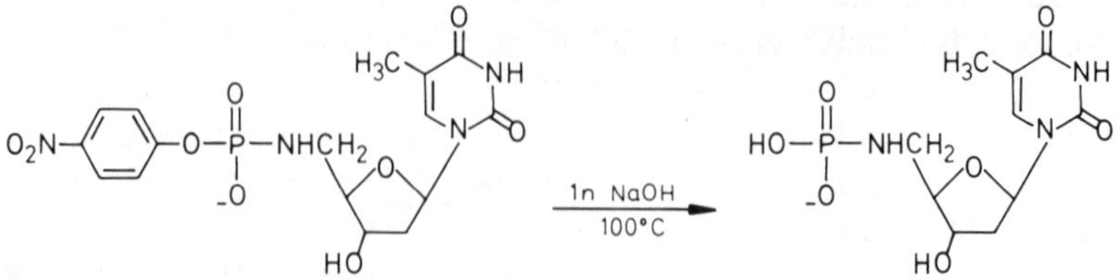

76 → (1n NaOH, 100°C) → 77

synthesized thymidylyl-(3′-5′)-5′-deoxy-5′-aminothymidine (**82**), the first analog of a dinucleoside monophosphate containing an internucleotide linkage with a P—N—C bond. Phosphorylation of 5′-*O*-tritylthymidine (**78**) by β,β,β-trichloroethylphosphoryl dichloride to **79**, reaction of the latter *in situ* with 5′-deoxy-5′-aminothymidine (**64**) and removal of the blocking groups from the neutral diesterphosphoramidate intermediate **80** yielded **82**. They also recognized the possibility of the use of 5′-deoxy-5′-

$Cl_3CCH_2P(O)Cl_2$

78 **79**

Trityl = $(C_6H_5)_3C$

azidothymidine in the above synthetic sequence that would allow the elongation of the synthesized dinucleoside monophosphate at the 5′-end after conversion of the 5′-azido into a 5′-amino group (48).

This basic concept was developed further and successfully applied to the chemical synthesis of oligothymidylic acids possessing P—N—C linkages (49, 50). As already pointed out, nucleoside *N*-phosphate esters are reasonably stable in aqueous solution above pH 7. Thus oligo and polynucleotides with phosphoramidate linkages should be stable enough to allow investigation of their behavior as templates in DNA, RNA, and protein synthesis. The enzymic synthesis of such polynucleotides requires the availability of nucleoside 5′-*N*-polyphosphates. Trowbridge et al. (51) and Letsinger et al. (52) independently discovered that 5′-deoxy-5′-aminonucleosides react with trimetaphosphate in aqueous solution to give 5′-*N*-triphosphates. By this approach adenosine 5′-*N*-triphosphate (**83**) (51) and thymidine 5′-*N*-triphosphate (**84**) (52) were prepared. Compound **84** is relatively stable in solution at pH 8 and above (52), but hydrolysis occurs rapidly at pH 6 yielding 5′-deoxy-5′-aminothymidine and trimetaphosphate (52). Nevertheless, it proved stable enough to allow its investigation as a substrate for DNA polymerase I from *E. coli* (52). Thymidine 5′-*N*-triphosphate (**84**) substituted for dTTP in the replication of X174 DNA and poly[d(A-T)·d(A-T)]. It is obvious that utilization of **84** by DNA polymerase I enables the enzymic synthesis of DNA with acid labile phosphoramidate linkages at defined positions. This might be of use in sequence analysis (52, 53).

Hexokinase accepted adenosine 5′-*N*-triphosphate (**83**) as a substrate in the phosphorylation of glucose. This enzymic reaction was employed on a

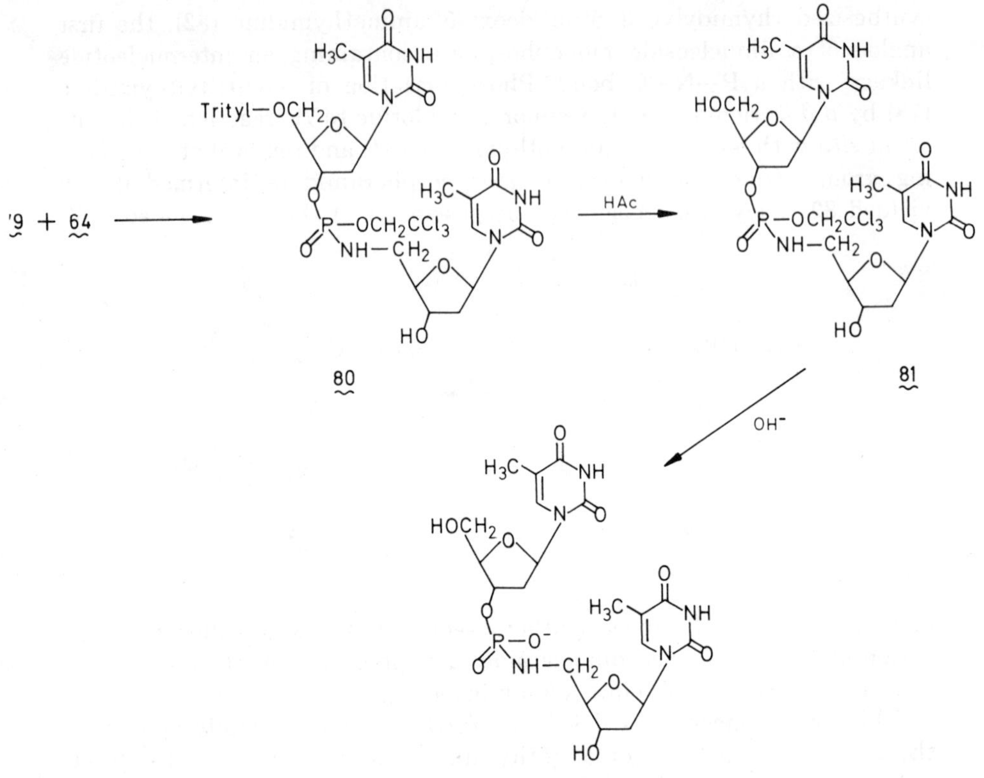

'9 + 64 → 80 —HAc→ 81 —OH⁻→ 82

Tritryl—OCH2 … P—OCH2CCl3 … NH—CH2 … HO

80

81

82

preparative scale for the synthesis of adenosine 5′-*N*-diphosphate (**85**) (54). Snake venom phosphodiesterase hydrolyzed **85** to 5′-deoxy-5′-amino-adenosine. In the absence of ADP, **85** was not a substrate for polynucleotide phosphorylase from *M. luteus* (54).

The reaction of triesterphosphites with azides represents a variation in the synthesis of diesterphosphoramidates. Thus 5′-deoxy-5′-azido-3′-*O*-

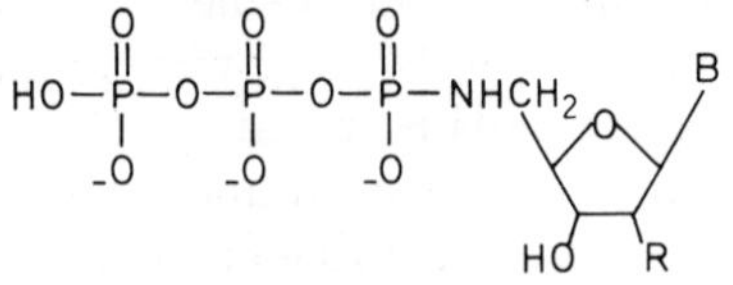

83, 84

83, R = OH B = ADENINE
84, R = H B = THYMINE

acetylthymidine (**86**) gave the 5′-*N*-phosphoric acid diphenyl ester (**87**) upon reaction with triphenylphosphite. Alkaline hydrolysis of **87** afforded thymidine 5′-*N*-phosphate phenylester (**88**) (55). The relative stability of

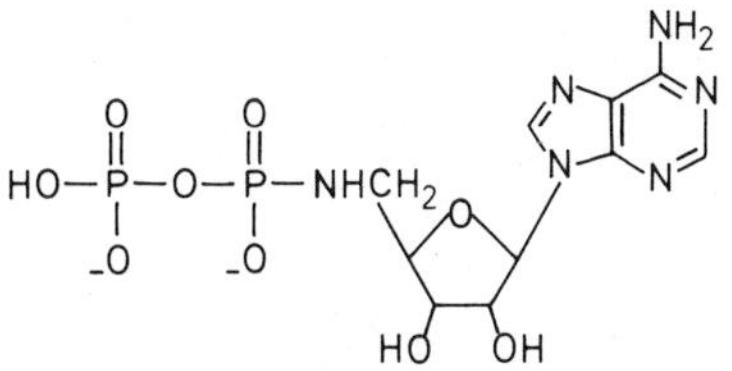

85

esterphosphoramidates obviously commended itself for the synthesis of analogs of adenosine 3′,5′-cyclic phosphate with P—N—C bonds (56). 5′-Deoxy-5′-aminoadenosine (**63**) or its *N*-substituted derivatives (**89–91**) were phosphorylated by di-(4-nitrophenyl)phosphorylchloride. The diesterphosphoramidates **92–95** were cyclized in a fast reaction in the

86 87 88

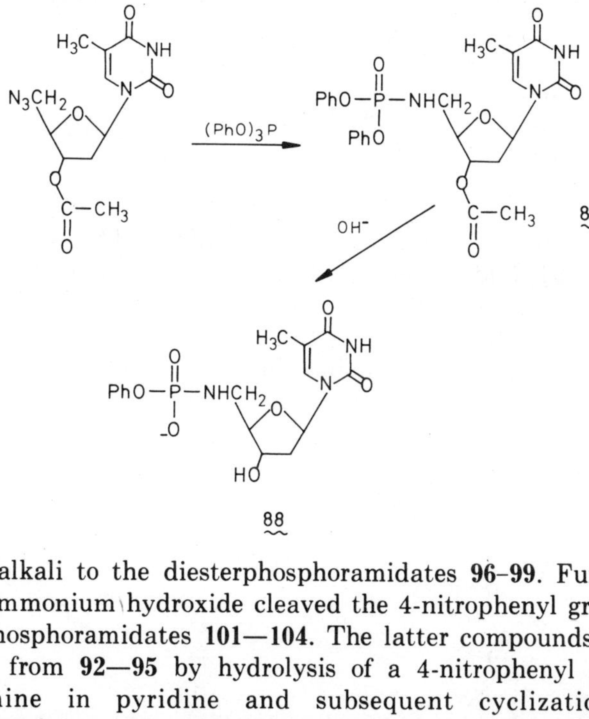

presence of alkali to the diesterphosphoramidates **96–99**. Further treatment with ammonium hydroxide cleaved the 4-nitrophenyl group to give the cyclic phosphoramidates **101—104**. The latter compounds could also be obtained from **92—95** by hydrolysis of a 4-nitrophenyl group with triethyl amine in pyridine and subsequent cyclization of the monoesterphosphoramidate **100** with potassium t-butoxide. Murayama et al., later discovered that the intermediates **96–99** were formed in a stereospecific reaction (57). Careful product analysis of the reaction leading to **99** revealed that two diastereomers of 5′-deoxy-5′-(octylamino) adenosine 3′-phosphoric acid-(4-nitrophenylester)-3′,5′-cycloamide (**99**) were actually formed. The absolute configuration of **99** could be assigned by ^{1}H-nmr spectroscopy. The formation of **99a** is largely favored over **99b**. It is interesting to note that 5′-deoxy-5′-amino-guanosine 5′-*N*-phosphoric acid di(4-nitrophenyl)-ester (**105**) (58) does not cyclize to 5′-deoxy-5′-

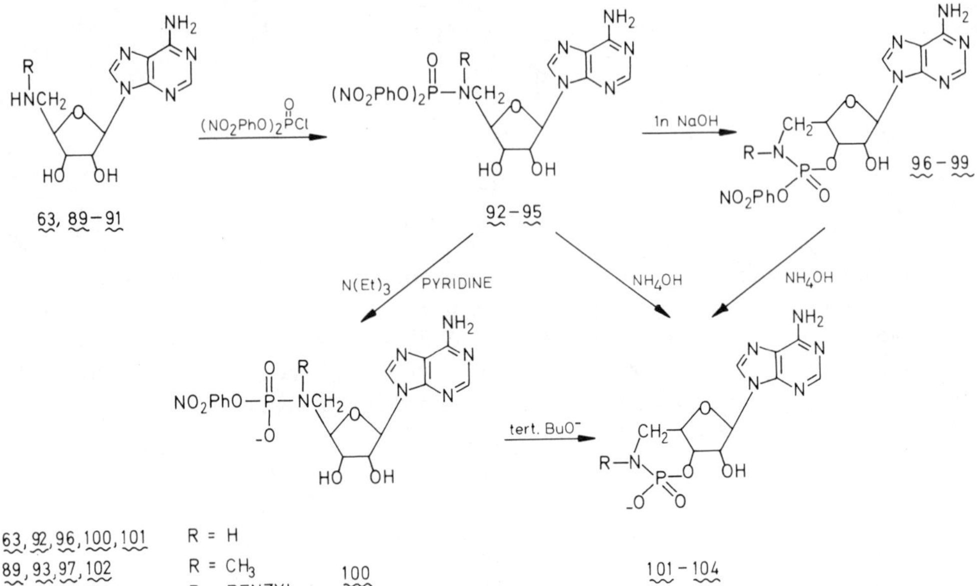

63, 92, 96, 100, 101 R = H
89, 93, 97, 102 R = CH_3
90, 94, 98, 103 R = BENZYL
91, 95, 99, 104 R = n-OCTYL

aminoguanosine 3′-5′-cyclic phosphoramidate. The phosphoramidate analogs **101–104** of adenosine 3′,5′-cyclic phosphate were unstable at pH 5; partial hydrolysis occurred at pH 7 and they were stable up to 5 hours at pH 9 (56).

The need for a greater stability in aqueous solutions at neutral pH prompted the synthesis of the respective thiophosphoramidates (**108, 109**) (59). Their synthesis closely followed the procedures for the preparation of **101–104**. The expectations were partially fulfilled. Compounds **108** and **109** were approximately 40% more stable than their oxygen congeners **102** and **104** within the pH range 5 to 7. The biological effects of the 5′-amido analogs of adenosine 3′,5′-cyclic phosphate have been investigated (60). Compound **101** stimulated cAMP-dependent protein kinase and competed with the parent compound in binding to cAMP-binding protein (60).

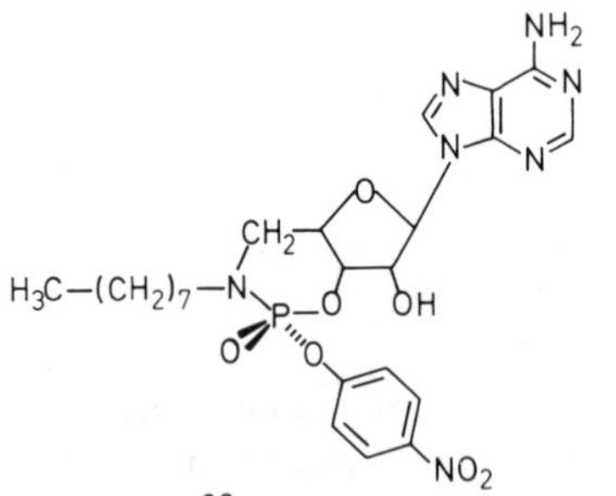

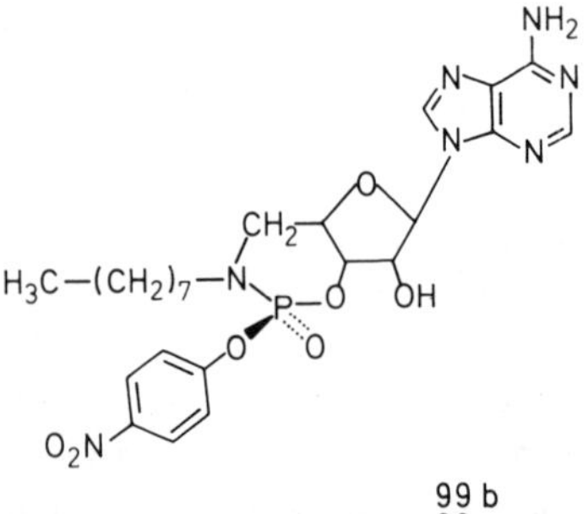

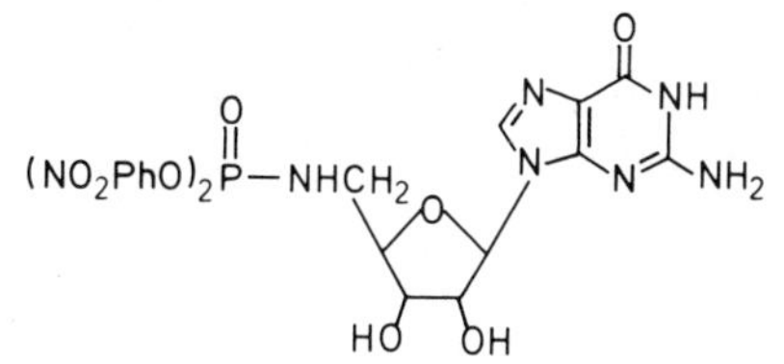

105

Phosphorylation of 3′-deoxy-3′-aminoadenosine (**110**) by phosphorus oxychloride in triethylphosphate gave 3′-deoxy-3′-aminoadenosine 5′-phosphate (**111**) which was cyclized to 3′-deoxy-3′-aminoadenosine 3′,5′-cyclic phosphate (**112**) by known procedures (61). Compound **112** was considerably more stable at pH 7 in aqueous solution than **101**. The analog

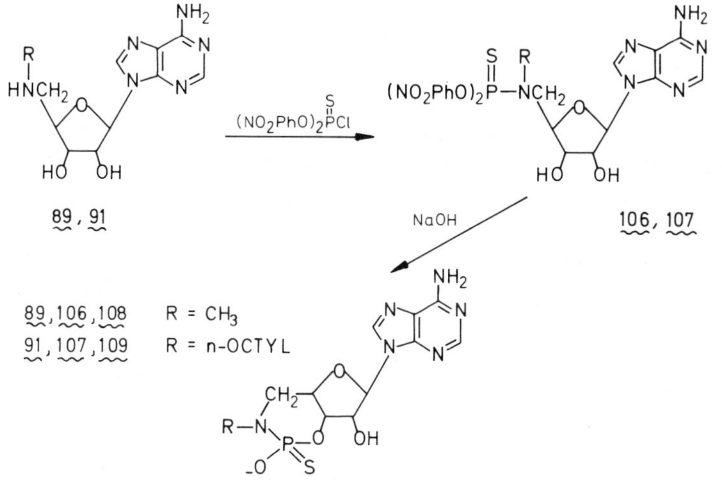

89, 91 → $(NO_2PhO)_2P(S)Cl$ → 106, 107 → NaOH → 108, 109

89, 106, 108 R = CH_3
91, 107, 109 R = n-OCTYL

112 stimulates cAMP dependent protein kinases from bovine brain and rabbit muscle respectively (61).

The chemical synthesis of 3′-deoxy-3′-aminoadenosine 3′-*N*,5′-cyclic phosphorothioate (**118**) turned out to be rather difficult (62). Phosphorylation of unprotected 3′-deoxy-3′-aminoadenosine (**110**) by thiophosphorylchloride in triethylphosphate resulted, not unexpectedly, in formation of 3′-deoxy-3′-aminoadenosine 3′,2′-cyclic phosphorothioate (**114**) (62). The formation of the side product **115** was explained by Morr and Ernst (62) to be due to the alkylation of **114** by triethyl phosphate with subsequent elimination of ethylmercaptane accompanied by hydrolysis of the 2′—O—P bond. This explanation, however, does not account for the formation of the *O*-ethyl ester. Compound **114** exists in the form of two diastereomers, which can be separated by anion-exchange chromatography. Phosphorylation 3′-deoxy-3′-(carbo-t-butoxy)aminoadenosine finally led to the desired 3′-deoxy-3′-aminoadenosine 3′,5′-cyclic phosphorothioate (**118**) (62). The diastereomers **118** exo and **118** endo are formed in a ratio of 80:20. They can be separated by anion-exchange chromatography (62).

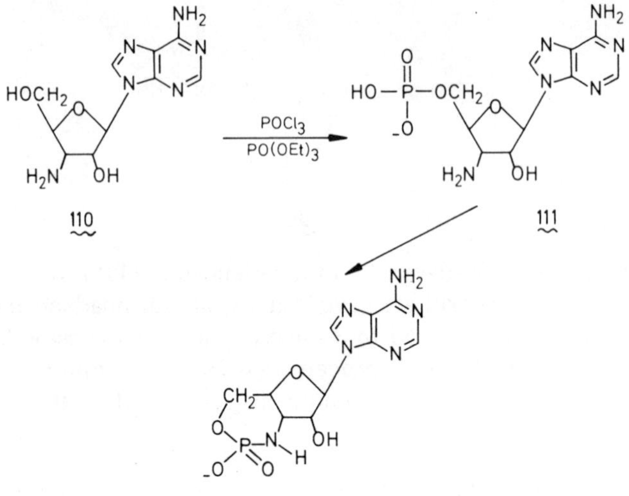

Thiophosphorylation of 5′,3′-dideoxy-5′,3′-diamino-adenosine (**119**) with thiophosphorylchloride proceeded as expected (62). The exo and endo isomer of **121** could be separated. Compound **121** proved to be stable at neutral pH (62).

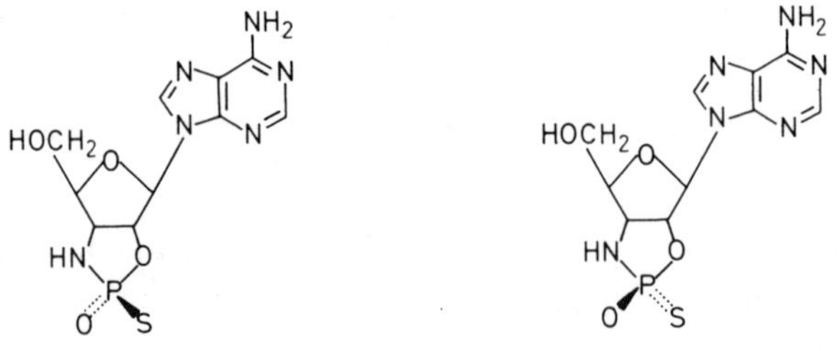

4.4 NUCLEOSIDE PHOSPHITES

Schofield and Todd (63) prepared nucleoside phosphites in good yield by reacting a suitably protected nucleoside with phosphorous acid in the presence of di-*p*-tolylcarbodiimide. Phosphorous acid, which resembles monoesterphosphoric acid with respect to its number of dissociation constants, did not need to be protected (this aspect is discussed in

110 $\xrightarrow{PSCl_3}$ 113 $\xrightarrow{NaOH}$ 114 → 115

113

114

115

$\xrightarrow{PSCl_3}$ 116

H^+

117

pH 10

EXO + ENDO

118

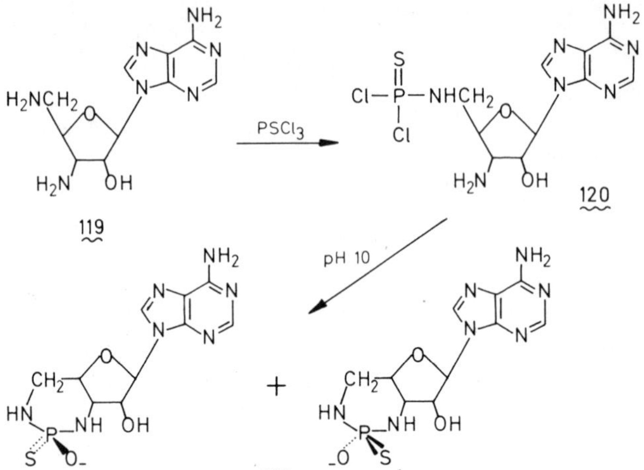

Chapter 6). On the contrary, the monoesters of phosphorous acid proved to be unreactive. When the procedure was applied to unprotected nucleosides, thymidine gave mainly the 3′-phosphite (**122**), but 2′-deoxyadenosine more of the 5′-phosphite (**123**). The purine nucleoside phosphites could be oxidized to the corresponding phosphates by treatment with permanganate (63). This method for the synthesis of nucleoside phosphites was taken up and reinvestigated by Holy and Sorm (64).

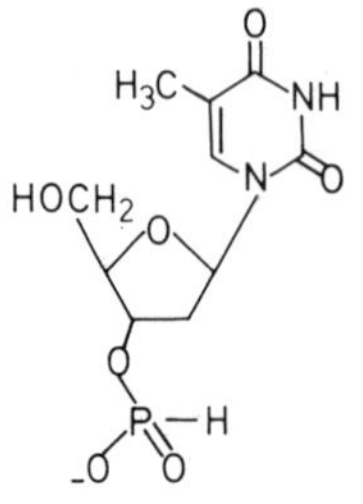

Transesterification reactions of triesters of phosphorous acid are well-known. Holy (65) applied this reaction to nucleosides and observed the general utility of this procedure for the synthesis of nucleoside phosphites. Transesterification with triphenylphosphite worked most efficiently under acid catalysis. Specific reaction, of course, requires suitably protected nucleosides (65). Work-up of the reaction mixture under neutral conditions allows the isolation of the intermediate **125**. Reaction of 5′-*O*-acetyl-6-azauridine with triphenylphosphite gave 5′-*O*-acetyl-6-azauridine 2′,(3′)-phosphite (**127**). Unprotected nucleosides, such as thymidine or adenosine, react with triphenylphosphite to give a mixture of isomeric nucleoside

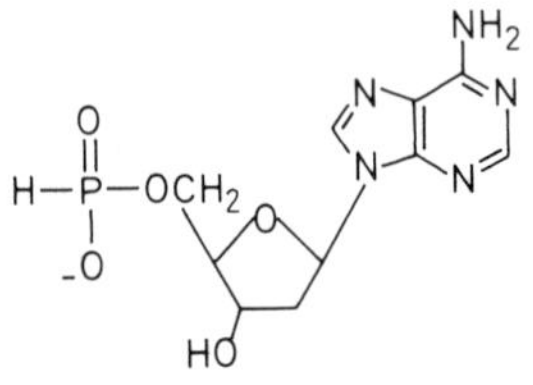

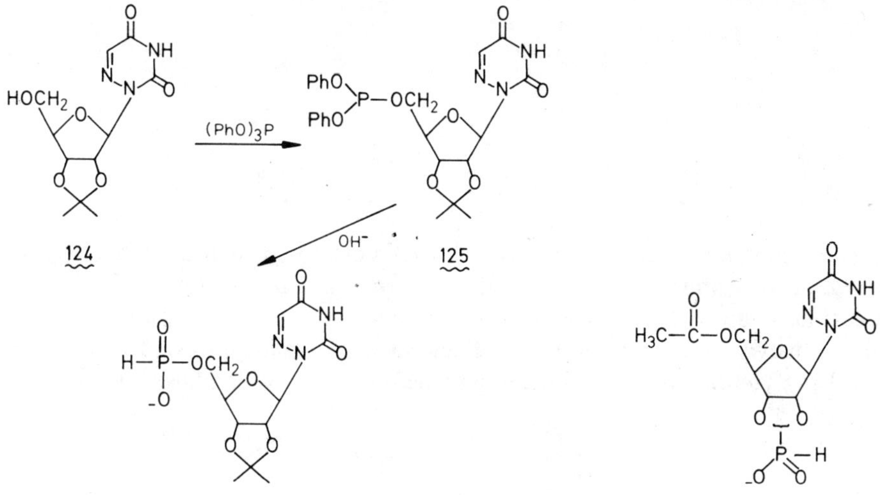

124 → (PhO)$_3$P → 125 → OH$^-$ → 126

127

phosphites (**129–131**). It should be emphasized that the transesterifications with triphenylphosphite give excellent yields of nucleoside phosphites (65). When unprotected nucleosides were allowed to react with stoichiometric amounts of triphenylphosphite, the principal product was the 2′,(3′)-phosphite (65). Even when an excess of triphenylphosphite was employed together with 'a 5′-protected nucleoside, the sole product obtained was the 2′,(3′)-phosphite. These results indicated that transesterification of ribonucleoside occurs preferentially through an intermediate

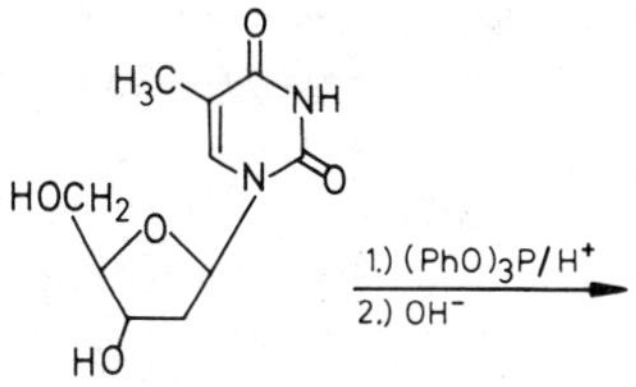

128 → 1.) (PhO)$_3$P/H$^+$ 2.) OH$^-$ → 129 + 130 + 131

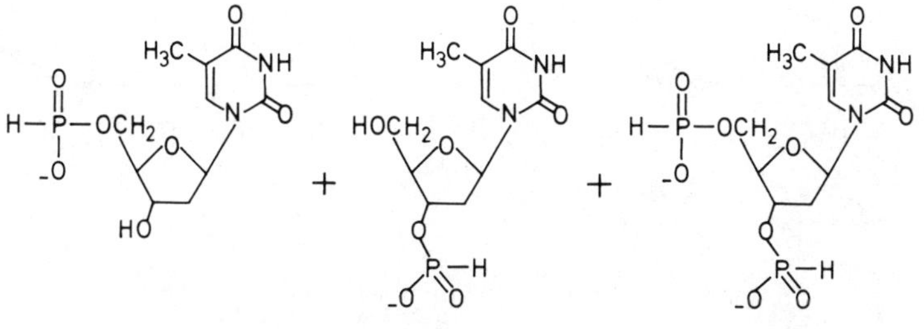

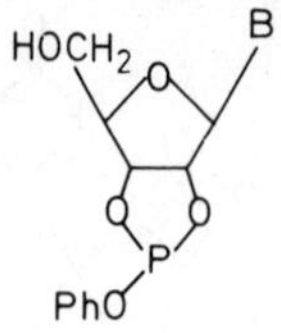

132

triesterphosphite **132** involving the 2′,3′-OH groups, which upon hydrolysis yields 2′,(3′)-phosphites. Consequently, reaction of unprotected ribonucleosides even with an excess of the less reactive triethylphosphite resulted in exclusive formation of ribonucleoside 2′,(3′)-phosphites (**133–139**) in excellent yields (66). The elegant synthesis of nucleoside phosphites by the

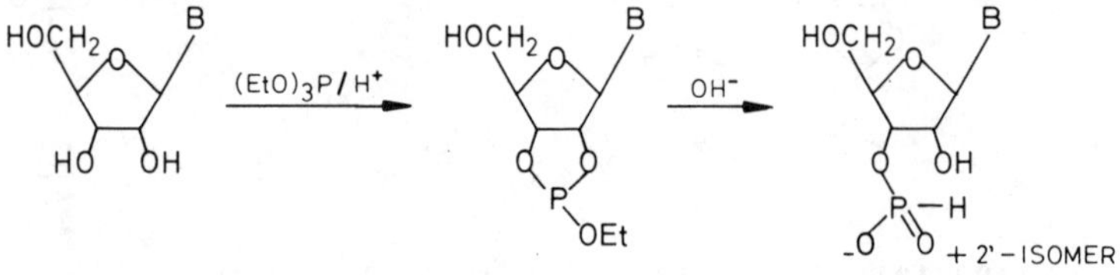

$(EtO)_3P/H^+$ — OH^- — + 2′-ISOMER

133–139

	B		B
133	Uracil	**137**	6-Azauracil
134	Cytosine	**138**	6-Azacytosine
135	Adenine	**139**	8-Azaguanine
136	Guanine		

above described transesterification procedure gained further importance when Holy et al., discovered that nucleoside phosphites react with certain α-haloketo compounds in a Perkov type reaction (67). For this reaction the authors proposed the mechanism outlined in Scheme 4.4. Compound **140** represents a reactive phosphodiester, which can participate in intra- or intermolecular phosphorylation reactions. Treatment of 6-azauridine 5′-phosphite (**144**) with hexachloroacetone leads to simultaneous formation of 6-azauridine 5′-phosphate, 6-azauridine 3′,5′-cyclic phosphate and

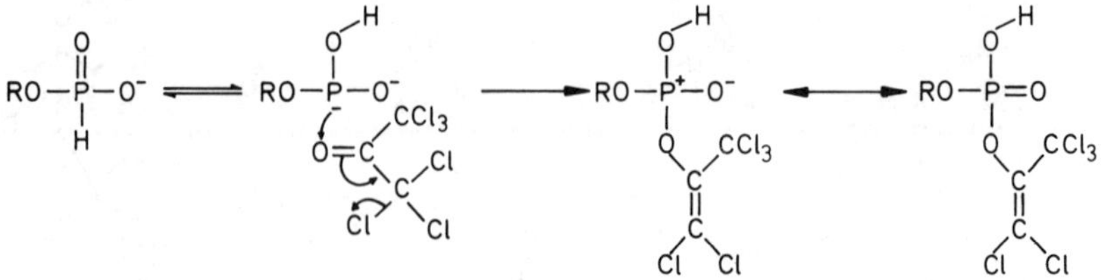

140

Scheme 4.4

P^1,P^2-di(6-azauridine-5′)pyrophosphate respectively (67). The reactive ester **143**, formed in the reaction of nucleoside 2′,(3′)-phosphites (**141**) with hexachloroacetone, gives in a nearly quantitative intramolecular phosphorylation the nucleoside 2′,3′-cyclic phosphate (**145**). The reaction of ribonucleosides with triethylphosphite, and subsequent treatment with hexachloroacetone of the resulting nucleoside phosphites represents a straightforward method for the synthesis of nucleoside 2′,3′-cyclic phosphates (68). This procedure has been successfully employed in numerous cases.

141, 142 + 2′-ISOMER — Cl_3C-CO-CCl_3 → 143, 144 + 2′-ISOMER → 145, 146

141, 143, 145 B = 6-AZAURACIL
142, 144, 146 B = 8-AZAGUANINE

4.5 NUCLEOSIDE PHOSPHONATES

Nucleoside phosphonates have not yet been isolated from biological sources. Esters of nucleoside phosphonates are easily available by an Arbuzov reaction between trialkylphosphites and deoxyiodonucleosides (69–71). Hydrolysis of nucleoside phosphonic esters could not be achieved because of a base-catalyzed elimination of the heterocyclic ring as outlined below (**147–149**). This difficulty was circumvented by Rammler et al., who

147 → 148 + 149

performed the Arbuzov reaction with triallylphosphite and 5′-deoxy-5′-iodo-3′-*O*-acetylthymidine (**150**) and obtained diallyl-5′-deoxy-3′-*O*-acetylthymidine 5′-phosphonate (**151**) (71). Removal of the allyl ester groups was achieved by catalytic hydrogenation. The correct chemical name of **152** is 5′-deoxy 5′-(dihydroxyphosphinyl)thymidine. To emphasize the structural analogy between **152** and thymidine 5′-phosphate, the descriptive nomenclature thymidine 5′-phosphonate was chosen. 5′-Deoxy-2′-deoxyuridine 5′-phosphonate was obtained in an analogous procedure (72). 5′-Deoxy-thymidine 5′-phosphonylpyrophosphate (**153**) was prepared from the imidazolidate of **152** by reaction with inorganic pyrophosphate (71). Compound **153** was synthesized with the expectation that it would

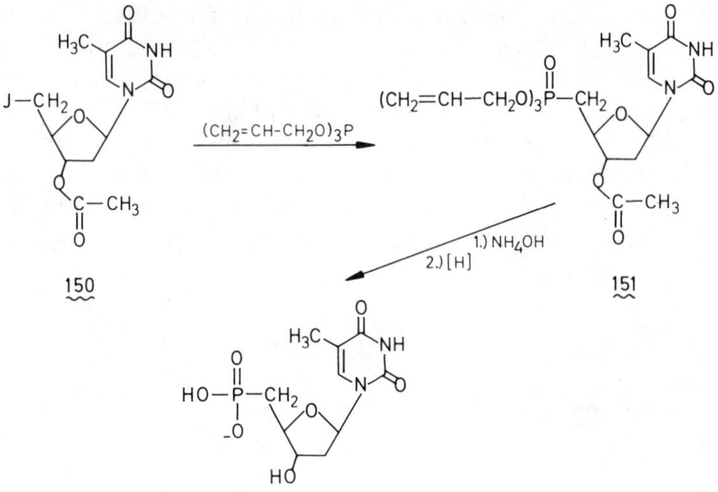

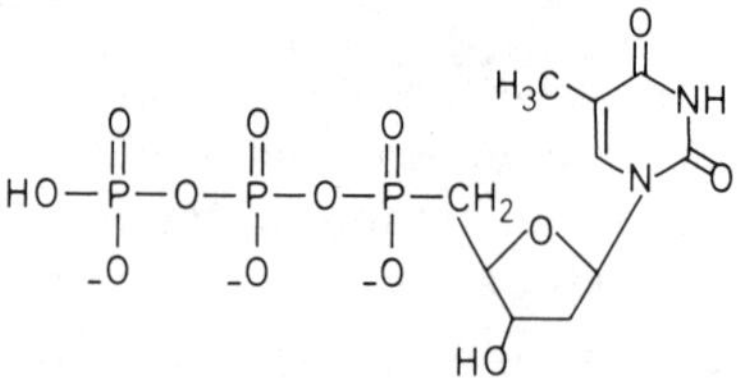

substitute for dTTP in the synthesis of synthetic deoxypolynucleotides by DNA polymerase I from *E. coli.* Such synthetic polynucleotides would possess many interesting features, in particular nonhydrolyzable phosphonate ester bonds. However, **153** turned out to be neither a substrate nor an inhibitor of DNA polymerase I from *E. coli* (71). Analogs of oligothymidylic acids prepared by chemical polymerization of 5′-deoxythymidine 5′-phosphonate (**152**) (73) were shown to be competitive inhibitors of *Microccocal* nuclease (74).

Holy developed an independent synthetic route to nucleoside phosphonates (75). 5′-Deoxy-5′-iodo-2′,3′-*O*-isopropylideneuridine (**154**) was reacted with tri-(2-benzyloxyethyl)phosphite (**155**) to give the diesterphosphonate **156**. Catalytic hydrogenation of **156** furnished the 2-hydroxyethyl ester **157** that was iodinated to the respective 2-iodoethyl ester **158**. Base-catalyzed β-elimination and acid hydrolysis converted **158** to 5′-deoxyuridine 5′-phosphonate (**159**). The overall yield of **159** in this reaction sequence was respectable 16%.

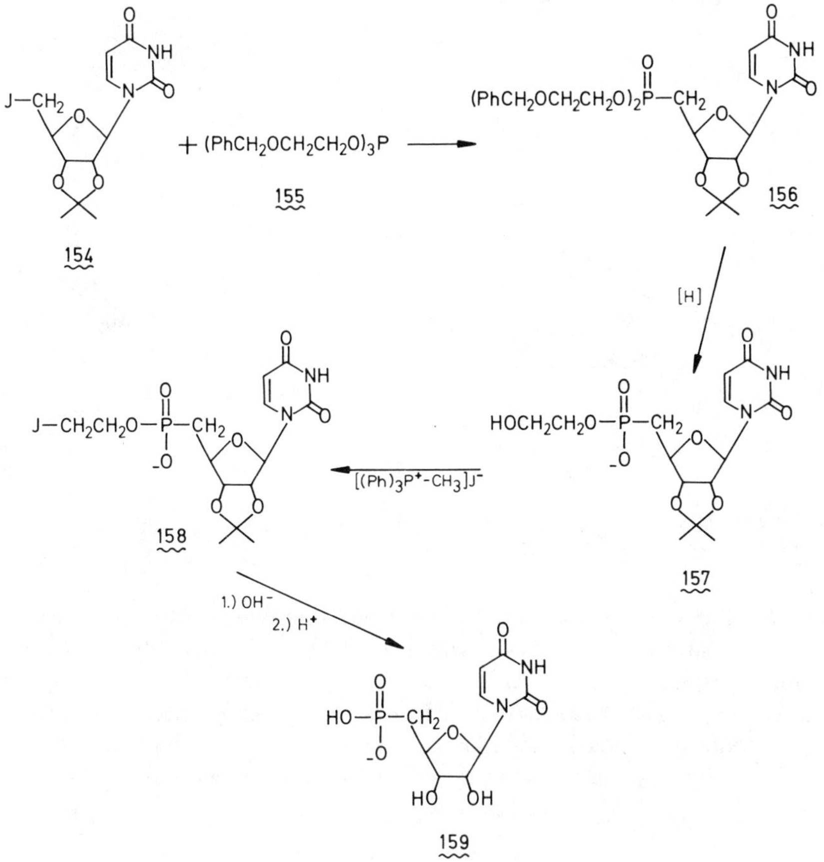

The failure of **153** to substitute for dTTP in an enzymic reaction was explained by the alteration of the spatial relationship between the α,β—P—O—P bond, where catalysis has to take place, and the nucleoside portion of the molecule (71). The shortening due to the omittance of the oxygen of the P—O—C(5′) bond should be offset by the insertion of a methylene bridge. These considerations led Moffatt et al. (76) to synthesize 5′-deoxy-5′-(dihydroxyphosphinylmethyl)nucleosides, isosteric phosphonate analogs of nucleoside 5′-phosphates. This ambitious synthetic task was approached by reaction of a protected nucleoside 5′-aldehyde (**160**, **161**) (77) with the Wittig reagent diphenyl triphenylphoranylidenemethylphosphonate (**162**). Subsequent catalytic hydrogenolysis of the α,β-unsaturated diesterphosphonates (**163**, **164**) proceeded smoothly to the 6′-deoxynucleoside 6′-phosphonic acid diphenyl esters **165**

160, 163, 165, 167, 169, 171 B = URACIL
161, 164, 166, 168, 170, 172 B = ADENINE

and **166**. Compounds **165** and **166** gave the phenylester phosphonates **167** and **168** on alkaline hydrolysis, which could then be hydrolyzed by snake venom phosphodiesterase to the 6′-deoxyhomonucleoside 6′-phosphonic acids **169** and **170**, respectively. More suitable for preparative purposes was the transesterification of **165** and **166** by reaction with sodium benzoxide. Hydrogenolysis of the resulting dibenzylester phosphonates (**171**, **172**)

gave the 6′-deoxyhomonucleoside 6′-phosphonates in essentially quantitative yields (76). The trivial name 6′-deoxyhomonucleoside 6′-phosphonate was favored by Jones and Moffatt to stress the structural relationship to the nucleoside 5′-phosphates (76). It should be noted in this context that 6′-deoxyhomoadenosine 6′-phosphonic acid ethyl ester was obtained in a rather elaborate synthetic attempt by Montgomery et al. (78).

Synthesis of the isosteric phosphonate analogs of nucleoside 3′-phosphates proved to be difficult, because of the instability of 3′-ketonucleosides under basic conditions (79) and the lack of reactivity of less basic reagents such as **162**. Condensation of 1,2:5,6-di-*O*-isopropylidene-α-D-ribohexofuranose-3-ulose (**173**) with tetraethyl methylenediphosphonate in the presence of *n*-butyllithium afforded the vinylphosphonate **174** in high yield. Compound **174** was catalytically hydrogenated to the D-allophosphonate **175**. In five subsequent steps, **175** was transformed into 3-deoxy-3-(diethoxyphosphinylmethyl)-5-*O*-benzoyl-2-*O*-acetyl-D-ribofuranosylchloride (**180**). Condensation of **180** with the mercury salts of 6-benzamidopurine, N^4-acetylcytosine, thymidine, uracil, and 6-chloropurine, followed by alkaline hydrolysis yielded the 3′-deoxy-3′-(dihydroxyphosphinylmethyl) nucleosides (**181–185**). The β-conformation of **181–185** was confirmed by ORD and nmr spectroscopy. The facile hydrolysis of the diethylesters of the phosphonates was of course because of the presence of the vicinal 2′-OH group (80). The crystal structure of 3′-deoxy-3′-(dihydroxyphosphinylmethyl)adenosine (**181**) was determined by X-ray diffraction (81) and was found to be very similar to that of adenosine 3′-phosphate (82).

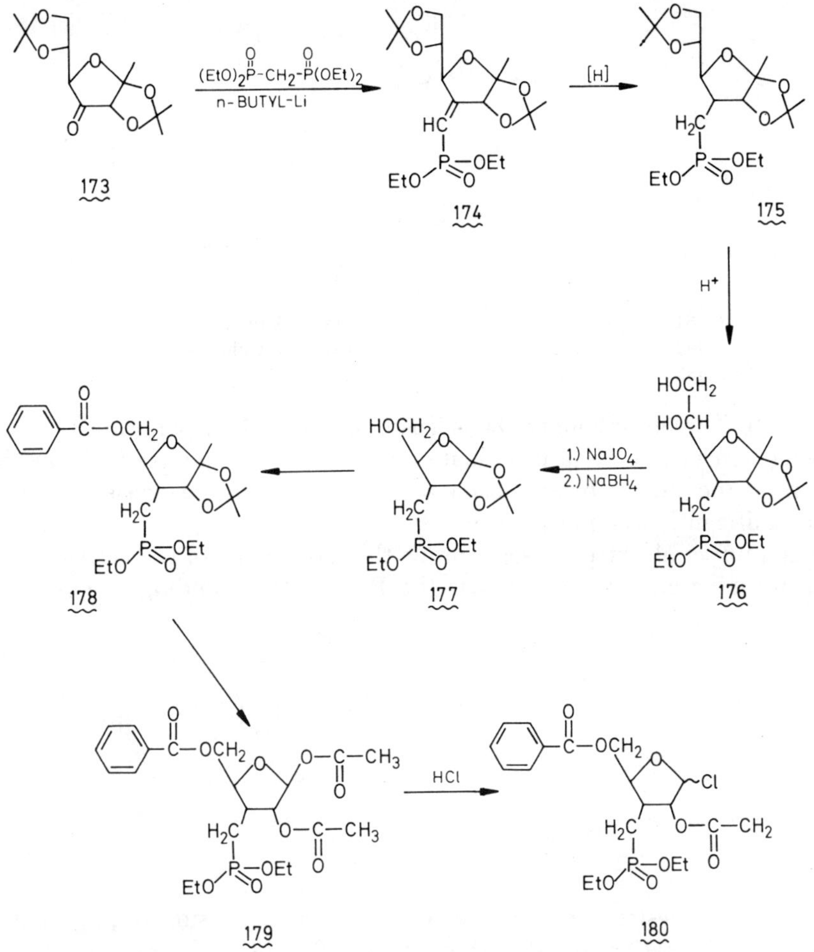

The procedure of Moffatt et al., for the synthesis of 6′-deoxyhomouridine 6′-phosphonates was adopted for the preparation of 1-(5,6-dideoxy-β-D-ribo-hexofuranosyl-6-phosphonic acid)-1,2,4-triazole-3-carboxamide

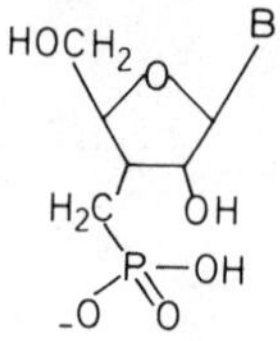

181–185

	B		B
181	Adenine	**184**	Uracil
182	Cytosine	**185**	6-Chloropurine
183	Thymine		

(**186**) (83). The phosphonate **186** is isosteric with 1-β-D-ribofuranosyl-1,2,4-triazole-3-carboxamide 5′-phosphate (ribavirin 5′-phosphate) and like the parent compound inhibited inosine 5′-phosphate dehydrogenase, although with significantly lower efficiency (83).

Hampton et al., attempted to probe the topography of the active site of AMP-utilizing enzymes at or near the P—O—C(5′) binding site. To this

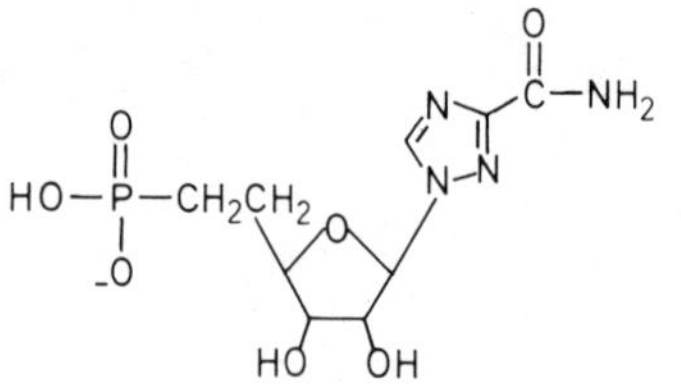

186 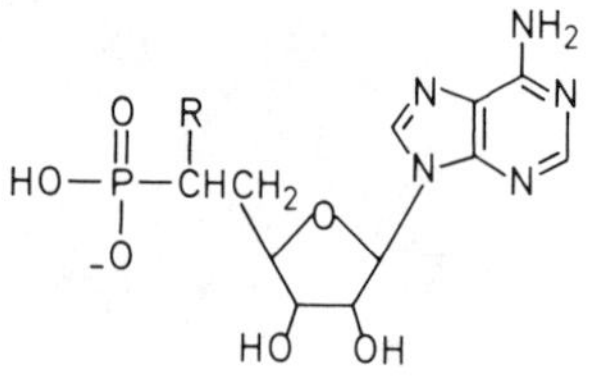 **187**

end they synthesized analogs of 6′-deoxyhomoadenosine 6′-phosphonic acid bearing substituents at carbon atom C(6′) (**187**). Hydroboration-oxidation of diphenyl [9-(5′,6′-dideoxy-2′,3′-*O*-isopropylidene-β-D-ribohex-5′-enofuranosyl)-adenine] 6′-phosphonate (**164**), a compound already described by Moffatt et al. (76), gave the 6′-hydroxy derivative **188** (84). Enzymic cleavage of the phenyl ester by snake venom phosphodiesterase and acid hydrolysis led to 5′-deoxy-5′-(*C*-dihydroxyphosphinyl)hydroxymethyl-adenosine (**189**) (6′-hydroxy-6′-homoadenosine 6′-phosphonate) (83). Oxidation of the organoborane intermediate seems to require the concomitant removal of one phenoxy group because the organoborane intermediate of the dibenzyl ester analog of **164**, which is much more stable in alkali, could not be oxidized (84). 6′-Hydroxy-6′-homoadenosine 6′-phosphonate (**189**) exists in the form of two epimers (84). Both epimers were found to be substrates for rabbit muscle adenosine 5′-phosphate kinase, but only one of the 6′-hydroxy-6′-homoadenosine 6′-phosphorylphosphonates produced, functioned as a substrate for pyruvate kinase. The epimers of **189** were found to be substrates for AMP aminohydrolases and pig muscle AMP kinase with low K_m- and V_{max}-values. Their high affinities to the aforementioned enzymes indicates that the AMP-binding sites tolerate a substituent the size of a hydroxymethyl group in place of the *O*(5′)-oxygen (84).

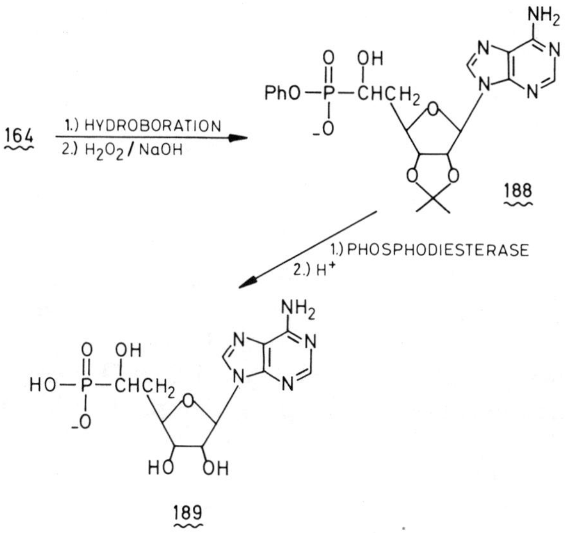

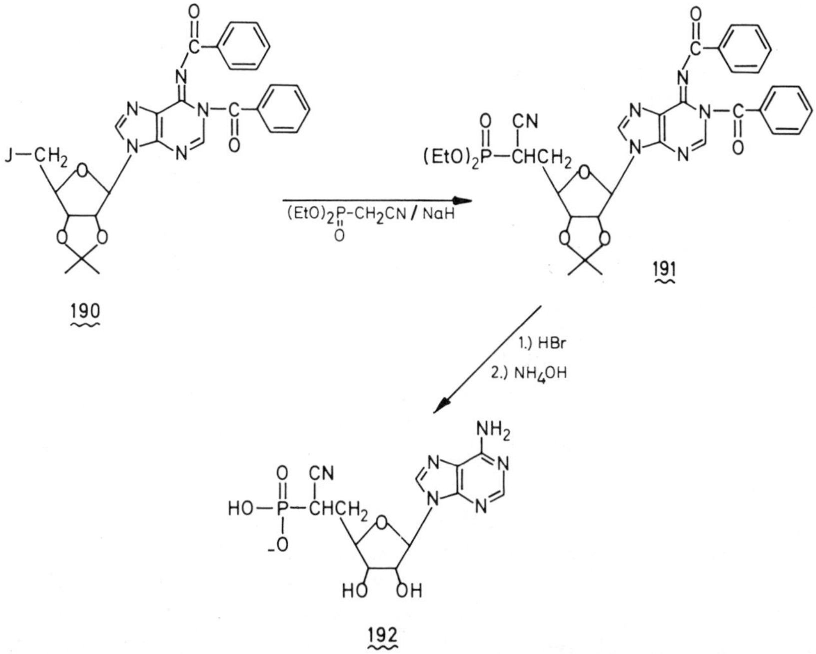

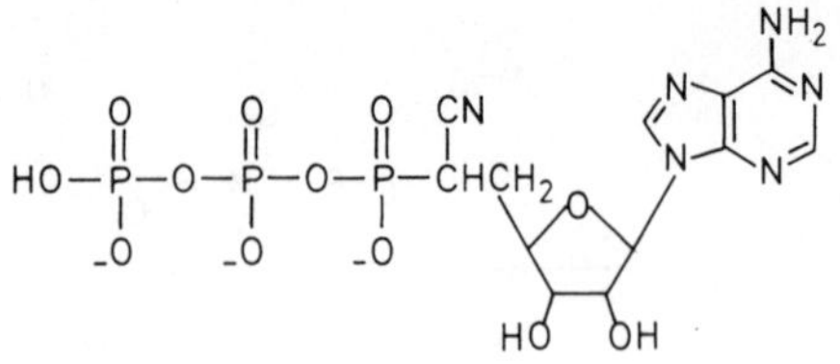

193

For the synthesis of derivatives of 6′-deoxyhomoadenosine-6′-phosphonates carrying substituents others than hydroxyl at carbon C(6′) a different approach was followed (84). Alkylation of diethyloxycyanomethyl phosphate by 5′-deoxy-5′-iodo-N^1,N^6-dibenzoyl-2′,3′-*O*-isopropylideneadenosine (**190**) in the presence of sodium hydride resulted in formation of 6′-cyanohomoadenosine 5′-phosphonate derivative **191**. Removal of the blocking groups from **191** was achieved by treatment with hydrogen

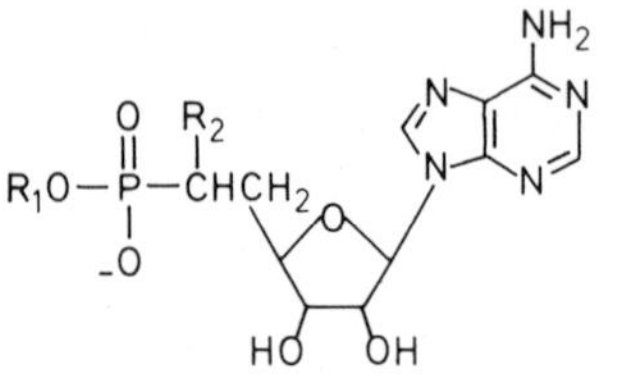

194–202

	R_2	R_1
194	$OCOCH_3$	H
195	CH_2NH_2	H
196	$CH_2NHCOCH_3$	H
197	$CH_2NHCOC_2H_5$	H
198	$CH_2NHCOC_6H_6$	H
199	$CH_2NHCOOC_2H_5$	H
200	$CH_2NHCOOC_3H_7$	H
201	$CH_2NHCOCH_3$	$P(O)O-O-P(O)(OH)O^{2-}$
202	$CH_2NHCOC_6H_5$	$P(O)O-O-P(O)(OH)O^{2-}$

bromide followed by hydrolysis with ammonia. 5′-Deoxy-5′-(dihydroxyphosphinyl)cyanomethyladenosine (**192**) was shown by ^{1}H-nmr spectroscopy to exist in the form of two epimers in a ratio of 70:30 (85). Conversion of **192** to 5′-deoxy-5′-(hydroxypyrophosphoroxyphosphinyl)cyanomethyladenosine (**193**) involved conventional procedures (85). A number of homologs of **192** and **193** were obtained by similar methods (86). The derivatives of 6′-deoxyhomoadenosine 6′-phosphonates **192, 194–202** were shown to be substrates and/or competitive inhibitors of AMP kinases from rabbit or pig muscle with affinities equal to or greater than that of AMP itself (85). The substrate and inhibitor properties of these analogs suggest the existence of a lipophilic region at the nucleotide binding site on the enzyme in the vicinity of the P—O—C(5′) part of the nucleotide (85, 87). With the AMP kinases, only one epimer of **192** was a phosphoryl acceptor and only one epimer of **193** was a phosphoryl donor in the AMP kinase reaction. Donor and acceptor epimers were of opposite configuration at carbon atom C(6′) (85). The action of AMP aminohydrolase on **192** seemed to indicate the presence of equal amounts of both epimers, an observation which conflicts with an analysis by ^{1}H-nmr spectroscopy (85).

Methylphosphonic acid esters of various nucleosides have been prepared. The first report was by Myers and Simon (88) who synthesized 5′-adenylyl methylphosphonate (**203**) and 5′-adenylyl chloromethylphosphonate (**204**) respectively. 6-Azauridine 5′-methylphosphonate (**205**) and 6-azauridine-2′,(3′)-methylphosphonate were synthesized by routine procedures (89). The expectations that **205** and **206** should be resistant toward phosphatases were fulfilled. This also holds for 5-bromo-2′-deoxyuridine 5′-methylphosphonate (**207**) described by Wigler and Lozzio (90). Com-

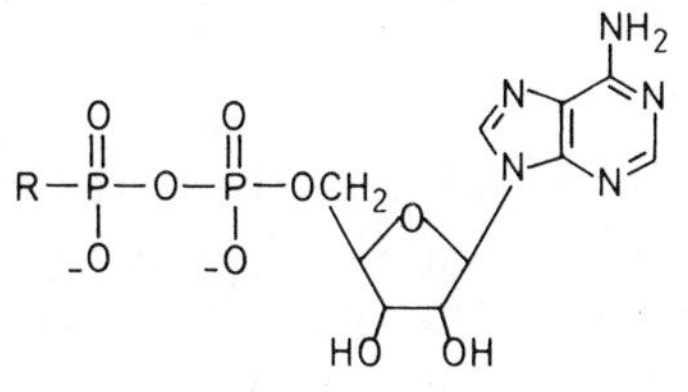

203, 204

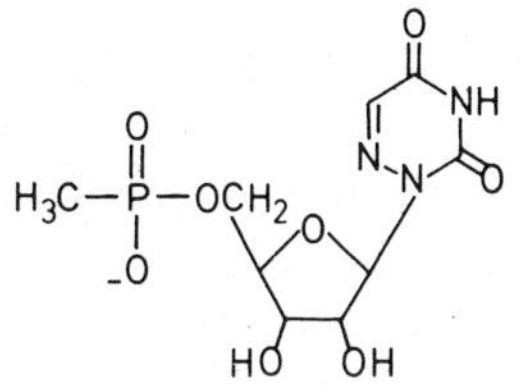

205

203, R = CH_3

204, R = CH_2Cl

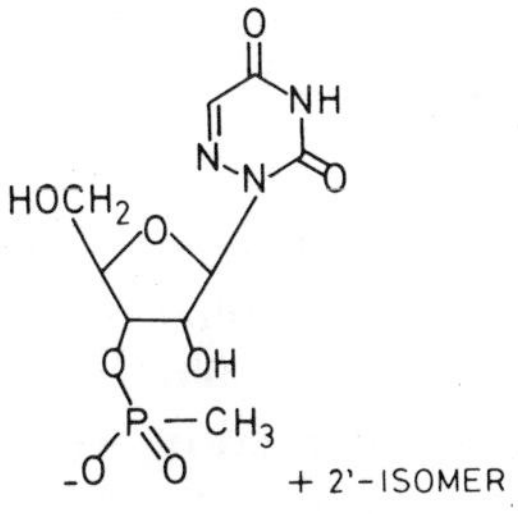

+ 2′-ISOMER

206

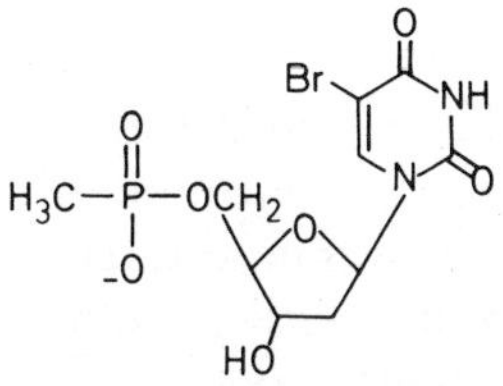

207

pound **207** was found to penetrate mammalian cells and to exhibit delayed cytotoxicity (90). The fate of **207** within cells and the mechanism of interference with cellular metabolism is not well understood (90).

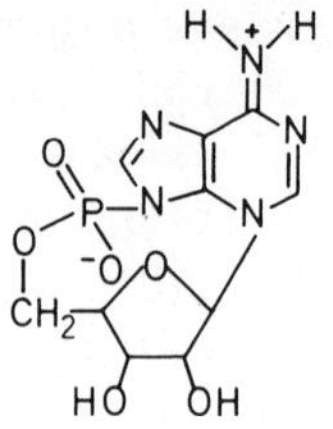

208

On phosphorylation of 3-(β-D-ribofuranosyl)adenine by means of phosphorus oxychloride in triethylphosphate Uchic (91) noticed the formation of an odd-side product, to which he assigned the structure 3-(β-D-ribofuranosyl)adenine 9,5′(P)-cyclic phosphonate (**208**). Compound **208** is similar in some respects to nucleoside 5′-phosphorimidazolidates. With concentrated ammonia **208** yielded 3-(β-D-ribofuranosyl)adenine 5′-phosphoramidate (91).

4.6 MISCELLANEOUS

Several analogs of nucleoside 5′-di- and -triphosphates substituted at the terminal phosphate moieties have been synthesized. One type of substitution is based on the early observation by Wittmann (92) that nucleoside 5′-fluorophosphates (**209**, **210**), formed in a reaction between nucleoside

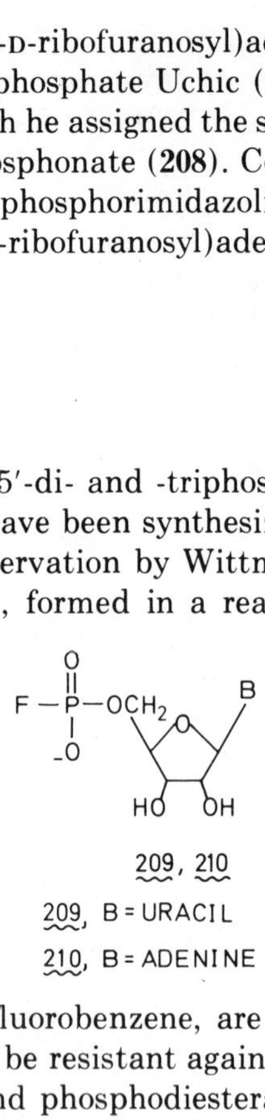

209, **210**

209, B = URACIL

210, B = ADENINE

5′-phosphates and 2,4-dinitrofluorobenzene, are stable compounds. Compounds like **209** were found to be resistant against phosphatases, but were hydrolyzed by nucleotidase and phosphodiesterase (92, 93). The reaction of nucleotides with 2,4-dinitro-fluorobenzene was studied in greater detail by Johnson et al. (94).

Much more interesting from a biochemical point of view was the synthesis of P^3-fluoro P^1-5′-adenosine triphosphate (**214**) by Haley and Yount (95). Preparation of **214** involved activation of fluorophosphate (**211**) by diphenylphosphorylchloride (**212**) and subsequent reaction of the P^1-fluoro P^2-diphenylpyrophosphate (**213**) with adenosine 5′-diphosphate. γ-Fluoro ATP (**214**) is relatively stable in solution at pH 7.4. It binds the divalent cations Ca^{2+}, Mg^{2+}, and Mn^{2+} with remarkably lesser affinity than ATP (95). Compound **214** was not a substrate for myosin, heavy meromyosin, or hexokinase respectively, but competitively inhibited the hydrolysis of ATP by heavy meromyosin (95). Like ATP, **214** protected myosin against heat inactivation (95). Contrary to expectations, **214** did not function as an irreversible inhibitor of the above enzymes (95). The authors pointed to the potential usefulness of **214** for studies of its interactions with proteins by means of ^{19}F-nmr spectroscopy. The preparation of P^3-fluoro P^1-5′-guanosine 5′-triphosphate (**215**) and P^2-fluoro P^1-5′-guanosine diphosphate (**216**) following the procedure of Haley et al. (95) was described by Eckstein (96). The attachment of the fluorine atom to the terminal phosphates in **215** and **216** was affirmed by ^{31}P-nmr spectroscopy. Eckstein et al., reported on futile attempts to synthesize **215** or **216** respectively by reaction of nucleotides with 2,4-dinitro-fluorobenzene according to Wittmann (92). P^3-Fluoro P^2-5′-guanosine triphosphate (**215**) is a potent inhibitor of the ribosome dependent elongation factor G GTPase (96).

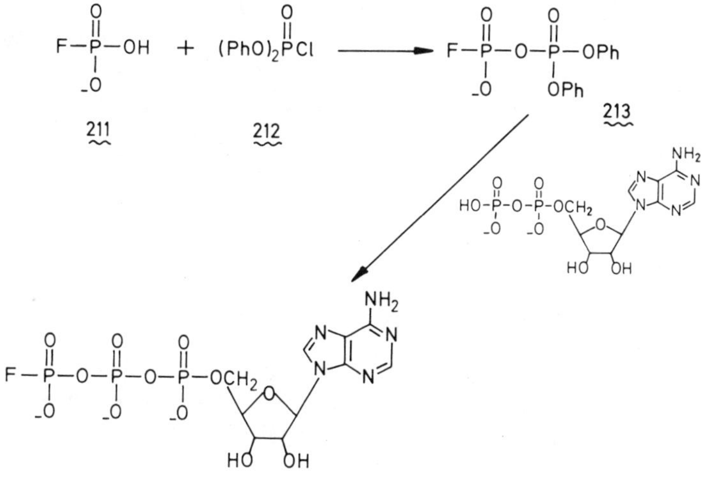

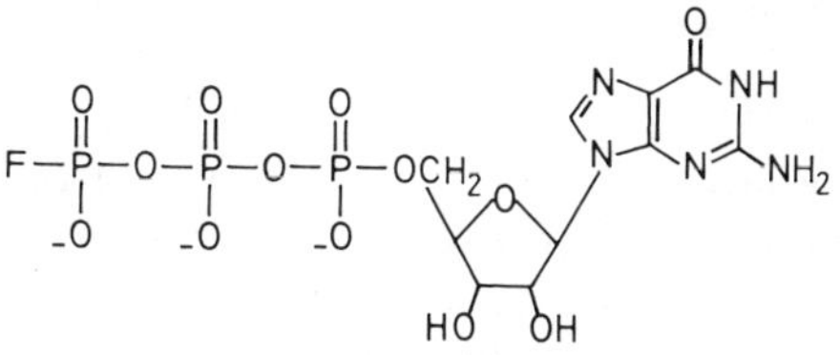

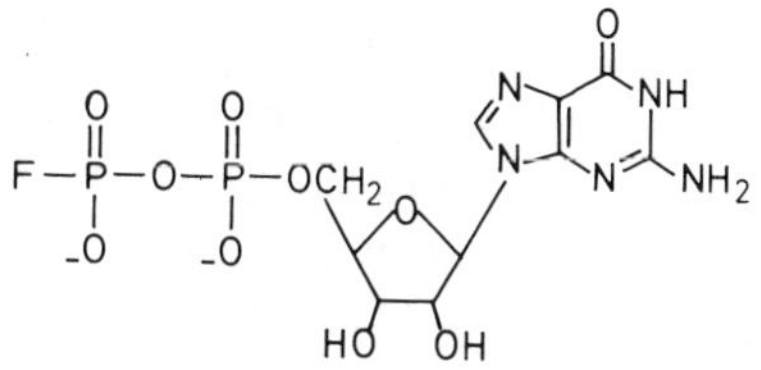

Eckstein et al. (96, 97) reported the synthesis of a series of GTP and ATP analogs, which were esterified at their terminal phosphate residues. The P^3-alkyl P^1-5′-nucleoside triphosphates (**217–221**) were obtained by

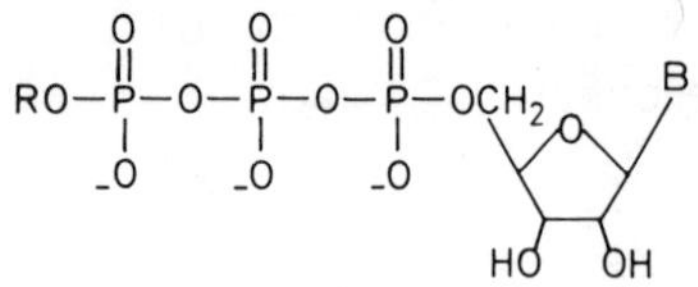

217–221

	R	B		R	B
217	CH_3	Guanine	**220**	C_4H_9	Guanine
218	C_2H_5	Guanine	**221**	CH_3	Adenine
219	C_3H_7	Guanine			

reaction of alcohols with the anhydrides formed between the corresponding nucleoside 5′-triphosphate and diphenylphosphorylchloride. P^3-Phenyl P^1-5′-adenosine triphosphate as well as P^3-phenyl P^1-5′-guanosine triphosphate were available by reaction of phenylphosphoroimidazolidate with the respective nucleoside 5′-diphosphate (96). The position of esterification was verified by ^{31}P-nmr spectroscopy. P^3-Methyl P^1-5′-adenosine triphosphate (**221**) and similarly P^3-fluoro P^1-5′-adenosine triphosphate (**214**) are substrates for DNA-dependent RNA polymerase from *E. coli* possessing rather low affinities for the transcriptional complex (97). A similar behavior was observed for P^3-anilido P^1-5′-adenosine triphosphate (**222**) (98). The synthesis of **222** was achieved by activation of ATP with water soluble carbodiimide followed by reaction with excess aniline (99).

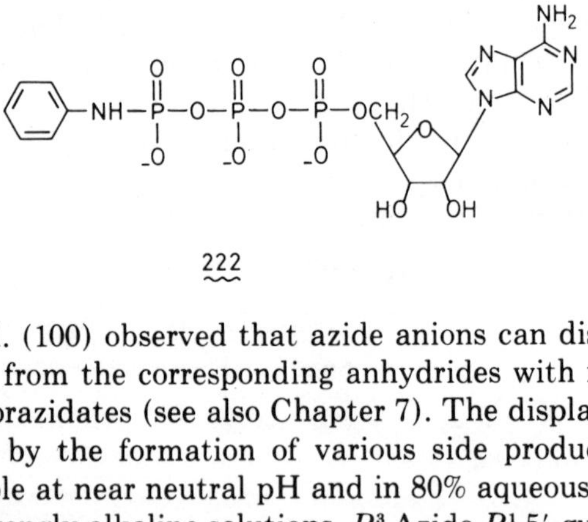

222

Chladek et al. (100) observed that azide anions can displace diphenylphosphoric acid from the corresponding anhydrides with nucleotides thus forming phosphorazidates (see also Chapter 7). The displacement reaction is accompanied by the formation of various side products. Compounds **223–230** are stable at near neutral pH and in 80% aqueous acetic acid, but decompose in strongly alkaline solutions. P^3-Azido P^1-5′-guanosine triphosphate (**225**) is a competitive inhibitor of the ribosome-dependent elongation factor G GTPase (100).

Anionic displacement of diphenylphosphoric acid from P^2-diphenyl P^1-5′-guanosine diphosphate by hypophosphate afforded guanosine 5′-phosphohypophosphate (**231**) (101). GTP-dependent ribosomal poly phenylalanine synthesis was competitively inhibited by this GTP analog (101). Adenosine 5′-phosphohypophosphate (**232**) was prepared in close analogy to the

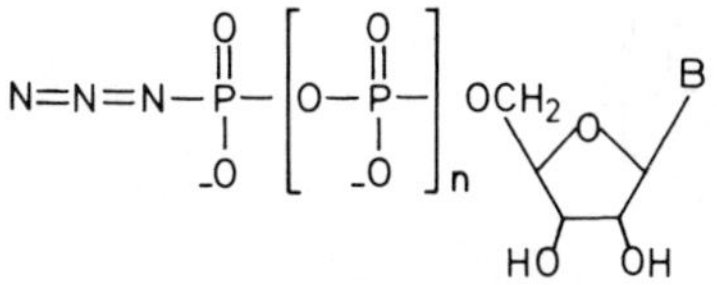

223 - 230

	n	B		n	B
223	0	Guanine	**227**	1	Adenine
224	1	Guanine	**228**	2	Adenine
225	2	Guanine	**229**	3	Adenine
226	0	Adenine	**230**	4	Adenine

synthesis of **231** (102). The ATP analog **232** was neither a substrate for valine tRNA-synthetase nor hexokinase, but proved to be a competitive inhibitor for both enzymes (103). It is interesting to note that **232** displayed a higher affinity for the enzymes as did the substrate ATP. Phosphorylation of 2′,3′-*O*-isopropylideneadenosine by hypophosphoric acid in the presence

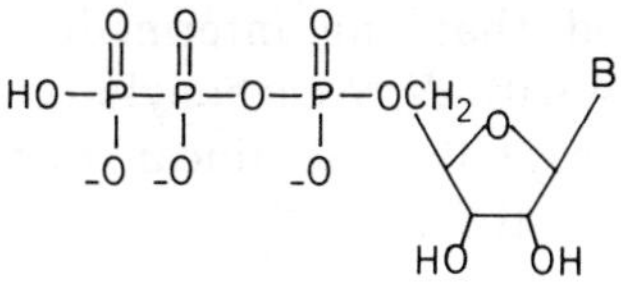

231, 232

231, B = GUANINE
232, B = ADENINE

of dicyclohexyl carbodiimide afforded adenosine 5′-hypophosphate after removal of the blocking group (104). Adenosine 5′-hypophosphate is a competitive inhibitor of AMP kinase with a greater affinity for the enzyme than the substrate ADP (105). The hypophosphate analog of ADP has very little affinity for pyruvate kinase (105).

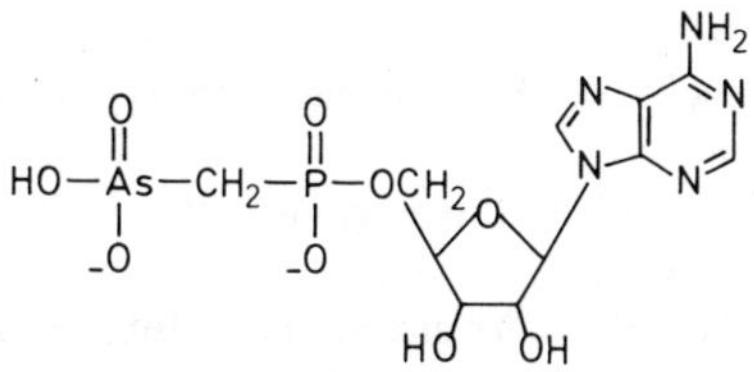

233

An analog of ADP in which the terminal phosphate group has been replaced by an arsonomethyl group was described by Webster et al. (106). Anhydrides between phosphoric acid and **233** formed in an enzymic phosphate transfer would not be stable. As a consequence one would expect the formation of inorganic phosphate in cases where **233** is selected by an en-

234 $\xrightarrow{H_2PO_4^-\,[NH(But)_3]^+}$ 235

zyme as an acceptor for phosphate transfer. However, **233** proved to be only a very poor substrate for 3-phosphoglycerate kinase and pyruvate kinase respectively (106). No substrate function of **233** was detected with AMP kinase and creatine kinase (106).

A branched isomer of ATP, so-called pseudo ATP (**235**), was formed in low yield besides ADP as the major product on reaction of adenosine 5′-phosphordiimidazolidate (**234**) with mono tri-*n*-butylammonium phosphate (107). The branched ATP (**235**) neither replaced ATP in aminoacylation of tRNA by synthetase nor as a substrate for hexokinase (107). It was furthermore demonstrated that the intermediate of the reaction of a nucleoside 5′-diphosphate with *N,N*-carbonyldiimidazole affords on treatment with inorganic phosphate the linear rather than the branched nucleoside 5′-triphosphates (107).

REFERENCES

1. T. C. Myers, K. Nakamura, and J. W. Flesher, *J. Am. Chem. Soc.* **85,** 3292–3295 (1963).
2. T. C. Myers, K. Nakamura, and A. B. Danielzadeh, *J. Org. Chem.* **30,** 1517–1520 (1965).
3. J. W. B. Hershey and R. E. Monro, *J. Mol. Biol.* **18,** 68–76 (1966).
4. T. C. Myers, *Chem. Abstr.* **64,** 15973f (1966).
5. A. M. Spiegel, R. W. Downs, Jr., and G. D. Aurbach, *Biochem. Biophys. Res. Commun.* **76,** 758–764 (1977).
6. R. G. Yount, D. Babcock, W. Ballantyne, and D. Ojala, *Biochemistry* **10,** 2484–2489 (1971).
7. F. Eckstein, M. Kettler, and A. Parmeggiani, *Biochem. Biophys. Res. Commun.* **45,** 1151–1158 (1971).
8. M. Larsen, R. Willett, and R. G. Yount, *Science* **166,** 1510–1511 (1969).
9. R. G. Yount, D. Ojala, and D. Babcock, *Biochemistry* **10,** 2490–2496 (1971).
10. F. Krug, I. Parikh, G. Illiano, and P. Cuatrecasas, *J. Biol. Chem.* **248,** 1203–1206 (1973).
11. R. G. Yount, *Adv. in Enzymol.* **43,** 1–56 (1975).
12. G. N. Bennett, R. G. Gough, and P. T. Gilham, *Biochemistry* **15,** 4623–4628 (1976).
13. F. Eckstein, *J. Am. Chem. Soc.* **88,** 4292–4294 (1966).
14. F. Eckstein, *J. Am. Chem. Soc.* **92,** 4718–4723 (1970).

15. A. W. Murray and M. R. Atkinson, *Biochemistry* **7,** 4023–4029 (1968).
16. D. Dunaway-Mariano, *Tetrahedron* **32,** 2991–2996 (1976).
17. F. Eckstein, H. H. Schulz, H. Rüterjans, W. Haar, and W. Maurer, *Biochemistry* **11,** 3507–3512 (1972).
18. W. Saenger and F. Eckstein, *J. Am. Chem. Soc.* **92,** 4712–4718 (1970).
19. F. Eckstein, *FEBS-Lett.* **2,** 85–86 (1968).
20. D. A. Usher, D. J. Richardson, and F. Eckstein, *Nature* **228,** 663–665 (1970).
21. D. A. Usher, E. S. Erdenreich, and F. Eckstein, *Proc. Natl. Acad. Sci. (US)* **69,** 115–118 (1972).
22. F. Eckstein, *Angew. Chem.* **87,** 179–212 (1975).
23. F. Eckstein, L. P. Simonson, and H. P. Bär, *Biochemistry* **13,** 3806–3810 (1974).
24. F. Eckstein and H. Sternbach, *Biochim. Biophys. Acta* **146,** 618–619 (1967).
25. F. Eckstein and H. Gindl, *Biochim. Biophys. Acta* **149,** 35–40 (1967).
26. F. Eckstein and R. S. Goody, *Biochemistry* **15,** 1685–1691 (1976).
27. R. S. Goody and F. Eckstein, *J. Am. Chem. Soc.* **93,** 6252–6257 (1971).
28. M. Cohn and T. R. Hughes, *J. Biol. Chem.* **235,** 3250–3253 (1960).
29. C. H. Schwalbe, R. S. Goody, and W. Saenger, *Acta Crystallogr. Sect.* **B29,** 2264–2272 (1973).
30. R. S. Goody, F. Eckstein, and R. H. Schirmer, *Biochim. Biophys. Acta* **276,** 155–161 (1972).
31. F. Eckstein, V. W. Armstrong, and H. Sternbach, *Proc. Natl. Acad. Sci. (US)* **73,** 2987–2990 (1976).
32., P. M. J. Burgers and F. Eckstein, *Proc. Natl. Acad. Sci. (US)* **75,** 4798–4800 (1978).
33. F. Eckstein, H. Sternbach, and F. v. d. Haar, *Biochemistry* **16,** 3429–3432 (1977).
34. E. K. Jaffe and M. Cohn, *J. Biol. Chem.* **253,** 4823–4825 (1978).
35. E. K. Jaffe and M. Cohn, *Biochemistry* **17,** 652–657 (1978).
36. R. D. Cornelius and W. W. Cleland, *Biochemistry* **17,** 3279–3286 (1978).
37. F. Cramer, *Progr. Nucl. Acid. Res. Mol. Biol.* **11,** 391–421 (1971).
38. S. Akerfeldt, *Acta. Chem. Scand.* **16,** 1897–1907 (1962).
39. J. Nagyvary, S. Chladek, and J. Roe, *Biochem. Biophys. Res. Commun.* **39,** 878–882 (1970).
40. S. Chladek and J. Nagyvary, *J. Am. Chem. Soc.* **94,** 2079–2085 (1972).
41. E. Bradbury and J. Nagyvary, *Nucleic Acids Res.* **3,** 2437–2442 (1976).
42. D. K. Lavallee and C. L. Coulter, *J. Am. Chem. Soc.* **95,** 576–581 (1973).
43. A. M. Michelson, *J. Chem. Soc.* **1962,** 979–982.
44. A. Hampton, L. W. Brox, and M. Bayer, *Biochemistry* **8,** 2303–2311 (1969).
45. A. Stütz and K. H. Scheit, *Eur. J. Biochem.* **50,** 343–349 (1975).
46. K. H. Scheit and A. Stütz, *J. Carbohydr. Nucleosides Nucleotides* **1,** 485–490 (1974).
47. K. H. Scheit and A. Stütz, unpublished results.
48. B. Jastorff and H. Hettler, *Chem. Ber.* **102,** 4119–4127 (1969).
49. R. L. Letsinger and W. S. Mungall, *J. Org. Chem.* **35,** 3800–3803 (1970).
50. W. S. Mungall, G. L. Greene, G. A. Heavner, and R. L. Letsinger, *J. Org. Chem.* **40,** 1659–1662 (1975).

51. D. B. Trowbridge, D. M. Yamamoto, and G. L. Kenyon, *J. Am. Chem. Soc.* **94,** 3816–3824 (1972).
52. R. L. Letsinger, J. S. Wilkes, and L. B. Dumas, *J. Am. Chem. Soc.* **94,** 292–293 (1972).
53. R. L. Letsinger, J. S. Wilkes, and L. B. Dumas, *Biochemistry* **15,** 2810–2816 (1976).
54. J. S. Wilkes, B. Hapke, and R. L. Letsinger, *Biochem. Biophys. Res. Commun.* **53,** 917–922 (1973).
55. W. Freist, K. Schattka, F. Cramer, and B. Jastorff, *Chem. Ber.* **105,** 991–999 (1972).
56. A. Murayama, B. Jastorff, F. Cramer, and H. Hettler, *J. Org. Chem.* **36,** 3029–3033 (1971).
57. A. Murayama, B. Jastorff, H. Hettler, and F. Cramer, *Chem. Ber.* **106,** 3127–3131 (1973).
58. K. Schattka and B. Jastorff, *Chem. Ber.* **107,** 3042–3052 (1974).
59. B. Jastorff and T. Krebs, *Chem. Ber.* **105,** 3192–3202 (1972).
60. B. Jastorff and H. P. Bär, *Eur. J. Biochem.* **37,** 497–504 (1973).
61. M. Morr, M. R. Kula, G. Roesler, and B. Jastorff, *Angew. Chem.* **86,** 308, (1974).
62. M. Morr and L. Ernst, *Chem. Ber.* **111,** 2152–2172 (1978).
63. J. A. Schofield and A. Todd, *J. Chem. Soc.* **1961,** 2316–2320.
64. A. Holy, J. Smrt and F. Sorm, Coll. Czech. Chem. Commun. **30,** 1635–1641 (1965).
65. A. Holy and F. Sorm, *Coll. Czech. Chem. Commun.* **31,** 1544–1561 (1966).
66. A. Holy and F. Sorm, *Coll. Czech. Chem. Commun.* **31,** 1562–1568 (1966).
67. A. Holy, J. Smrt, and F. Sorm, *Coll. Czech. Chem. Commun.* **30,** 3309–3319 (1965).
68. A. Holy and J. Smrt, *Coll. Czech. Chem. Commun.* **31,** 1528–1534 (1966).
69. J. R. Parikh, M. E. Wolff, and A. Burger, *J. Am. Chem. Soc.* **79,** 2778–2781 (1957).
70. M. E. Wolff and A. Burger, *J. Am. Pharm. Assoc.* **48,** 56–59 (1959).
71. D. H. Rammler, L. Yengoyan, A. V. Paul, and P. C. Bax, *Biochemistry* **6,** 1828–1837 (1967).
72. L. Yengoyan and D. H. Rammler, *Biochemistry* **5,** 3629–3638 (1966).
73. P. C. Bax and D. H. Rammler, unpublished results.
74. D. H. Rammler, A. Bagdasarian, and F. Morris, *Biochemistry* **11,** 9–12 (1972).
75. A. Holy, *Tetrahedron Lett.* **10,** 881–884 (1967).
76. G. H. Jones and J. G. Moffatt, *J. Am. Chem. Soc.* **90,** 5337–5338 (1968).
77. K. E. Pfitzner and J. G. Moffatt, *J. Am. Chem. Soc.* **85,** 3027–3028 (1963).
78. J. A. Montgomery and K. Hewson, *Chem. Commun.* **1969,** 15–16.
79. A. F. Cook and J. G. Moffatt, *J. Am. Chem. Soc.* **89,** 2697–2705 (1967).
80. H. P. Albrecht, G. H. Jones, and J. G. Moffatt, *J. Am. Chem. Soc.* **92,** 5511–5513 (1970).
81. S. M. Hecht and M. Sundaralingam, *J. Am. Chem. Soc.* **94,** 4314–4319 (1972).
82. M. Sundaralingam, *Acta Crystallogr.* **21,** 495–506 (1966).

83. M. Fuertes, J. T. Witkowski, D. G. Streeter, and R. K. Robins, *J. Med. Chem.* **17,** 642–645 (1974).
84. A. Hampton, F. Perini, and P. J. Harper, *Biochemistry* **12,** 1730–1736 (1973).
85. A. Hampton, T. Sasaki, and B. Paul, *J. Am. Chem. Soc.* **95,** 4404–4414 (1973).
86. A. Hampton, T. Sasaki, F. Perini, L. A. Slotin, and F. Kappler, *J. Med. Chem.* **19,** 1029–1033 (1976).
87. A. Hampton, L. A. Slotin, F. Kappler, T. Sasaki, and F. Perini, *J. Med. Chem.* **19,** 1371–1377 (1976).
88. T. C. Myers and L. N. Simon, *J. Org. Chem.* **30,** 443–446 (1965).
89. A. Holy, *Coll. Czech. Chem. Commun.* **32,** 3713–3718 (1967).
90. P. W. Wigler and C. B. Lozzio, *J. Med. Chem.* **15,** 1020–1024 (1972).
91. J. T. Uchic, *Tetrahedron Lett.* **43,** 3775–3778 (1977).
92. R. Wittmann, *Chem. Ber.* **96,** 771–779 (1963).
93. Z. Kucerova and J. Skoda, *Biochim. Biophys. Acta* **247,** 194–196 (1971).
94. P. W. Johnson, R. von Tigerstrom, and M. Smith, *Nucleic Acids Res.* **2,** 1745–1749 (1975).
95. B. Haley and R. G. Yount, *Biochemistry* **11,** 2863–2871 (1972).
96. F. Eckstein, W. Bruns, and A. Parmeggiani, *Biochemistry* **14,** 5225–5232 (1975).
97. V. Armstrong and F. Eckstein, *Eur. J. Biochem.* **70,** 33–38 (1976).
98. M. A. Grachev, and E. F. Zaychikov, *FEBS-Lett.* **49,** 163–166 (1974).
99. G. T. Bakina, W. Zarytova, and D. G. Knorre, *Bioorgan. Chem.* **1,** 611–615 (1975).
100. S. Chladek, K. Quiggle, G. Chinali, J. Kohut III, and J. Ofengand, *Biochemistry* **16,** 4312–4319 (1977).
101. P. Remy, M. L. Engel, G. Dirheimer, J. P. Ebel, and M. Revel, *J. Mol. Biol.* **48,** 173–176 (1970).
102. P. Remy, G. Dirheimer, and J. P. Ebel, *Biochim. Biophys. Acta* **136,** 99–107(1967).
103. P. Remy, J. Setondji, G. Dirheimer, and J. P. Ebel, *Biochim. Biophys. Acta* **204,** 31–38 (1970).
104. J. Setondji, P. Remy, G. Dirheimer, and J. P. Ebel, *Biochim. Biophys. Acta* **224,** 136–143 (1970).
105. J. Setondji, P. Remy, J. P. Ebel, and G. Dirheimer, *Biochim. Biophys. Acta* **232,** 585–594 (1971).
106. D. Webster, M. J. Sparkes, and H. B. F. Dixon, *Biochem. J.* **169,** 239–244 (1978).
107. J. W. Kozarich, A. C. Chinault, and S. M. Hecht, *Biochemistry* **12,** 4458–4463 (1973).

FIVE

Nucleotides with Altered Sugar Parts

This chapter deals with two principal types of nucleotide analogs. The major class comprises those analogs that are chemically substituted at the OH groups or hydrogen atoms of the ribose or deoxyribose moieties. It is not always obvious whether a nucleotide analog with, for example, a 2′-chloro substituent instead of either a 2′-OH group, or a 2′-hydrogen atom, should be classified as an analog of a ribo- or 2′-deoxyribonucleotide. In some, but not all cases, this question can be decided on the basis of the analogs biochemical properties. Nucleotide analogs containing sugar OH-groups substituted by amino groups also belong to this class. The *N*-phosphoryl derivatives of deoxy-aminonucleosides have been already treated in Chapter 4. The deoxyaminonucleosides are the only members of this class that are of natural origin. They are produced as antibiotics by certain microorganisms. For the sake of completion, the only exception known so far, 9-(4′-fluoro-5′-*O*-sulfamoylpentofuranosyl)adenine (1), a nucleoside antibiotic containing a fluoro sugar, should be mentioned.

The second class of analogs includes nucleotides with configurations and/or chemical structures different from D-ribose or 2′-deoxy-D-ribose. Many nucleosides of the second class are antibiotics produced by microorganisms. Examples are 9-β-D-arabinosylnucleosides (**1**), 9-β-D-xylofuranosylnucleosides (**2**), and 9-β-D-lyxofuranosylnucleosides (**3**).

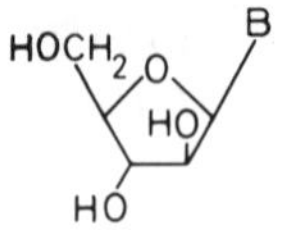

1 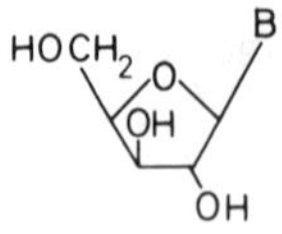 2 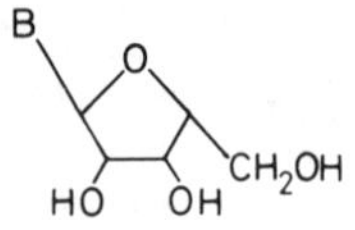 3

Their biological functions require uptake by the recipient cell and conversion to phosphorylated derivatives that interfere with cellular metabolism. The chemical synthesis of nucleotides derived from **1**, **2**, and **3** was a prerequisite for the elucidation of the various pathways effected by the metabolites of the above antibiotics, some of which have gained enormous therapeutic importance. The biological function of the nucleoside antibiotics 6-amino-9-(β-D-psicofuranosyl)adenine (**4**) (2) and 6-amino-9-(β-D-psicofuranoseenyl)adenine (**5**) (3) very likely does not involve phosphoryla-

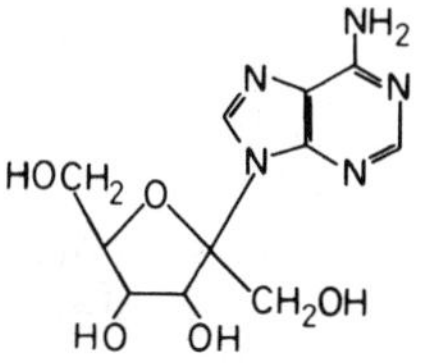

4

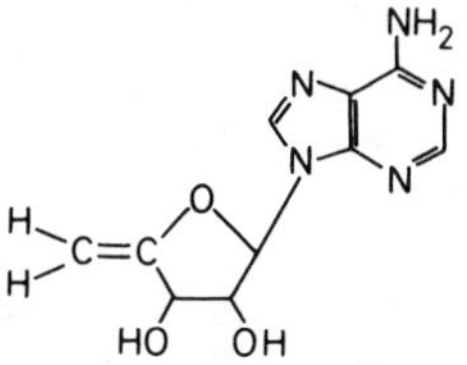

5

tion to corresponding nucleotides within the cell. All nucleotides that have so far been isolated from natural sources possess D-ribose moieties with the β-configuration about the glycosidic bonds. Nevertheless, various nucleotide analogs have been synthesized containing L-ribose residues and/or α-configurated glycosidic bonds with a view to investigating how enzymes behave *in vitro* toward unnatural anomeric analogs of their normal substrates.

5.1 ARABINO-, XYLO-, AND LYXO-NUCLEOTIDE ANALOGS

At first 9-(β-D-arabinofuranosyl)adenine was obtained by chemical synthesis (4, 5) before it was discovered as an antibiotic in cultures of *Streptomyces antibioticus* (6). The fermentation process now seems to be the most economical for the preparation of this compound. Chemical synthesis of 9-(β-D-arabinofuranosyl)adenine 5′-phosphate (**6**) by phosphorylation of arabinoadenosine with 2-cyanoethylphosphate (7, 8) or phosphorus oxy-

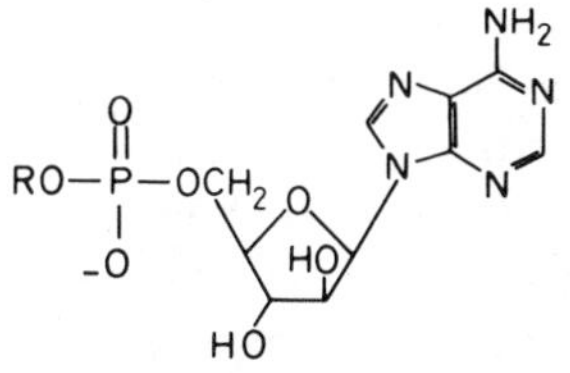

6, R = H

7, R = P(O)(OH)O⁻

6 , 7

chloride in triethylphosphate (9) has been described. Because of the lack of the easily protected cis diol system in arabinoadenosine, phosphorylation by 2-cyanoethyl phosphate in the presence of dicyclohexylcarbodiimide proved to be inefficient. Phosphorylation by means of phosphorus oxy-

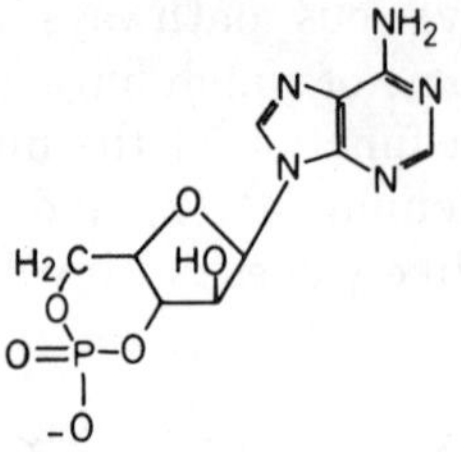

8

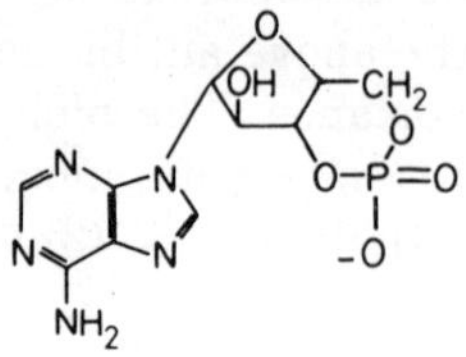

9

chloride and, as reported later, also by pyrophosphorylchloride in *m*-cresol (10) turned out to be more satisfactory. The morpholidate (7) or the imidazolidate (10) of **6** was employed in the preparation of arabinoadenosine 5′-triphosphate (**7**). The synthesis of arabinoadenosine 3′,5′-cyclic phosphate (**8**) proved to be problematic since direct cyclization of **6** led to the exclusive formation of **9** (11). This behavior is common to all arabinonucleoside 5′-phosphates (12). The first successful synthesis of **8** involved the preparation of a 2′-*O*-substituted derivative of arabinoadenosine by total synthesis, which was then phosphorylated and cyclized to the 3′,5′-cyclic phosphate (13). *N*-Oxidation of arabinoadenosine 5′-phosphate (**6**) furnished **10**, deamination by means of nitrous acid gave 9-(β-D-arabinofuranosyl)hypoxanthine 5′-phosphate (**11**) (14). The formation of

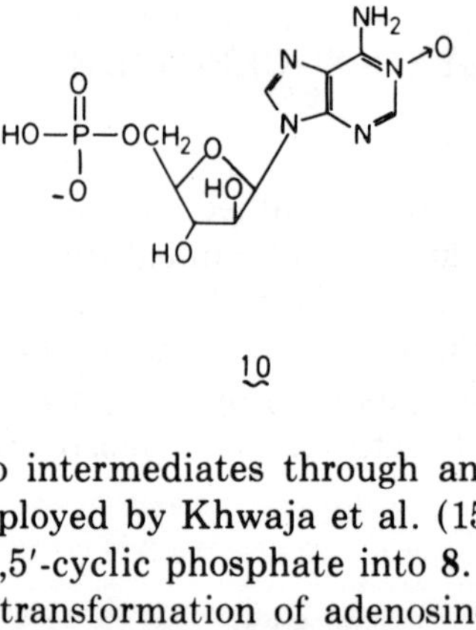

10

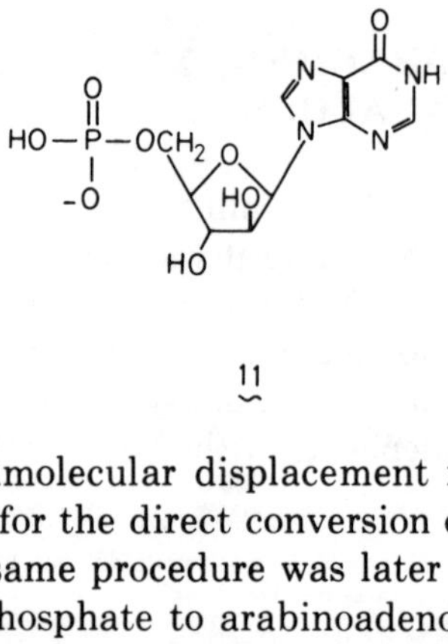

11

anhydro intermediates through an intramolecular displacement reaction was employed by Khwaja et al. (15, 16) for the direct conversion of adenosine 3′,5′-cyclic phosphate into **8**. The same procedure was later utilized for the transformation of adenosine 5′-phosphate to arabinoadenosine 5′-phosphate (**6**) (17). Although this synthetic scheme proceeds through many steps, the yields of the individual steps are reasonably good. It is interesting to note that the 8,2′-*O*-anhydroadenosine 5′-phosphate (**16**) itself is rather inert toward nucleophilic attack by SH^--anions. Only after acylation can the anhydro bridge in **17** be attacked to give the 8-mercapto derivative **18**.

The enzymatic conversion of arabinoadenosine 5′-phosphate to the corresponding 5′-triphosphate **7** by means of phosphoenol pyruvate synthetase from *E. coli* was developed by Johnson et al. (18). According to these authors this procedure should prove useful for small scale preparations.

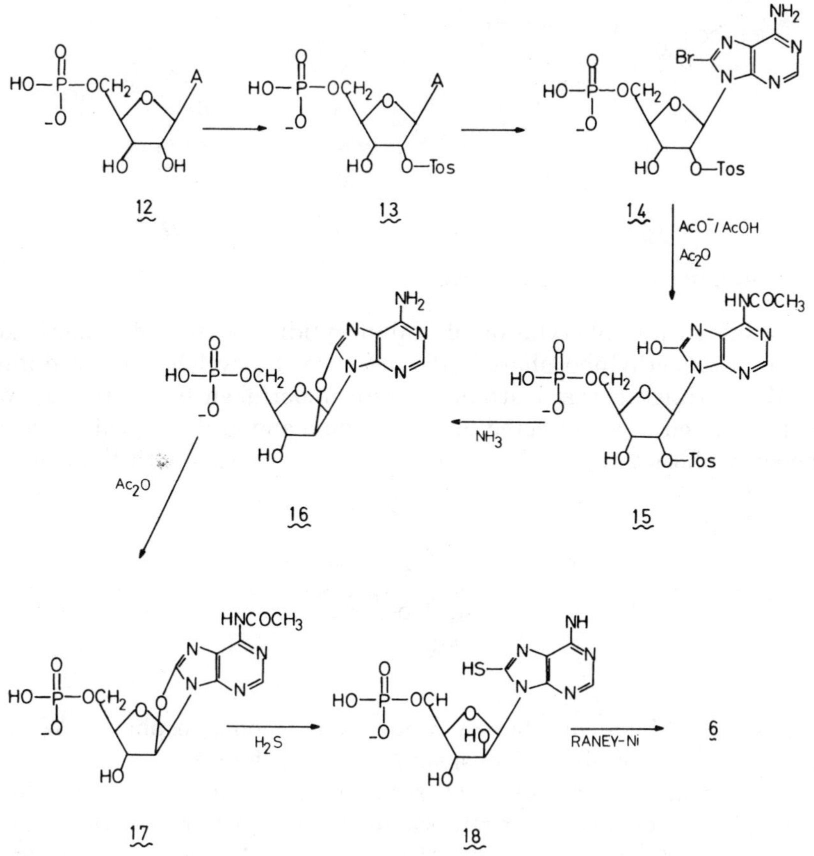

A= ADENINE Tos = TOSYLATE

1-(β-D-Arabinofuranosyl)uracil and 1-(β-D-arabinofuranosyl)thymine were isolated from sponges (19), but not 1-(β-D-arabinofuranosyl)cytosine which is a purely synthetic nucleoside. It is, however, the latter and not the natural compounds that exhibits antimetabolic properties. The first synthesis of 1-(β-D-arabinofuranosyl)cytosine 5′-phosphate (**19**) was accomplished enzymatically (20). Rather unsatisfactory chemical prepara-

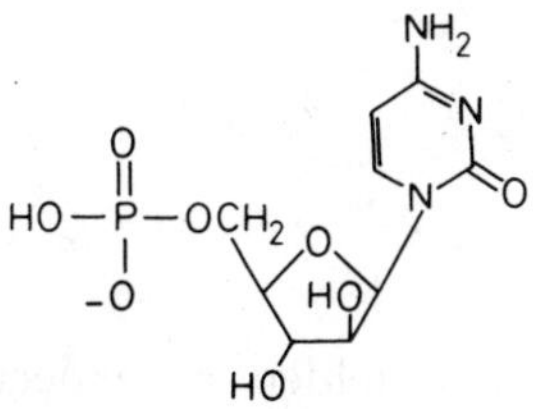

19

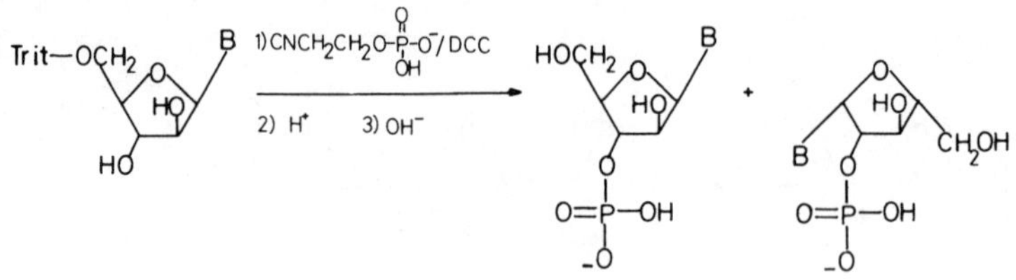

20 21 22

B = CYTOSINE Trit = $(C_6H_5)_3C$

tions involved phosphorylation of arabinocytidine by polyphosphoric acid (21) or 2-cyanoethylphosphate in the presence of dicyclohexylcarbodiimide (8). More straightforward attempts were undertaken by Smrt (22), who phosphorylated 2′,3′-protected arabinouridine and arabinocytidine with 2-cyanoethylphosphate. At the same time, in connection with the chemical

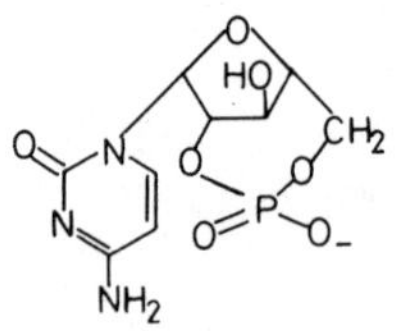

23

synthesis of a large number of dinucleoside monophosphates, Wechter described the synthesis of **19** by a similar approach (23).

The phosphorylation of 5′-*O*-tritylarabinocytidine (**20**) with 2-cyanoethylphosphate gave, after removal of the blocking groups, the isomeric 3′-(**21**) and 2′-phosphate (**22**) in a ratio of 3:1. Cyclization of

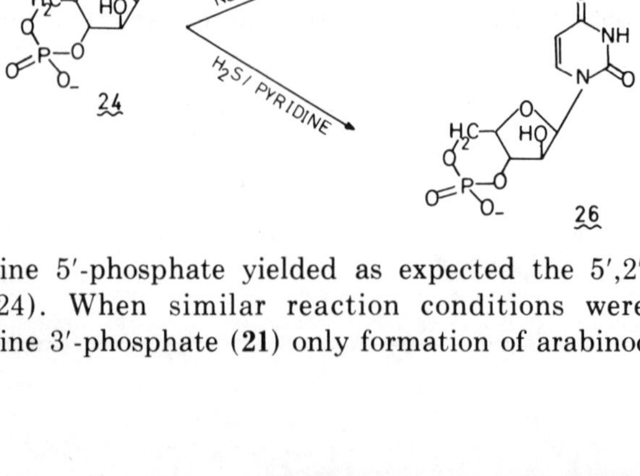

arabinocytidine 5′-phosphate yielded as expected the 5′,2′-cyclic phosphate **23** (24). When similar reaction conditions were applied to arabinocytidine 3′-phosphate (**21**) only formation of arabinocytidine 3′,5′-

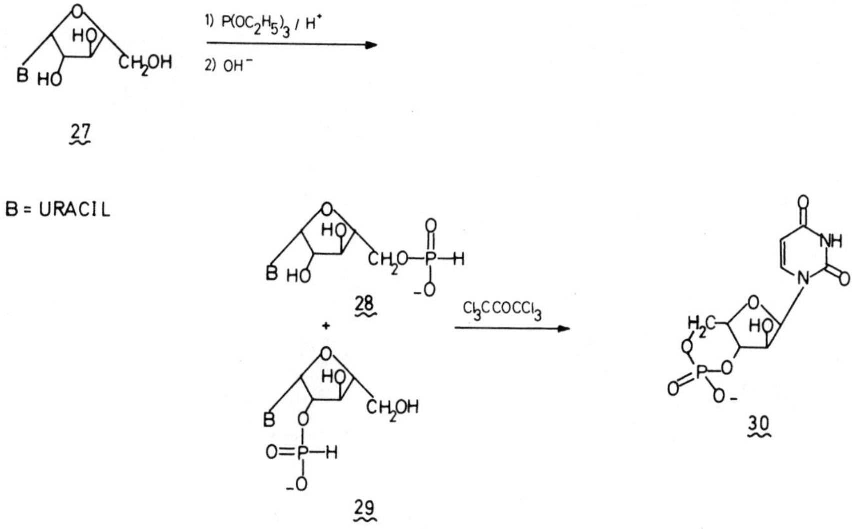

1) $P(OC_2H_5)_3$ / H^+
2) OH^-

27

B = URACIL

28

29

$Cl_3CCOCCl_3$

30

cyclic phosphate (**24**) was feasible for steric reasons (25, 26). Deamination of **24** afforded arabinouridine 3′,5′-cyclic phosphate (**25**) (25), and treatment with H_2S in pyridine yielded arabino-4-thiouridine 3′,5′-cyclic phosphate (**26**) (25). Holy and Sorm (27) investigated the reaction of arabinouridine (**27**) with triethyl phosphite. This reaction proceeded in low yields

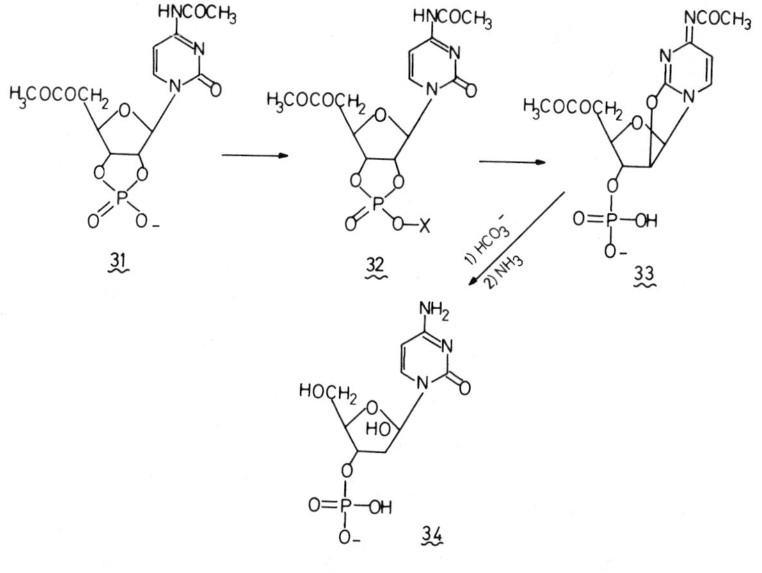

31

32

33

1) HCO_3^-
2) NH_3

34

giving a mixture of isomeric phosphites **28** and **29**. Treatment of the mixture of phosphites **28** and **29** with hexachloroacetone leads to a cyclic phosphodiester, which the authors tentatively assigned to be the 3′,5′-cyclic phosphate **30**. It seems not unlikely, especially considering the

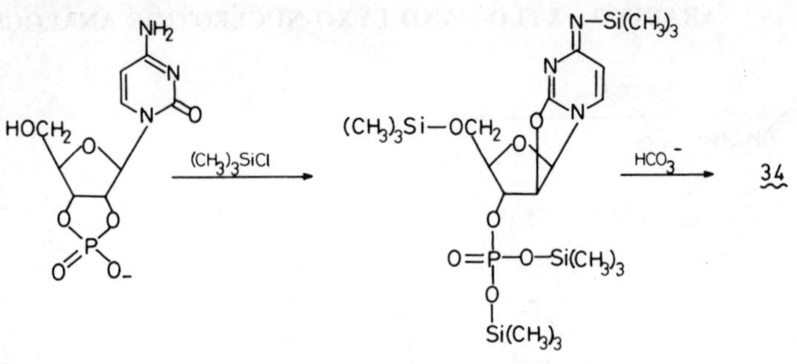

$(CH_3)_3SiCl$ HCO_3^- 34

35 36

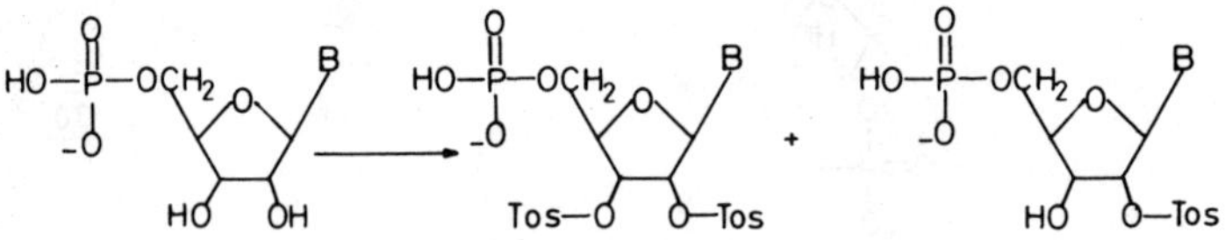

37 38 39

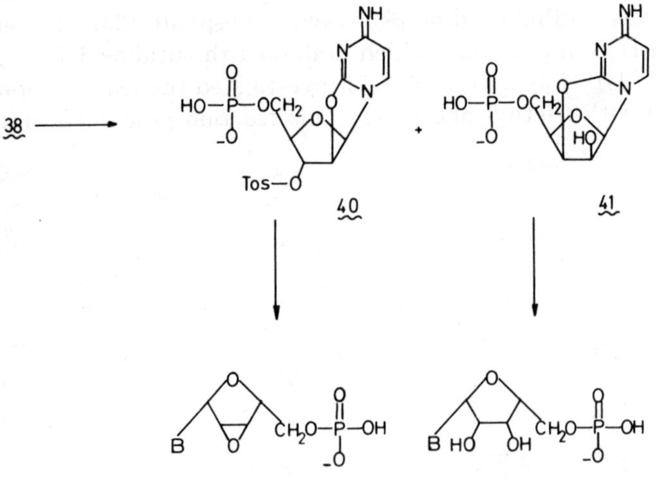

38 → 40 + 41

40 → 42

41 → 43

39 → 44 → 19

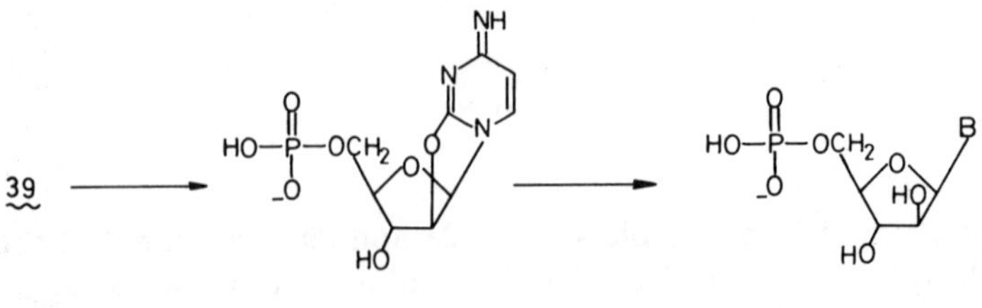

B = CYTOSINE
Tos = TOSYLATE

behavior of arabinouridine 5′-phosphate in the presence of an activating agent (12) that compound **30** is instead the 2′,5′-cyclic phosphate.

Upon reaction with methane sulfonylchloride, *p*-toluenesulfonyl chloride or diphenylphosphoryl chloride N^4-acetylcytidine 2′,3′-cyclic phosphate (**31**) forms cyclic phosphate ester anhydrides (**32**), which undergo an intramolecular attack by the 2-keto oxygen leading to the 2′,2-*O*-anhydro derivative **33**. Mild alkaline hydrolysis followed by removal of the N^4-acetyl group gave arabinocytidine 3′-phosphate in 45% yield (28). This reaction is certainly a useful procedure for the preparation of **34** from readily available **31**. An even simpler way from cytidine 2′,3′-cyclic phosphate (**35**) to **34** involves treatment of the former with trimethylsilyl chloride. The yield of **34** is of the order of 80% (29). Conversion of cytidine 5′-phosphate to the arabino derivative is achieved via the 2′—*O*—*p*-toluenesulfonate of **37** (30). As outlined in the synthetic scheme below, the composition of the reaction mixture, which one obtains is rather complex. Nevertheless, the overall yield of **19** formed in this reaction approaches 60%. Side products of this transformation were 2′,3′-anhydro-1-(β-D-lyxofuranosyl)cytosine 5′-phosphate (**42**) and 1-(β-D-lyxofuranosyl)-cytosine 5′-phosphate (**43**).

Nagyvary et al. (31) discovered that uridylic acid residues engaged in an internucleotidic bond can be converted into arabino derivatives without fission of the phosphodiester bond. Activation of the phosphodiester bond of 5′-*O*-acetyluridilyl[2′(3′)-5′]-2′,3′-*O*-isopropylideneuridine (**45**) by

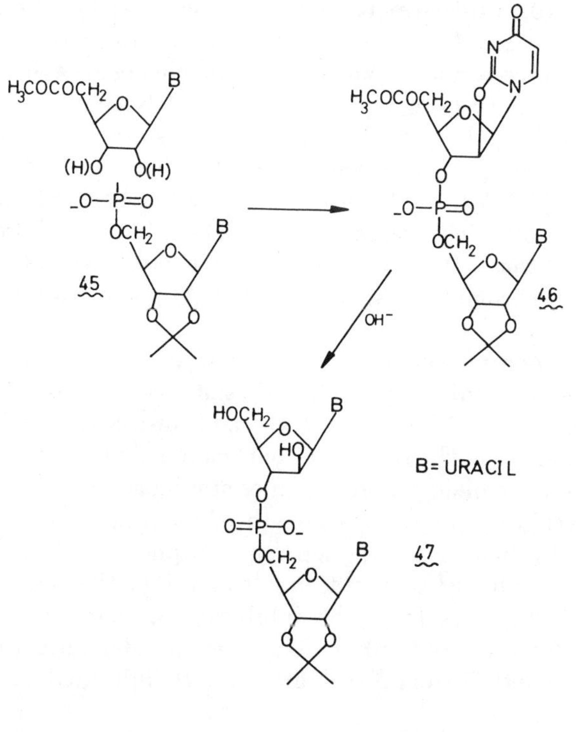

B= URACIL

various reagents such as diphenylphosphoryl chloride, *p*-toluenesulfonyl chloride, methane sulfonyl chloride, and tetraethyl pyrophosphate leads to formation of a 2′,2-*O*-anhydro compound **46** via a hypothetical cyclic triester intermediate. The intramolecular rearrangement **45** to **46** is carried out at 90°. Alkaline hydrolysis of **46** afforded arabinouridilyl(3′-5′)-2′,3′-*O*-isopropylideneuridine (**47**), the overall yield being approximately 80%. The internucleotidic linkage in **47**, as expected, is completely stable to alkaline hydrolysis. This procedure was successfully applied to convert oligouridiylic acid to polyarabinouridylic acid (32).

Arabinoadenosine strongly inhibits bacterial growth (33). The inhibition requires phosphorylation of the nucleoside analog within the cell, and the main target for the antimetabolic function seems to be DNA synthesis. Of greater importance is the inhibition of tumor growth by arabinoadenosine (33). The effective metabolite of this analog in the mammalian cell appears to be arabinoadenosine 5′-triphosphate (ara ATP), and experimental evidence suggests that the molecular basis for the inhibitory effect of arabinoadenosine could be the inhibition of mammalian DNA polymerase by ara ATP (34, 35, 36). It is noteworthy that neither DNA polymerase I nor DNA-dependent RNA polymerase from *E. coli* were inhibited by ara ATP (7, 35, 36). However, ara ATP interferes with other metabolic processes in cells. In mammalian cells, arabinonucleotides inhibit ribonucleoside diphosphate reductase (36). Ribonucleotide triphosphate reductase from *Lactobacillus leichmannii* was shown to be inhibited by ara ATP (10). Furthermore, bacterial adenylate cyclase is susceptible to inhibition by ara ATP (37). Whether this also holds for mammalian adenylate cyclases is not yet known. Arabinoadenosine exhibits considerable antiviral activity (33), especially against DNA viruses. The fact that adenosine rather than 2′-deoxyadenosine suppresses the action of arabinoadenosine as well as the incorporation of arabinoadenosine into both DNA and RNA (33, 38) raises doubts as to whether DNA polymerase really is the principal target for araATP. Indeed, Rose and Jacob described the very effective inhibition of polyadenylation of mRNA catalyzed by a chromatin associated poly (rA) polymerase in the presence of ara ATP (39).

The crystal and molecular structure of arabinoadenosine as determined by X-ray diffraction displays *C*(3′)-endo sugar conformation, gauche-trans conformation of the C(4′)—C(5′) bond and anti conformation about the glycosidic bond (40). The structural parameters do not provide an obvious explanation for the biological function of arabinoadenosine.

Arabinocytidine inhibits the growth of bacteria, various viruses, and tumors (41). Its biological function requires phosphorylation, and it now seems to be established with reasonable certainty that the cytotoxicity of arabinocytidine results from the inhibition of mammalian DNA polymerases by ara CTP (42–45). Reichard et al., demonstrated that DNA polymerases II and III from *E. coli* are similarly inhibited by ara CTP (46).

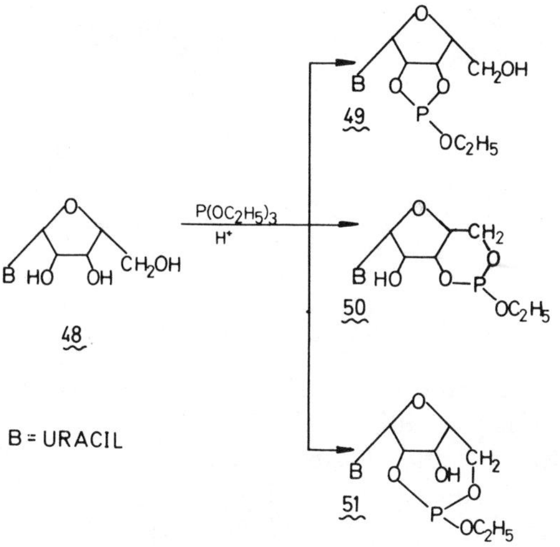

DNA-dependent RNA polymerase from *E. coli* is not affected by ara CTP (36). Phosphorylation of arabinocytidine in mammalian cells was studied by Momparler et al. (45, 47). These authors reported the strong inhibition of deoxycytidine kinase as well as deoxyadenosine kinase by ara CTP (45, 47).

Holy and Sorm (27) in an interesting study probed the possible transesterification reaction that 1-(β-D-lyxofuranosyl)uracil can undergo with triethyl phosphite. The β-D-lyxofuranoside **48** can potentially form three different kinds of cyclic phosphites (**49–51**). Alkaline hydrolysis of the reaction mixture leads to formation of the 2′-(**52**), 3′-(**53**), and 5′-phosphite (**54**) of 1-β-D-lyxofuranosyluracil. Compounds **53** and **54** may be presumed to have originated from the intermediate cyclic phosphite **49**, and compounds **52** and **53** from intermediate **50**. Although the presence of all possible phosphites in the reaction mixture is compatible with the formation of all three intermediary cyclic phosphites **49–51**, the authors, by not very convincing reasoning, exclude formation of species **51** (27).

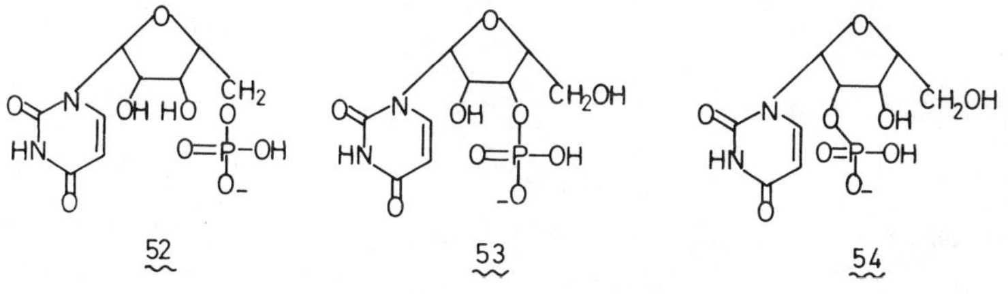

1-(5′-*O*-Benzoyl-β-D-lyxofuranosyl)uracil (**55**) reacted with triethyl phosphite to give the isomeric mixture of 2′- and 3′-phosphites (**53**, **54**), and treatment with hexachloroacetone converted this mixture into 1-(β-D-lyxofuranosyl)uracil 2′,3′-cyclic phosphate (**56**) (27). The exclusive formation of **56** without concomitant cyclization of **53** to the corresponding 3′,5′-cyclic phosphate seems plausible.

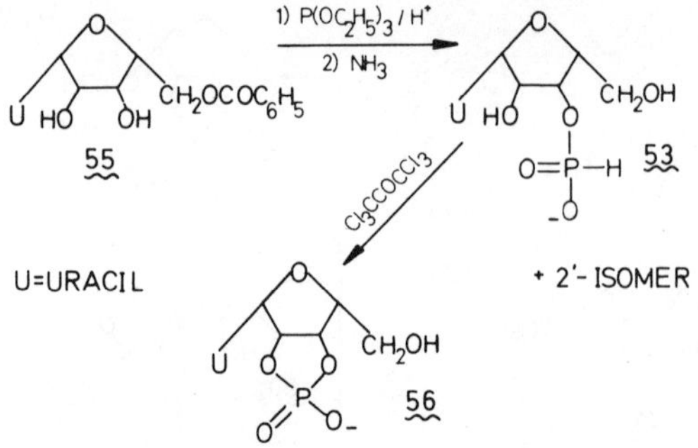

Protection of the 2′- and 3′-OH functions as in 1-(2′,3′-*O*-ethoxymethylene-β-D-lyxofuranosyl)uracil (**57**) allows selective phosphitylation at the 5′-position (**58**) (27). Reaction of **58** with hexachloroacetone leads to formation of 1-(β-D-lyxofuranosyl)uracil 3′,5′-cyclic phosphate (**59**) as well as 1-(β-D-lyxofuranosyl)uracil 2′,5′-cyclic phosphate (**60**) (27).

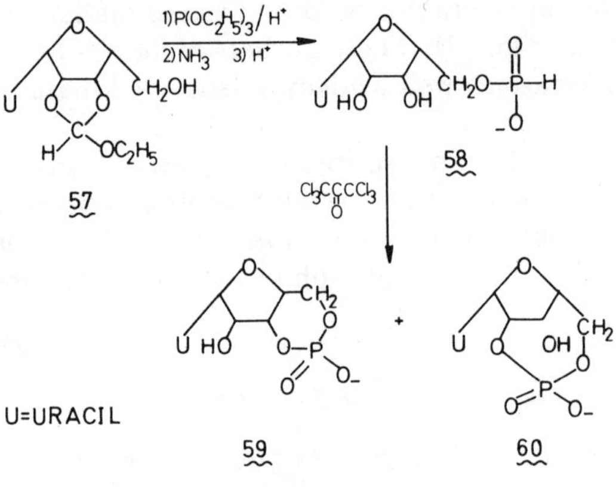

Transesterification of 1-(β-D-xylofuranosyl)uracil (**61**) with triethyl phosphite via the cyclic intermediate **62** gave a mixture of phosphites **63** and **64**, which could not be separated (27). Oxidative cyclization of the mixture of compounds **63** and **64** by treatment with hexachloroacetone leads to the exclusive formation of 1-(β-D-xylofuranosyl)uracil 3′,5′-cyclic phosphate (**65**) (27).

Deamination of 9-(β-D-5′,3′-*O*-isopropylidenexylofuranosyl)-6-amino-2-chloropurine (**66**) followed by aminolysis afforded 9-(β-D-xylofuranosyl)guanine (**68**) (48). Enzymatic phosphorylation of **68** to the corresponding 5′-phosphate (**69**) by nucleoside phosphotransferase from *Pseudomonas trifolii* was achieved in high yield (48). The chemical synthesis of 9-(β-D-xylofuranosyl)adenine 3′,5′-cyclic phosphate as reported by Hubert-Habart and Goodman (11) started with the total synthesis of the suitably protected nucleoside **72**. 9-(β-D-Xylofuranosyl)adenine inhibits the growth of tumors and viruses (33). In mammalian cells, the nucleoside is metabolized to the corresponding 5′-triphosphate (xylo ATP). Xylo ATP is an inhibitor of 5-phosphoribosyl-1-pyrophosphate formation (48, 49). Since the latter compound is required for the biosynthesis of pyrimidine and purine nucleotides, this observation would explain why xylofuranosyladenosine inhibits both RNA and DNA synthesis.

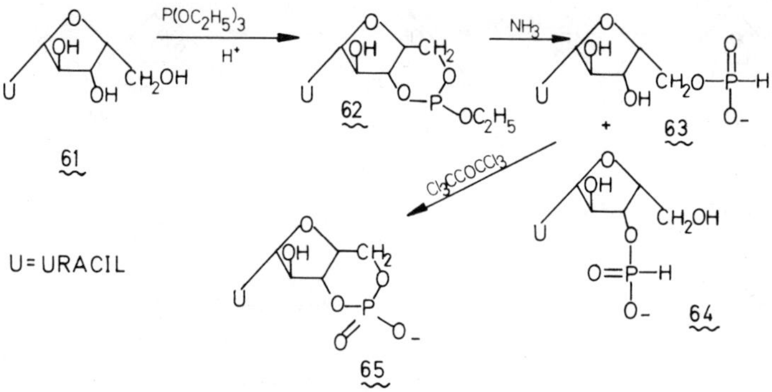

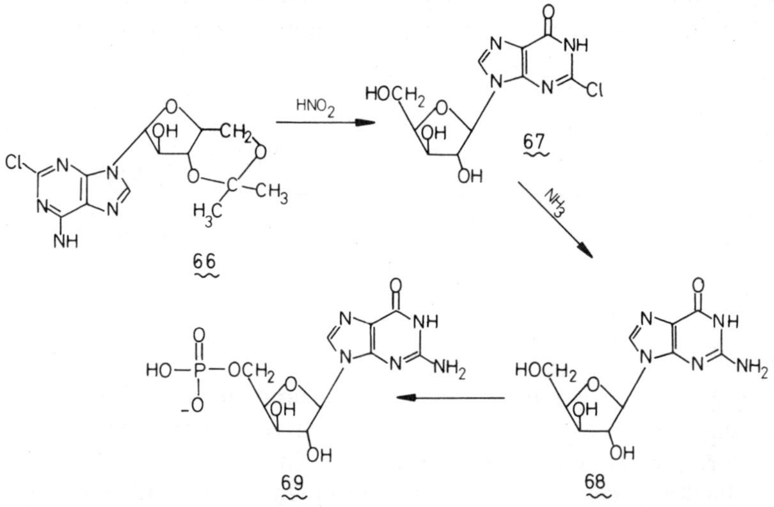

Although 9-(β-D-lyxofuranosyl)adenine seems to be metabolized in mammalian cells to the corresponding nucleotides, it displays only marginal cytotoxicity (33).

5.2 UNNATURAL ENANTIOMERIC AND ANOMERIC FORMS OF NUCLEOTIDES

The first report on the chemical synthesis of ribonucleotides containing L- instead of D-ribose was from Asai et al. (50). D,L-Adenosine (**75**), synthesized by the chloromercuri method, was converted to the 2′,3′-*O*-iso-

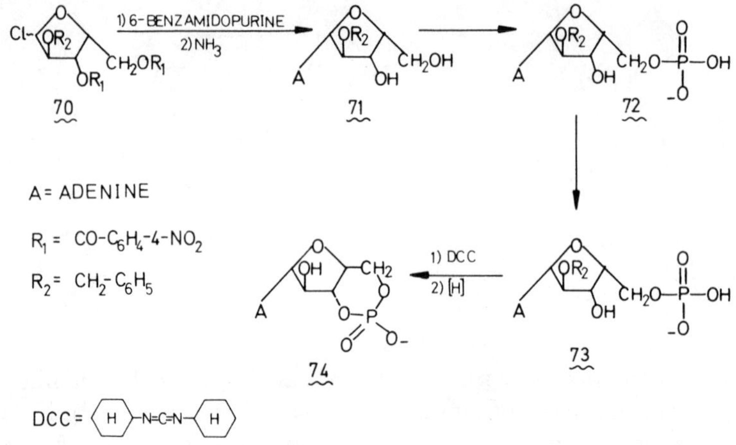

A = ADENINE

$R_1 = CO\text{-}C_6H_4\text{-}4\text{-}NO_2$

$R_2 = CH_2\text{-}C_6H_5$

propylidene derivative (**76**) and subsequently phosphorylated with phenylphosphorodichloridate. Removal of the phenyl group afforded D,L-adenosine 5′-phosphate (**77**). D,L-Inosine (**78**), obtained from D,L-adenosine by deamination, was phosphorylated by means of tetra-*p*-nitrophenylpyrophosphate. Removal of the blocking groups gave D,L-inosine 5′-phosphate (**80**) (50). Enzymatic hydrolysis of **77** and **80** by crude snake venom led to the exclusive cleavage of the respective D-enantiomers. Thus the pure L-forms of **77** and **80** remained (50). On treatment of D,L-adenosine with a bacterial cell suspension the D-isomer was degraded and L-adenosine could be recovered together with L-inosine, the latter resulting from partial deamination of L-adenosine (50). L-Adenosine was unable to support the growth of mammalian cells dependent on exogenous purine source (50), and L-adenosine was found to be nontoxic (50). Both, L- and D-enantiomers of adenosine possess identical physical and chemical properties with the exception of the Cotton-effects in their ORD spectra, which are identical in amplitude but of opposite sign (50).

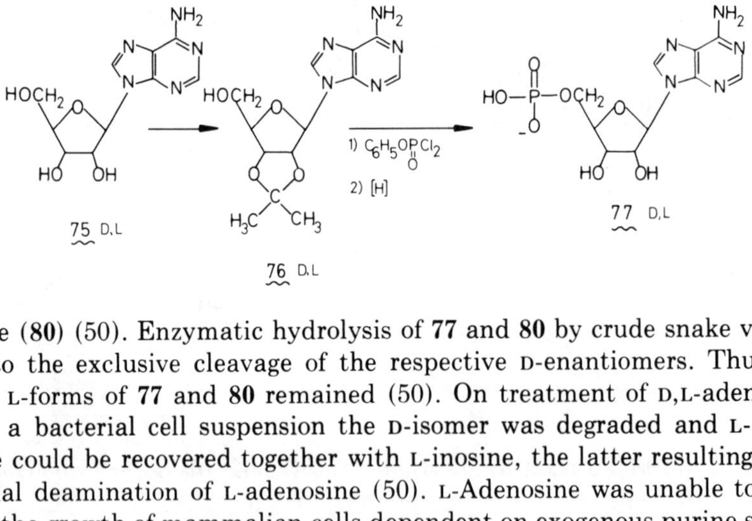

Condensation of 5′-*O*-diphenylphosphoryl-2′,3′,-di-*O*-benzoyl-L-ribofuranosylbromide (**81**) with O^2,O^4-di-(trimethylsilyl)uracil (**82**) gave a mixture of the α- and β-anomers of 1-(5′-*O*-diphenylphosphoryl-2′,3′-*O*-dibenzoyl-L-ribofuranosyl)uracil (**83, 84**), which could be separated. The diphenyl ester groups were removed by catalytic hydrogenation and the

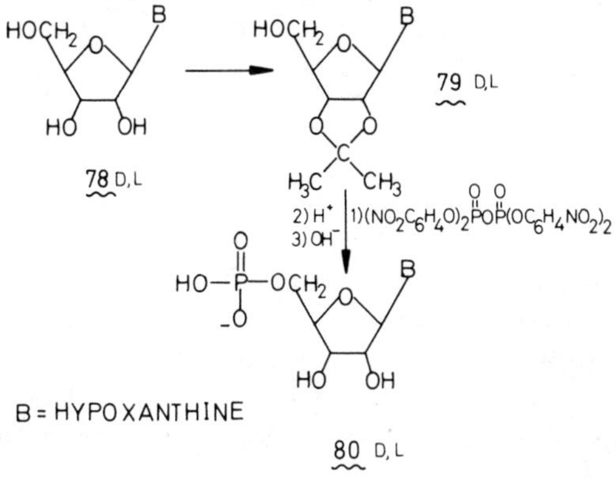

HOCH2 B 78 D,L → 79 D,L H3C CH3 1)(NO2C6H4O)2POP(OC6H4NO2)2 2) H+ 3) OH− 80 D,L

B = HYPOXANTHINE

benzoyl groups by alkaline hydrolysis to give α-L-uridine 5′-phosphate (**85**) and β-L-uridine 5′-phosphate (**86**) (51). Employing the same procedure with halosugar derivatives of either L or D configuration the synthesis of α-L-ribofuranosylthymine (**87**), β-L-ribofuranosylthymine 5′-phosphate (**88**), α-D-ribofuranosylcytosine 5′-phosphate, α-L-ribofuranosylcytosine 5′-phos-

81 + 82 → 83 + 84

R = CO-C6H5

83 1) [H] 2) OH− → 85

84 1) [H] 2) OH− → 86

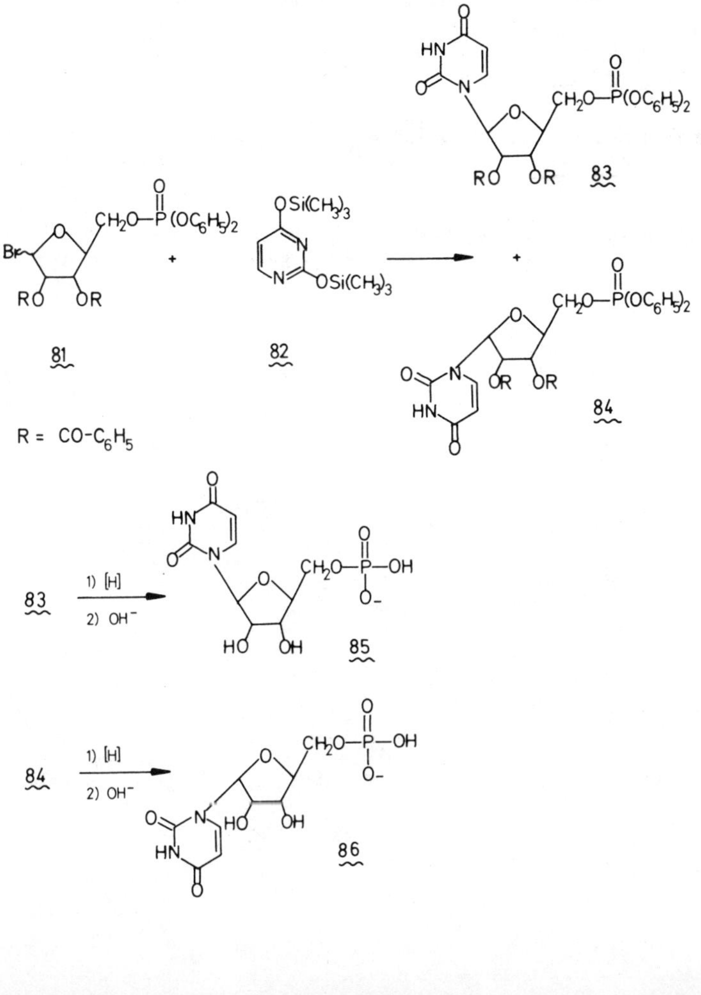

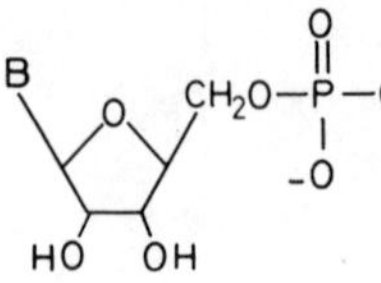

87,88, B=THYMINE

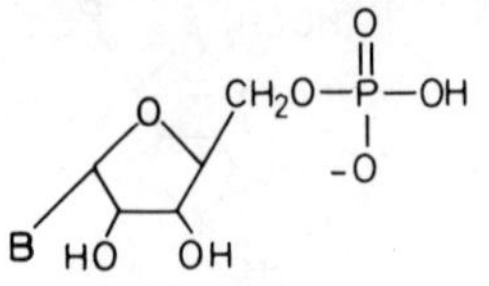

89,90, B=CYTOSINE

88,89

87,90

phate (**89**), and β-L-ribofuranosylcytosine 5′-phosphate (**90**) was achieved. Reaction of **81** with adenine in accordance with the method described by Leonard (52) led to 9-(β-L-ribofuranosyl)adenine 5′-phosphate (**91**) and 3-(β-L-ribofuranosyl)adenine 5′-phosphate (51) (**92**). Confirmation of the purity of the respective enantiomers came from nmr, optical rotation and

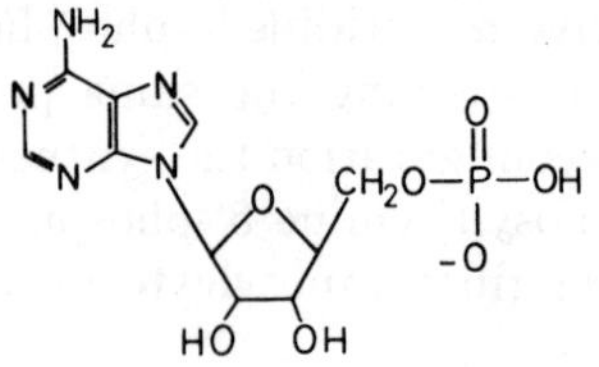

91

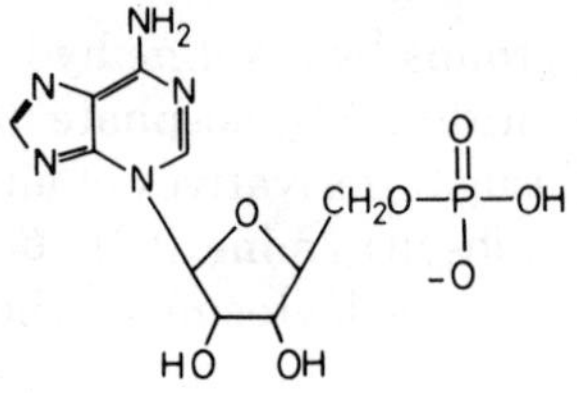

92

ORD measurements. Nuleotides with β,D and α,L configuration possessed positive Cotton effects, α,D and β,L nucleotides negative Cotton effects (51).

Reaction of 9-(α-L-lyxofuranosyl)adenine (**93**) with triethyl phosphite afforded three isomeric monophosphites **94–96** (53), which were separated

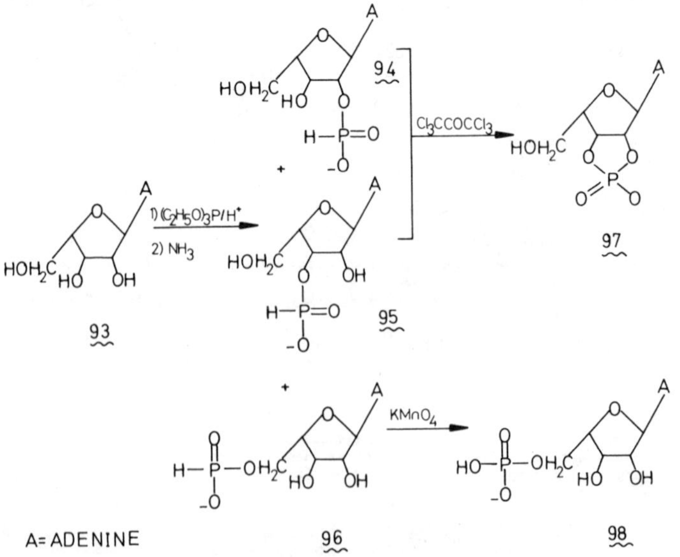

1) $(C_2H_5O)_3P/H^+$
2) NH_3

$Cl_3CCOCCl_3$

$KMnO_4$

A= ADENINE

93 94 95 96 97 98

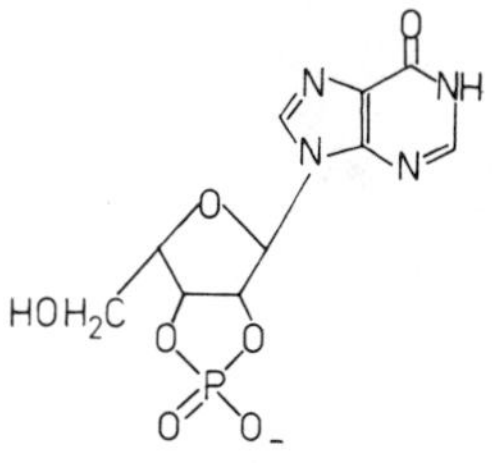

99

as borate complexes by anion exchange chromatography. The 2′- and 3′-phosphites (**94**, **95**) were combined and transformed to 9-(α-L-lyxofuranosyl)adenine 2′,3′-cyclic phosphate (**97**) by reaction with hexachloroacetone. The 5′-phosphite was oxidized by permanganate at pH 7 to give 9-(α-L-lyxofuranosyl)adenine 5′-phosphate (**98**). 9-(α-L-

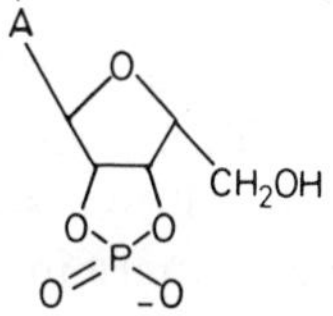

100 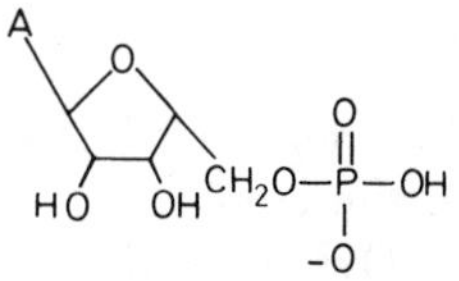 101 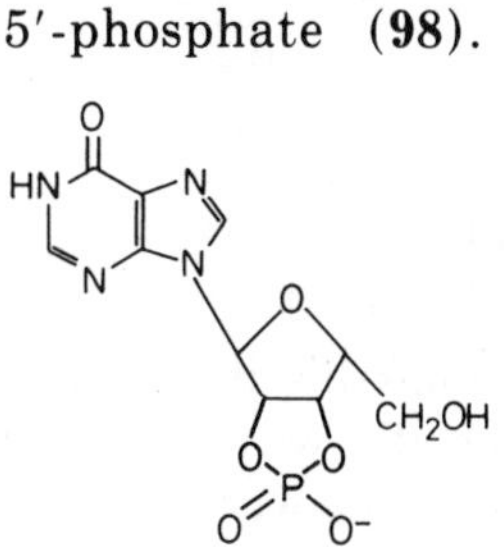 102

A = ADENINE

Lyxofuranosyl)hypoxanthine 2′,3′-cyclic phosphate (**99**) was obtained by deamination of **96** (53). By performing the same sequence of reactions with 9-(α-D-lyxofuranosyl)adenine instead of **93**, the following phosphorylated derivatives were synthesized: 9-(α-D-lyxofuranosyl)adenine 2′,3′-cyclic phosphate (**100**), 9-(α-D-lyxofuranosyl)adenine 5′-phosphate (**101**), and 9-(α-D-lyxofuranosyl)hypoxanthine 2′,3′-cyclic phosphate (**102**) (53). The behavior of compounds **97** and **100** as substrates for ribonuclease T_2 from *Aspergillus oryzae* was investigated (53). Only the L-enantiomer **97** was utilized as substrate. Whether **100** inhibited the enzyme was not established. Similarly, from compounds **99** and **102** only the L-enantiomer **99** served as substrate for ribonuclease T_2 (53). A comparison of the structures of an α-L-nucleoside (**103**) and a β-D-nucleoside (**104**) reveals that they differ only with respect to configuration of the 5′-hydroxymethyl

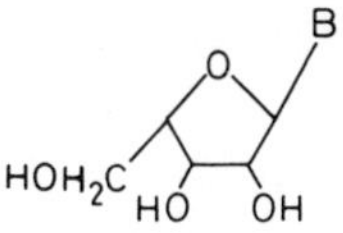

103 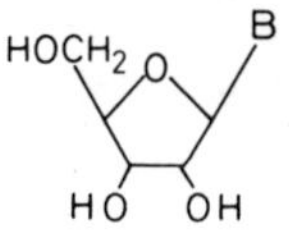 104

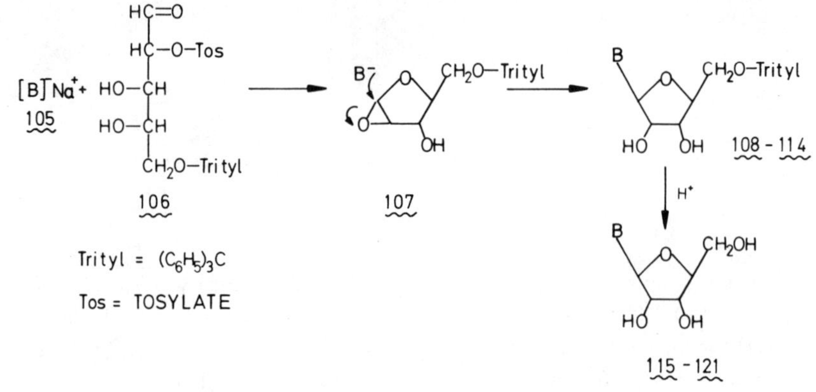

Trityl = $(C_6H_5)_3C$

Tos = TOSYLATE

	B		B
108, 115	Uracil	**112, 119**	Thymine
109, 116	Cytosine	**113, 120**	4-Methoxypyrimidine-2-one
110, 117	Adenine	**114, 121**	6-Azauracil
111, 118	Guanine		

group. From this it follows that neither ribonuclease T_1 nor ribonuclease T_2 are specific with respect to the configuration of this 5′-hydroxymethyl group and moreover that the 5′-OH group does not participate in the enzymatic reaction (53).

A variety of β-L-ribonucleosides have been made available by an intelligent synthetic approach reported by Holy and Sorm (54). Reaction of the sodium salts of purine or pyrimidine bases (**105**) with 2—O—p-toluenesulfonyl-5-*O*-trityl-L-arabinose (**106**) proceeds via an intermediary 1,2-oxirane (**107**), which is stereospecifically attacked by the anion of the heterocycle under forming of 5′-*O*-trityl-β-L-ribonucleosides (**108–114**). The above β-L-ribonucleosides (**115–121**) were subjected to transesterification with triethyl phosphite. The resulting isomeric mixtures of 2′- and 3′-phosphites gave the corresponding 2′,3′-cyclic phosphates on treatment with hexachloroacetone (54). Neither ribonuclease A, ribonuclease T_1, nor ribonuclease T_2 hydrolyzed any of the β-L-ribonucleoside 2′,3′-cyclic phosphates (**122–126**) (54).

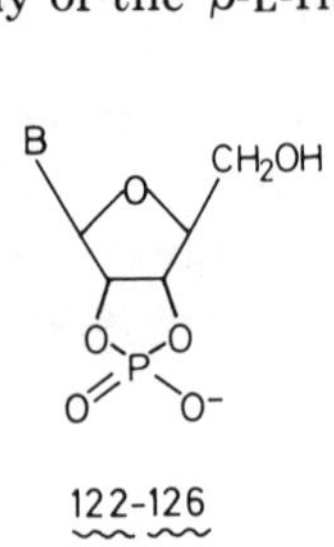

	B		B
122	Uracil	**125**	Guanine
123	Cytosine	**126**	Thymine
124	Adenine		

Phosphorylation of unprotected β-L-ribonucleosides by phosphorus oxychloride or of protected β-L-ribonucleosides by means of 2-cyanoethylphosphate provided the 5′-monophosphates (**127–130**), which were further processed to the corresponding 5′-di- and 5′-triphosphates by standard procedures (55). β-L-Ribonucleoside 5′-phosphates (**127–130**) are completely resistant against hydrolysis by 5′-nucleotidase from snake venom (55). Unspecific phosphomonoesterases like alkaline phosphatase from *E. coli* accept β-L-nucleotides as substrates. It is claimed that β-L-nucleoside 5′-diphosphates are recognized as substrates, albeit very poor ones, by

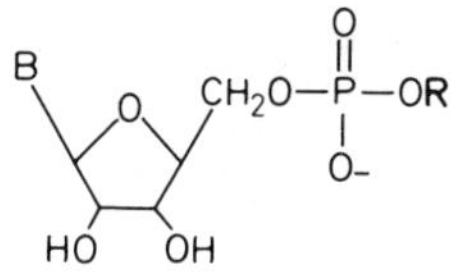

127-134

	R	B		R	B
127	H	Uracil	**131**	P(O)(OH)O⁻	Uracil
128	H	Cytosine	**132**	P(O)(OH)O⁻	Cytosine
129	H	Adenine	**133**	P(O)(OH)O⁻	Adenine
130	H	Thymine	**134**	P(O)O—O—P(O)(OH)O^{2-}	Adenine

polynucleotide phosphorylase from *E. coli* and *Streptomyces aureofaciens* (55, 56). β-L-Adenosine 5′-triphosphate (**134**) does not seem to be a substrate for DNA-dependent RNA polymerase from *E. coli*, but is a competitive inhibitor (55). Neither β-L-adenosine, β-L-cytidine, nor β-L-azauridine inhibit the growth of *E. coli* even at high concentrations (54).

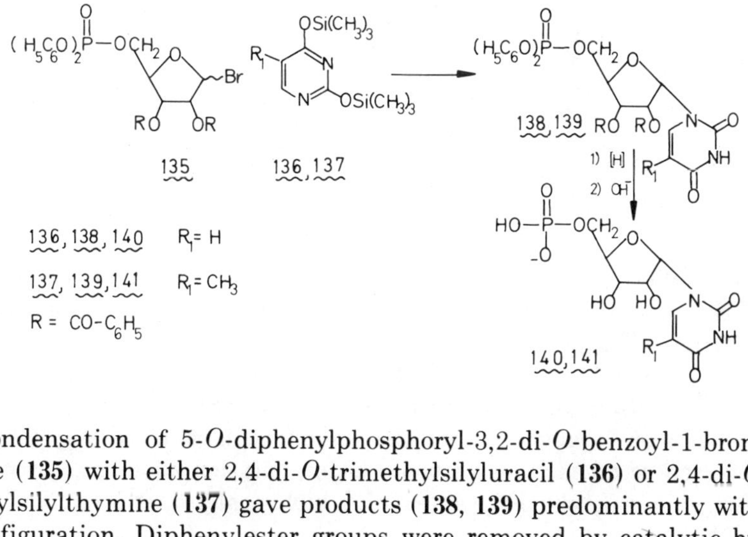

135   136, 137   138, 139   140, 141

1) [H] 2) OH^-

136, 138, 140 R_1 = H

137, 139, 141 R_1 = CH_3

R = $CO-C_6H_5$

Condensation of 5-*O*-diphenylphosphoryl-3,2-di-*O*-benzoyl-1-bromo-D-ribose (**135**) with either 2,4-di-*O*-trimethylsilyluracil (**136**) or 2,4-di-*O*-trimethylsilylthymine (**137**) gave products (**138, 139**) predominantly with the α-configuration. Diphenylester groups were removed by catalytic hydro-

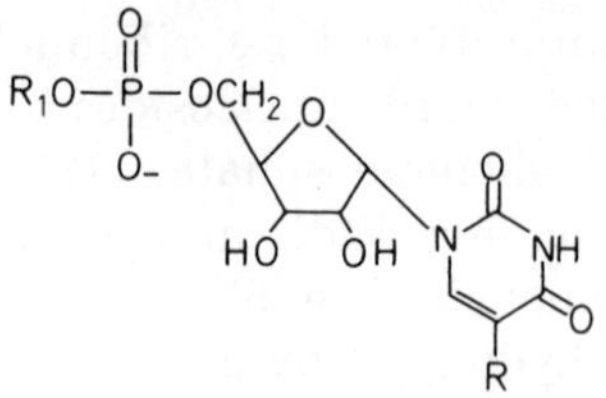

142–145

	R_1	R		R_1	R
142	$P(O)(OH)O^-$	H	**144**	$P(O)O—O—P(O)(OH)O^{2-}$	H
143	$P(O)(OH)O^-$	CH_3	**145**	$P(O)O—O—P(O)(OH)O^{2-}$	CH_3

genation, and benzoyl groups by alkaline hydrolysis to give α-D-uridine 5′-phosphate (**140**) and α-D-thymidine 5′-phosphate (**141**) (57). The 5′-diphosphates (**142, 143**) and 5′-triphosphates (**144, 145**) of α-D-uridine or α-D-thymidine were prepared by standard procedures (57). Holy described the synthesis of 2′-deoxy-α-D-uridine 5′-phosphate (**146**), 2′-deoxy-α-D-

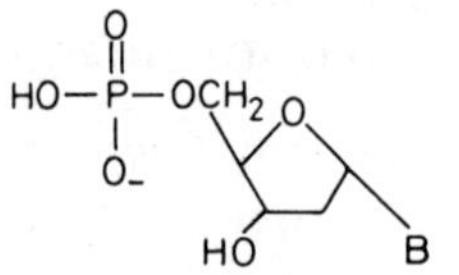

146, B = URACIL

147, B = CYTOSINE

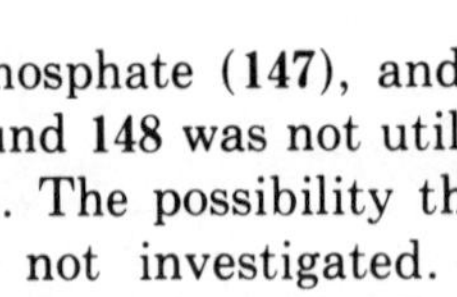

146, 147

148

cytidine 5′-phosphate (**147**), and α-D-uridine 2′,3′-cyclic phosphate (**148**) (58). Compound **148** was not utilized as a substrate by pancreatic ribonuclease A (58). The possibility that it might be an inhibitor of the latter enzyme was not investigated. The 5′-phosphates **146** and **147** were hydrolyzed by 5′-nucleotidase from snake venom (58).

In conclusion it appears that the unnatural L-enantiomers of nucleotides as well as nucleotides with the α-configuration about the glycosidic bond do not in general exhibit interesting biological properties.

5.3 SUBSTITUTION OF RIBOSE MOIETIES IN NUCLEOTIDES

The synthesis of DNA and RNA occurs by template-instructed enzymatic polymerization of 2′-deoxynucleoside 5′-triphosphates or nucleoside 5′-triphosphates respectively. The 3′-5′ nature of the internucleotidic bonds in both DNA and RNA requires substrates with free 3′-OH groups for extensive polymerization to take place. This situation initiated synthetic efforts aimed at the substitution of 3′-OH or 2′-OH groups in the above mentioned nucleotides by hydrogen, halogen, amino, and azido groups. Other synthetic work has been guided by the necessity to understand the role of the 2′-OH group in ribonucleotides with respect to its function in RNA molecules. Substitution of 2′-OH groups in nucleotides was therefore performed to determine whether the modified nucleotide behaved as a 2′-deoxy- or ribonucleotide analog.

The natural occurring antibiotic 3′-deoxyadenosine (cordycepin) (**149**) (59, 60) was shown to be a powerful inhibitor of RNA synthesis in mam-

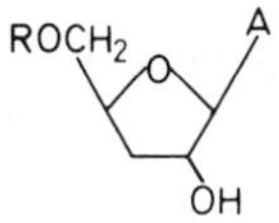

ROCH₂ … A … OH

149–**151**

149, R = H

150, R = P(O)(OH)O^-

151, R = P(O)O-O-P(O)O-O-P(O)(OH)O^{3-}

malian, tumor, and bacterial cells (61, 62). 3′-Deoxyadenosine is phosphorylated within cells to the corresponding 5′-mono-, 5′-di-, and 5′-triphosphate (61). The chemical synthesis of 3′-deoxyadenosine 5′-phosphate (**150**) and 5′-triphosphate (**151**) was reported more or less simultaneously by various groups (63, 64). 3′-Deoxyadenosine 5′-triphosphate (3′dATP) inhibits DNA-dependent RNA synthesis *in vitro* in a bacterial system most probably by chain termination (65, 66). Furthermore, Gumport et al. (64) demonstrated that 3′dATP competitively inhibited ATP-dependent DNA synthesis in *E. coli*. A possible explanation for this phenomenon may be the inhibition of primer RNA synthesis, necessary for initiation of replication. The incorporation of 3′dATP into the CCA-terminus of various tRNA molecules turned out to be a powerful tool for studies concerned with the process of aminoacylation of tRNA as well as ribosomal polypeptide synthesis (63, 67). 3′dATP participates in other metabolic processes of the cell either as substrate, effector, or inhibitor. Thus 3′dATP is a phosphate donor in the fructo kinase reaction (68) and 3′dATP can substitute for ATP as an effector of ribonucleoside diphosphate reductase from *E. coli* (69). Adenylate cyclase from *Brevibacterium liquefaciens* is strongly inhibited by 3′dATP (70).

ROCH$_2$ O B

OH

152-159

	R	B		R	B
152	H	Uracil	**156**	P(O)(OH)O$^-$	Cytosine
153	H	Cytosine	**157**	P(O)(OH)O$^-$	Guanine
154	H	Guanine	**158**	P(O)O—O—P(O)(OH)O^{2-}	Uracil
155	P(O)(OH)O$^-$	Uracil	**159**	P(O)O—O—P(O)(OH)O^{2-}	Cytosine

Novak and Sorm achieved the chemical synthesis of 3′-deoxyadenosine (**149**), 3′-deoxyuridine (**152**), 3′-deoxycytidine (**153**), and 3′-deoxyguanosine (**154**) by condensation of 5,2-*O*-di-toluyl-1-bromo-3′-deoxy-D-ribose with trimethylsilyl derivatives of the corresponding heterocyclic bases (71). Phosphorylation of the unprotected nucleosides by phosphorus oxychloride led to the respective 5′-phosphates (**150**, **155–157**). The imidazolidates of **155** and **156** served as precursors for the preparation of 3′-deoxyuridine 5′-triphosphate (**158**) and 3′-deoxycytidine 5′-triphosphate (**159**) (71). No reports have so far been made on the biochemical properties of **158** and **159**.

Hamel reported the synthesis of 3′-deoxyguanosine 5′-triphosphate (**164**) by a combination of chemical and enzymatic procedures (72). Phorphorylation of the starting material 3′-deoxyadenosine was performed by routine procedure and the product (**150**) was then deaminated to 5′-deoxyinosine 5′-phosphate (**160**). The conversion of **160** to **164** was achieved exclusively by enzymic processes. The enzymes IMP dehydrogenase and XMP aminase used in this procedure are easily isolated from *E. coli*, the other enzymes, PEP kinase as well as GMP kinase, being commercial preparations. This procedure might be also applicable to the transformation of other adenosine 5′-phosphate analogs to the corresponding guanosine 5′-phosphate derivatives. Hamel established that 3′dGTP was neither a substrate for DNA polymerase I from *E. coli* nor DNA-dependent RNA polymerase, but proved to be a potent inhibitor of the latter enzyme (72). Whether this means that 3′dGTP is different in its behavior to 3′dATP, which is a chain-terminating substrate for RNA polymerase, was not clearly stated. The function of 3′dGTP in protein biosynthesis has also been investigated (73, 74).

The biochemical properties of the antibiotic 3′-amino-3′-deoxyadenosine (**165**) (75, 76) differ to some extent from those of 3′-deoxyadenosine. The former inhibits the growth of various tumors, and some strains of yeast, but is nontoxic to bacteria (77, 78). 3′-Amino-3′-deoxyadenosine is phosphorylated by Ehrlich ascites cells to the corresponding 5′-mono (**166**), -di- (**167**), and -triphosphate (**168**) (79, 80). 3′-Amino-3′-deoxyaden-

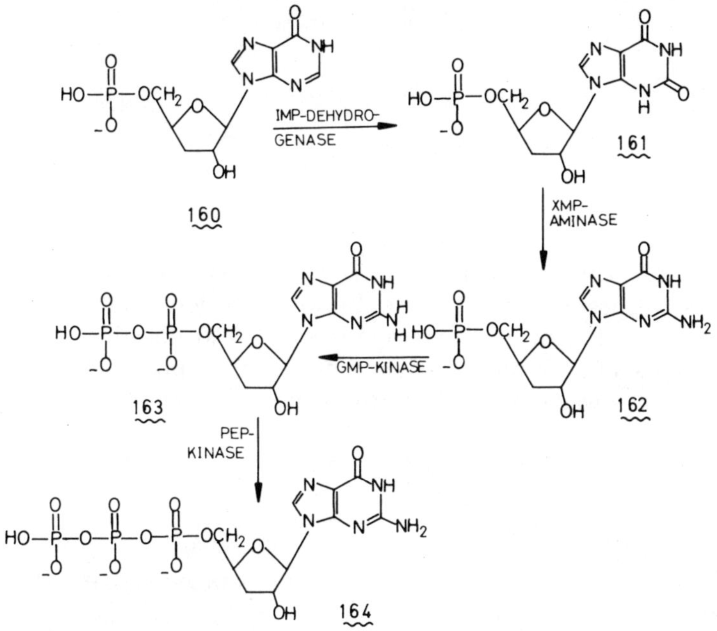

IMP-DEHYDROGENASE
XMP-AMINASE
GMP-KINASE
PEP-KINASE

160 → **161** → **162** → **163** → **164**

osine 5′-triphosphate (**168**) blocks RNA synthesis in Ehrlich ascites cells. It probably acts as a chain terminating substrate for DNA-dependent RNA polymerase (79, 80). Armstrong and Eckstein observed strong inhibition of DNA-dependent RNA polymerase from *E. coli* by **168** (81). 3′-Amino-3′-deoxyadenosine 5′-triphosphate was also found to act as

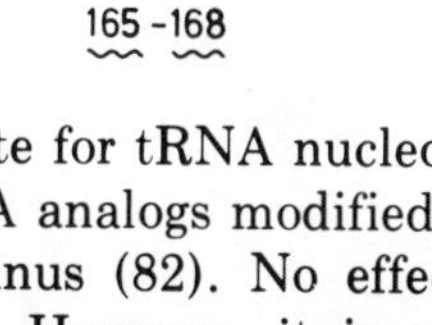

165–168

165, R = H

166, R = P(O)(OH)O^{-}

167, R = P(O)O–O–P(O)(OH)O^{2-}

168, R = P(O)O–O–P(O)O–O–P(O)(OH)O^{3-}

substrate for tRNA nucleotidyl transferase, thus allowing the preparation of tRNA analogs modified by a 3′-amino 3′-deoxyadenosine residue at the 3′-terminus (82). No effect of **168** on DNA synthesis could be demonstrated. However, it is a powerful inhibitor of adenylate cyclase from *Brevibacterium liquefaciens* (70).

2′,3′-Dideoxyribonucleosides are potential precursors of substrate analogs of DNA polymerases that cannot participate in chain elongation and hence should be inhibitors of DNA synthesis. The first synthesis of 2′,3′-dideoxythymidine 5′-phosphate (**169**) and the structurally related 2′,3′-dideoxyuridine 5′-phosphate (**170**) was reported by Pfitzner and Mof-

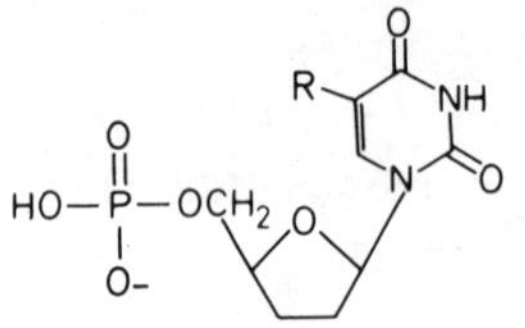

169, R= CH_3

170, R= H

169,170

fatt (83). 2′,3′-Dideoxyadenosine (84) (**171**) was phosphorylated to the respective 5′-phosphate (**172**) (88). Deoxynucleoside kinase from *E. coli* phosphorylated **172** to 2′,3′-dideoxyadenosine 5′-triphosphate (**173**) (85). The synthesis of 2′,3′-dideoxythymidine 5′-triphosphate (**178**) solely by chemical means was described by Russel and Moffatt (86). 2′,3′-Dideoxy-

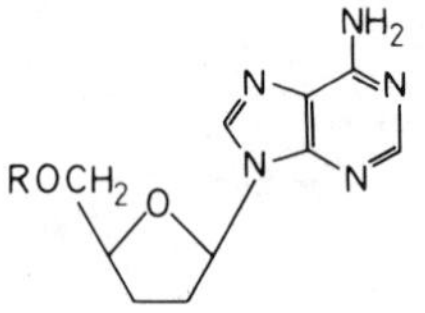

171, R= H

172, R= P(O)(OH)O^-

173, R= P(O)O-O-P(O)O-O-P(O)(OH)O^{3-}

171-173

3′-iodothymidine (87) was phosphorylated with 2-cyanoethylphosphate. The morpholidate of 2′,3′-dideoxy-3′-iodothymidine 5′-phosphate (**174**) served as an intermediate in the synthesis of the 5′-triphosphate **175**. The reaction of the morpholidate with tri-*n*-butylammonium pyrophosphate was accompanied by the formation of various side products. 2′,3′-

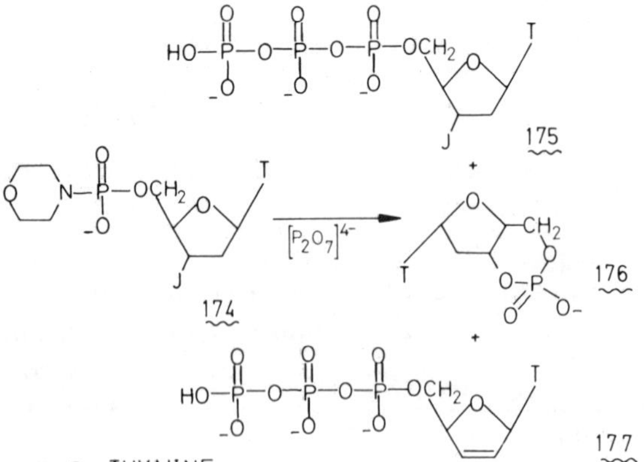

T = THYMINE

Didehydro-3′-deoxythymidine 5′-triphosphate (**177**) presumably arises because of dehydrohalogenation of **175** by excess tri-*n*-butylamine. Catalytic hydrogenation of both **175** and **177** leads to 2′,3′-dideoxythymidine 5′-triphosphate **178**). The mechanism of formation of the side

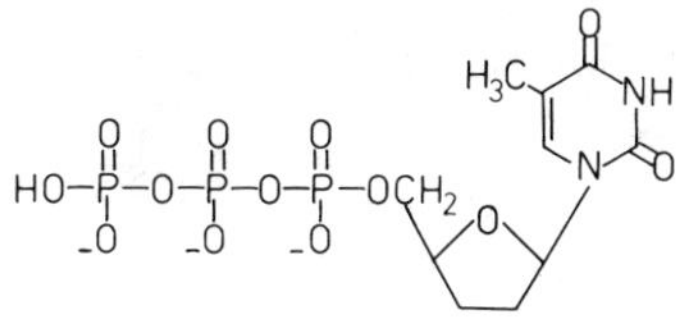

178

product 1-(2′-deoxy-β-D-*threo*-pentofuranosyl)thymidine 3′,5′-cyclic phosphate (**176**) is not quite clear. An attempt to circumvent the formation of side products by employing the phosphorylation of 2′,3′-dideoxy-3′-iodothymidine (**179**) with tris-imidazolylphosphinoxide failed. Reaction of 2′,3′-dideoxy-3′-iodothymidine 5′-phosphoroimidazolidate (**180**) with tri-*n*-butylammonium pyrophosphate led to the same pattern of products as in the previous reaction.

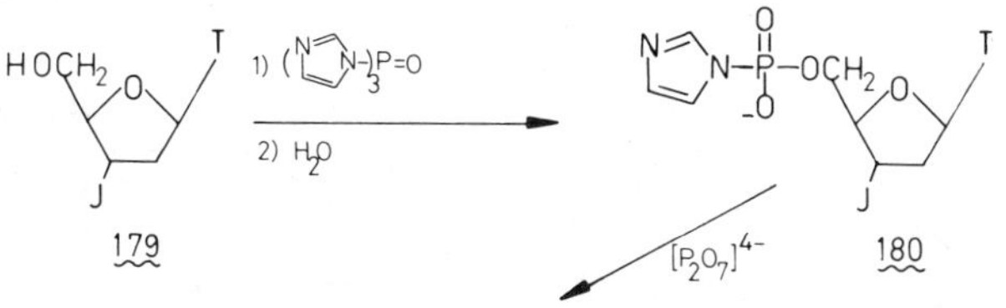

1) $(\text{Im})_3P{=}O$ 2) H_2O

179 180

$[P_2O_7]^{4-}$

175 + 176 + 177

T = THYMINE

2′,3′-Dideoxyadenosine 5′-triphosphate (88) and 2′,3′-dideoxythymidine 5′-triphosphate (89) were shown to be chain terminating substrates of DNA polymerase I from *E. coli*. Of the 2′,3′-dideoxynucleosides with the heterocyclic bases thymine, adenine, cytosine, and uracil only 2′,3′-dideoxyadenosine was toxic to bacterial cells (89, 90). None of them, however, had any effect on mammalian cells (89). It was suggested that in those cases where no biological effects were observed with the above compounds, the respective 2′,3′-dideoxynucleoside was not phosphorylated within a given cell.

On the other hand 2′,3′-dideoxy-3′-chlorothymidine and 2′,3′-dideoxy-3′-fluorothymidine (**181**) inhibit the growth of tumor cells, the latter being

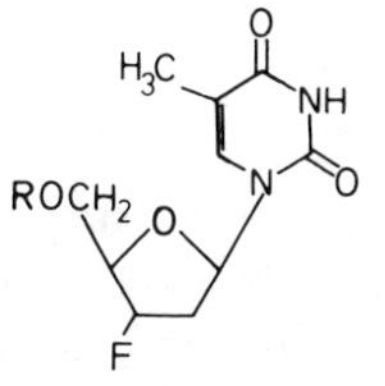

181, R = H

182, R = P(O)O-O-P(O)O-O-P(O)(OH)O^{3-}

181,182

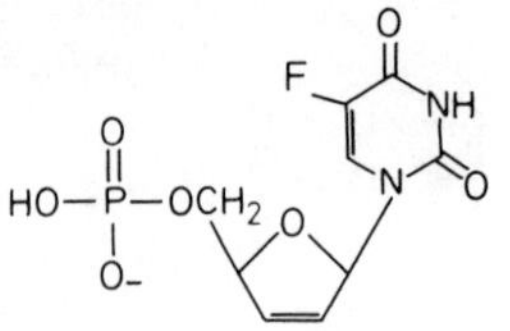

183

the more effective agent (90). Compound **181** inhibits DNA synthesis and it appears that it is phosphorylated within mammalian cells to 2′,3′-dideoxy-3′-fluorothymidine 5′-triphosphate (**182**), which presumably functions as a chain terminating substrate for DNA polymerase.

Khwaja and Heidelberger (91) synthesized 1-(2′,3′-dideoxy-2′,3′-didehydro-β-D-glycero-pentofuranosyl)5-fluorouracil and phosphorylated it by standard procedures to the corresponding 5′-triphosphate (**183**).

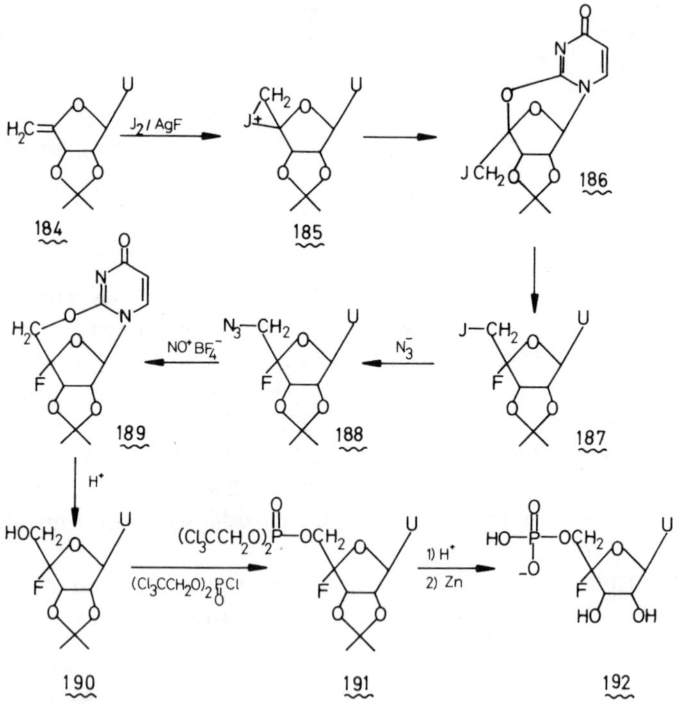

J2 / AgF
NO⁺BF4⁻
N3⁻
H⁺
(Cl3CCH2O)2PCl
1) H⁺
2) Zn
184
185
186
187
188
189
190
191
192

U = URACIL

The synthesis of 4′-fluorouridine derivatives is in itself a difficult synthetic task. Additional complication arises from the known acid lability of fluoro acetals, to which class 4′-fluorouridine belongs. Owen et al., reacted 1-(5-deoxy-2′,3′-*O*-isopropylidene-β-D-erythro-pent-4-enofuranosyl)uracil (**184**) with iodine in the presence of silver fluoride and obtained 5′-deoxy-4′-fluoro-5′-iodo-2′,3′-*O*-isopropylideneuridine (**187**) in a stereospecific reaction (92). Reaction of **187** with azide ions could be achieved only under forcing conditions. Treatment of **188** with nitrosyl tetrafluoroborate led to nitrosation of the azide moiety accompanied by loss of nitrogen and nitrous oxide. The intermediary 5′-carbonium species rearranges to the O^2,5′-cyclo nucleoside **189**. Controlled acid hydrolysis of the latter gave **190** which was phosphorylated with bis-(2,2,2-trichloroethyl) phosphorochloridate giving 4′-fluoro-2′,3′-*O*-isopropylidene 5′-*O*-bis-(2,2,2-trichloroethyl)phosphate (**191**). Acid hydrolysis of the ketal

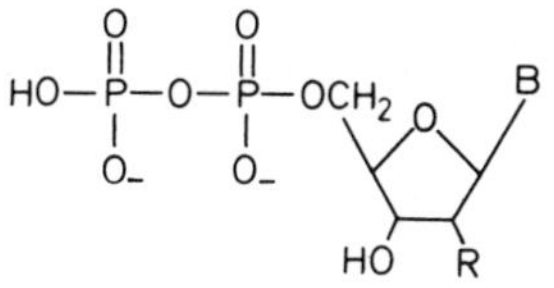

HO—P(=O)(O⁻)—O—P(=O)(O⁻)—OCH₂ [furanose ring: O, B; HO, R]

193-196

	B	R	Reference		B	R	Reference
193	Uracil	Cl	94	**195**	Uracil	F	95
194	Cytosine	Cl	94	**196**	Adenine	F	96

followed by *β*-elimination of the 2,2,2-trichloroethyl groups with the aid of zinc dust afforded 4′-fluorouridine 5′-phosphate (**192**). Compound **192** was found to be susceptible to both alkaline as well as acid hydrolysis, liberating uracil. Whether this limits the preparation of derivatives of biochemical interest remains to be seen.

The discovery that double-stranded polyribonucleotides such as poly (rI)·poly (rC) can stimulate interferon synthesis, and by virtue of this property, function as powerful antiviral agents stimulated synthetic efforts in the field of polyribonucleotide synthesis (93). Because of the high molecular weight of polyribonucleotide required for interferon stimulation, the method of choice was the polymerization of nucleoside 5′-diphosphates by polynucleotide phosphorylase or of nucleoside 5′-triphosphates by DNA-dependent RNA polymerase. An important objective of synthetic attempts was the preparation of polyribonucleotides with enhanced stability toward nucleolytic degradation. This property was provided by nucleoside 5′-diphosphates, in which the 2′-OH group was replaced by either chlorine or fluorine atoms. The synthesis of compounds **193–196** starting from the corresponding nucleosides was achieved by routine chemical procedures. These compounds were all polymerized by polynucleotide phosphorylase from various origins. The replacement of the 2′-OH group by iodine was not feasible because of the ease with which such compounds would undergo an intramolecular displacement reaction followed by rearrangement to arabinonucleotides. Indeed, even 2′-deoxy-2′-chloronucleotides at pH 8.9 and room temperature are already partially hydrolyzed via the respective O^2,2′-anhydro derivatives to the arabinonucleotides (97). In the case of 2′-deoxy-2′-chlorouridine 5′-phosphate or 5′-

diphosphate the O^2,2′-anhydronucleotides formed are relatively stable and are only slowly hydrolyzed to arabinonucleotides (97).

Biophysical studies on dinucleoside monophosphates containing 2′-deoxy-2′-fluorourididylic acid residues revealed that the 2′-deoxy-2′-fluororiboside apparently resembles ribose much more than 2′-deoxyribose (98).

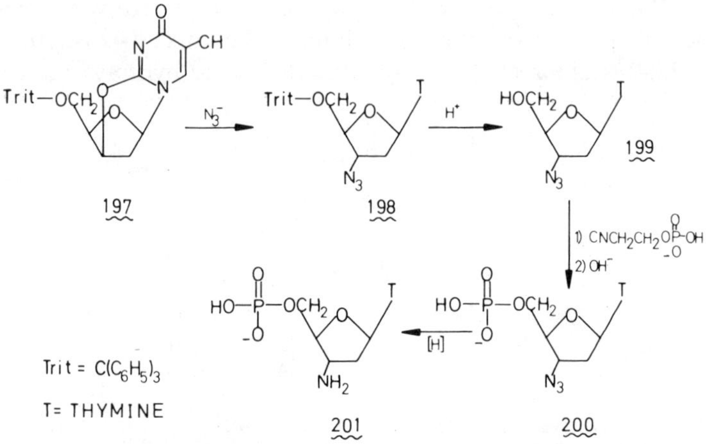

Trit = $C(C_6H_5)_3$

T= THYMINE

A useful and much used approach to the synthesis of deoxyaminonucleotides was first demonstrated in the preparation of 3′-deoxy-3′-aminothymidine 5′-phosphate (**201**) (99). The above approach was taken up by Wagner et al. (100) and applied to the synthesis of the 3′-(**204**) and 5′-phosphate (**205**) of 2′-deoxy-2′-aminouridine (100). The authors

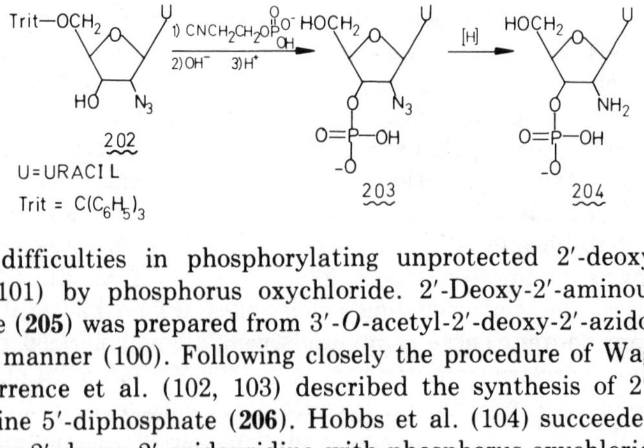

U=URACIL

Trit = $C(C_6H_5)_3$

reported difficulties in phosphorylating unprotected 2′-deoxy-2′-azido-uridine (101) by phosphorus oxychloride. 2′-Deoxy-2′-aminouridine 5′-phosphate (**205**) was prepared from 3′-*O*-acetyl-2′-deoxy-2′-azidouridine in a similar manner (100). Following closely the procedure of Wagner et al. (100), Torrence et al. (102, 103) described the synthesis of 2′-deoxy-2′-azidouridine 5′-diphosphate (**206**). Hobbs et al. (104) succeeded in phosphorylating 2′-deoxy-2′-azidouridine with phosphorus oxychloride without encountering major difficulties. Catalytic hydrogenation of **206** gave the 2′-deoxy-2′-amino derivative **207**. The conversion of 2′-deoxy-2′-azido-uridine 5′-phosphate to the diphosphate **206** involved routine procedures.

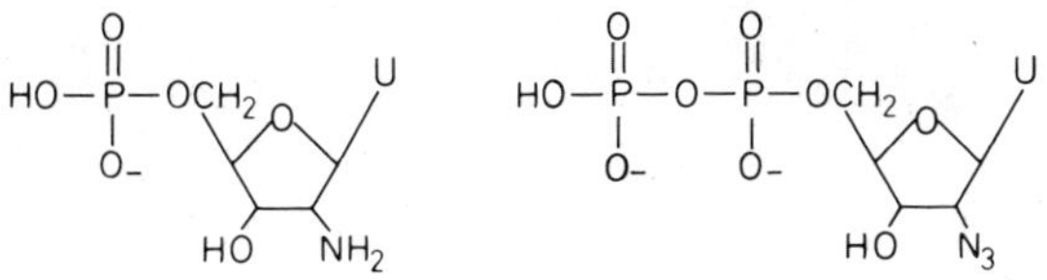

U = URACIL

205 **206**

The 5′-monophosphate (**208**) and 5′-diphosphate (**209**) of 2′-deoxy-2′-azidoadenosine were prepared (105) in close similarity to the method reported by Hobbs et al., for the uridine analogs (104). 2′-Deoxy-2′-azidocytidine (**212**) is available by chemical transformation of a corresponding uridine derivative (106). Routine procedures were employed for

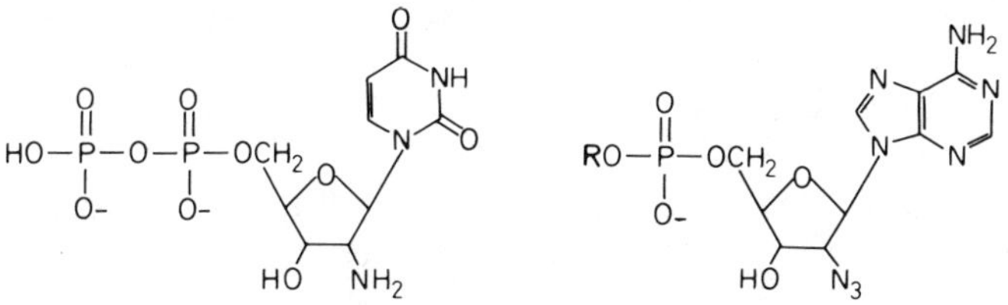

207 **208**, **209**

208, R = H

209, R = $P(O)(OH)O^-$

the phosphorylation of **212** to 2′-deoxy-2′-azidocytidine 5′-phosphate (**213**) and 5′-diphosphate (**214**) (109). Catalytic hydrogenation of **214** afforded 2′-deoxy-2′-aminocytidine 5′-diphosphate (**215**) (106). Compounds **206**, **207**, **209**, **214**, and **215** were found to be substrates of polynucleotide phosphorylases.

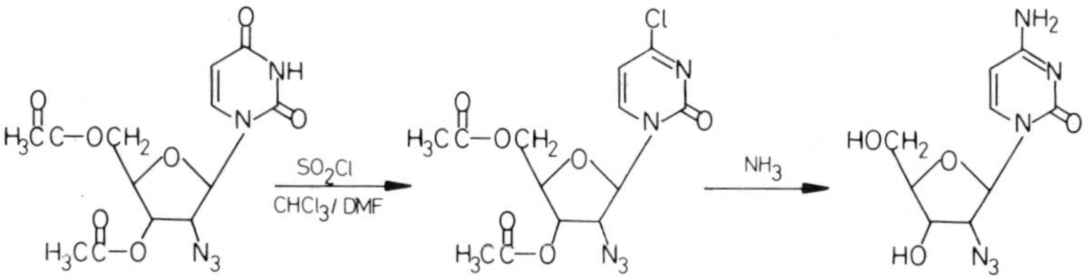

210 → (SO_2Cl, $CHCl_3$/DMF) → **211** → (NH_3) → **212**

Armstrong and Eckstein (81) investigated the substrate properties of 2′-amino-2′-deoxyadenosine 5′-triphosphate, 2′-amino-2′-deoxyuridine 5′-triphosphate, and 2′-chloro-2′-deoxyuridine 5′-triphosphate with DNA-dependent RNA polymerase from *E. coli*. These analogs were found to be weak competitive inhibitors, which is surprising, in view of the fact that 3′-amino-3′-deoxyadenosine 5′-triphosphate behaves as a powerful competitive inhibitor of transcription. It can not be ruled out, however, that the above analogs are substrates with very unfavorable kinetic parameters.

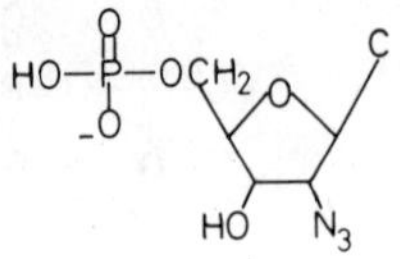

213

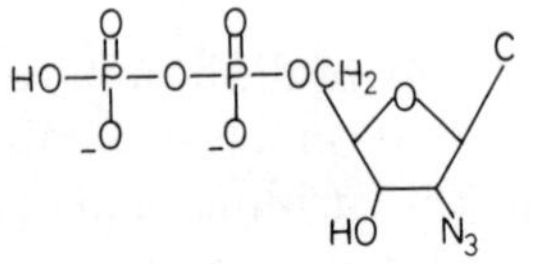

214

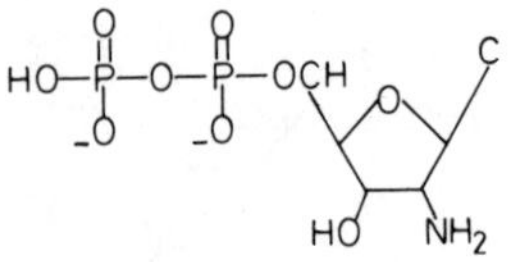

215

C = CYTOSINE

2′-*O*-Methylnucleotides were discovered as minor constituents of ribosomal RNA, 5 S RNA, and transfer RNA (107). The 2′-*O*-methyl derivatives of adenosine (108–111), guanosine (108), cytidine (111, 112), and uridine (111, 112) are most conveniently prepared by methylation of the respective nucleosides with diazomethane in organic solvents. In most

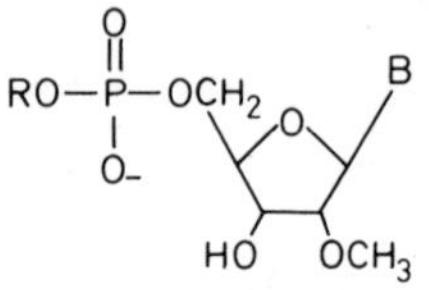

216-222

	R	B
216	H	Adenine
217	$P(O)(OH)O^-$	Adenine
218	$P(O)O—O—P(O)(OH)O^{2-}$	Adenine
219	H	Cytosine
220	$P(O)(OH)O^-$	Cytosine
221	$P(O)(OH)O^-$	Cytosine
222	$P(O)(OH)O^-$	Hypoxanthine

cases isomeric mixtures of *O*-methylnucleosides are obtained that can, however, be separated by chromatography. Phosphorylation of unprotected 2′-*O*-methyladenosine or 2′-*O*-methylcytidine with 2-cyanoethylphosphate occurred predominantly in the 5′-position (113, 114). 2′-*O*-Methyladenosine 5′-phosphate (**216**) was converted by means of rabbit muscle myokinase to the 5′-diphosphate (**217**) and 5′-triphosphate (**218**) (113). The morpholidate of 2′-*O*-methylcytidine 5′-phosphate (**219**) upon reaction with phosphate gave 2′-*O*-methylcytidine 5′-diphosphate (**220**) (114, 115). Synthesis of the cytosine analogs **219** and **220** by a combination of enzymic and chemical procedures was reported by Zmudzka et al. (116). Deamination of **220** with nitrous acid afforded the corresponding uracil derivative **221** and deamination of the adenine compound **222** gave the hypoxanthine analog **217** (118). An improved preparation of 2′-*O*-methylnucleoside 5′-diphosphates was reported by Pal and Schmidt (119). An original approach to the synthesis of 2′-*O*-methyladenosine 5′-diphosphate was contributed by Tazawa et al. (120). Methylation of adenosine 3′,5′-cyclic phosphate with methyl iodide at alkaline pH and subsequent enzymic hydrolysis of **224** by phosphodiesterase afforded **216**.

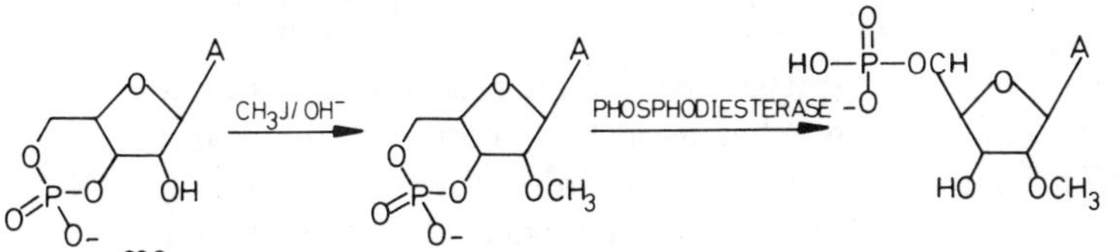

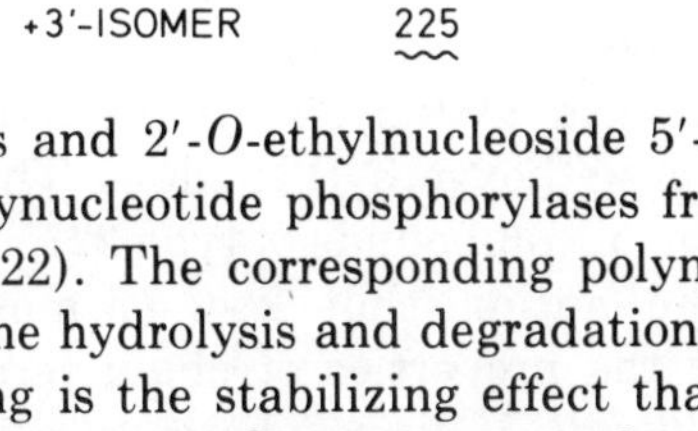

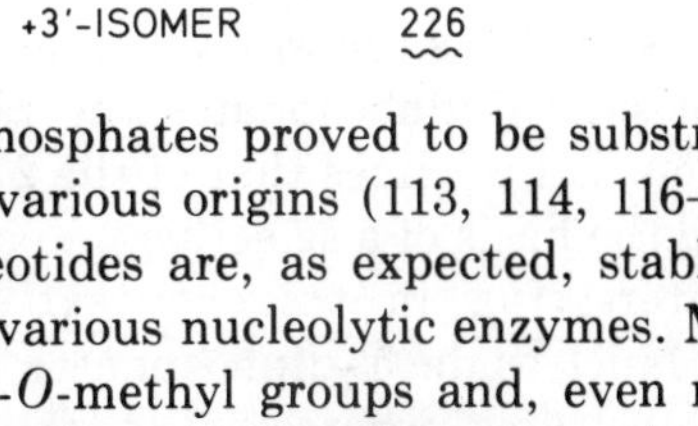

The preparation of 2′-*O*-ethyladenosine 5′-diphosphate (**225**) was first reported by Rottman et al. (118, 121). Alkylation of cytidine 5′-phosphate with diethyl sulfate in alkaline medium followed by deamination with nitrous acid gave an isomeric mixture of 2′,(3′)-*O*-ethyluridine 5′-phosphate which was chemically phosphorylated to the corresponding 5′-diphosphates (**226**) (122). Virtually all 2′-*O*-methylnucleoside 5′-diphosphates and 2′-*O*-ethylnucleoside 5′-diphosphates proved to be substrates of polynucleotide phosphorylases from various origins (113, 114, 116–118, 121, 122). The corresponding polynucleotides are, as expected, stable to alkaline hydrolysis and degradation by various nucleolytic enzymes. More striking is the stabilizing effect that 2′-*O*-methyl groups and, even more pronounced, 2′-*O*-ethyl groups have on the self complementary double-stranded structures of homo polynucleotides. A convincing explanation for this phenomenon has not yet been put forward. mRNA partially methylated at the 2′-OH groups still functions as a template in ribosomal protein synthesis, whereas fully methylated 2′-*O*-methylated mRNA does not (117). Single-stranded poly 2′-*O*-methyluridylate and poly 2′-*O*-methylcytidylate are templates for DNA-dependent RNA polymerase from *Pseudomonas putidas* in contrast to poly 2′-*O*-methyladenylate and poly 2′-*O*-methylinosinic acid (123). 2′-*O*-Methyladenosine 5′-triphosphate (**218**) is a substrate for DNA-dependent RNA polymerase from *Pseudomonas putidas* (124) with unfavorable kinetic parameters. It could not be distinguished whether the reduced V_{max} of synthesis in the presence of **218** is due to either reduced affinity or a smaller rate constant k_{cat} of the catalytic step. Most of the incorporated 2′-*O*-methyladenosine residues were found at the 3′-ends of synthesized RNA molecules, indicating that the rate of chain elongation after introduction of a 2′-*O*-methyladenosine moiety is drastically reduced. An interesting effect on transcription is observed with 3′-*O*-methyladenosine 5′-triphosphate as substrate for DNA-dependent RNA polymerase from *E. coli.* (125). At moderate ionic strength and temperature the analog seems to function as a chain terminating substrate, whereas at higher ionic strength and temperature it has the properties of a competitive inhibitor with $K_i = 4 \cdot 10^{-5}$ *M* (125).

Gilham et al. (126) discovered that 2′-*O*-substituted derivatives of nucleoside 5′-diphosphates were utilized by polynucleotide phosphorylase under proper experimental conditions in a single-step phosphorylation of a oligonucleotide primer with a free 3′-OH group. A variety of different approaches have been reported, from which the following have been selected for presentation. Bennett and Gilham (127) introduced the 2′-*O*-(α-methoxyethyl) substituted nucleoside 5′-diphosphates (**227**) as single

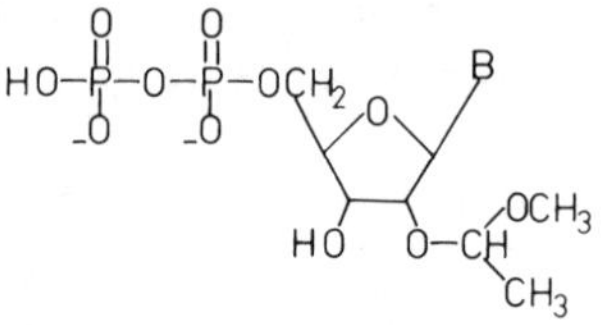

227

addition substrates. Reaction of nucleoside 5′-diphosphates with methylvinyl ether leads to preferential formation of 3′-*O*-(α-methoxyethyl) derivatives. It was, however, observed that in 2′,3′-*O*-di(α-methoxyethyl) nucleoside 5′-diphosphates the rate of acid hydrolysis for the 3′-*O*-substituent is two times that of the 2′-*O*-(α-methoxyethyl) residues. This provided the basis of a selective preparation of compounds of type **227**.

Interesting in this context is the notion that 2′-*O*-(α-methoxyethyl)nucleoside 5′-diphosphates have the properties of chain terminating substrates, whereas 2′-*O*-ethylnucleoside 5′-diphosphates in contrast, allow chain elongation to take place. 2′-*O*-Ortho-nitrobenzyl nucleoside 5′-diphosphates (**228–230**) were employed by Ikehara et al. (128) as single-step addition substrates for polynucleotide phosphorylase. The 2′-*O*-ortho-

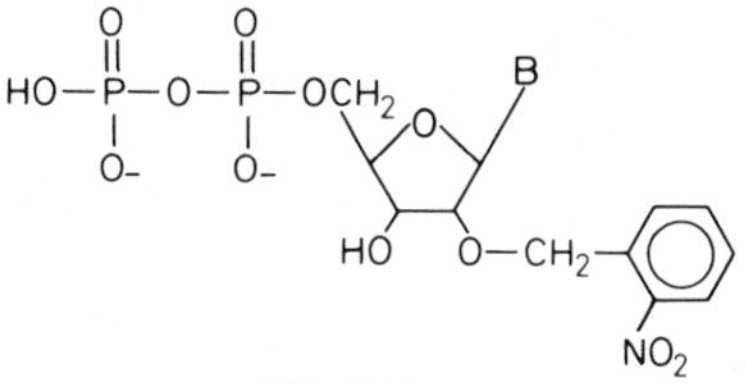

228, B = URACIL
229, B = CYTOSINE
230, B = ADENINE

228–230

nitrobenzyl groups can be removed photochemically after enzymic reaction has been accomplished. An interesting approach to the single-step addition with the aid of polynucleotide phosphorylase employs 2′(3′)-*O*-dihydrocinnamoylnucleoside 5′-diphosphates (**231**) (129). The dihydrocinnamoyl group can be removed from the oligonucleotide product

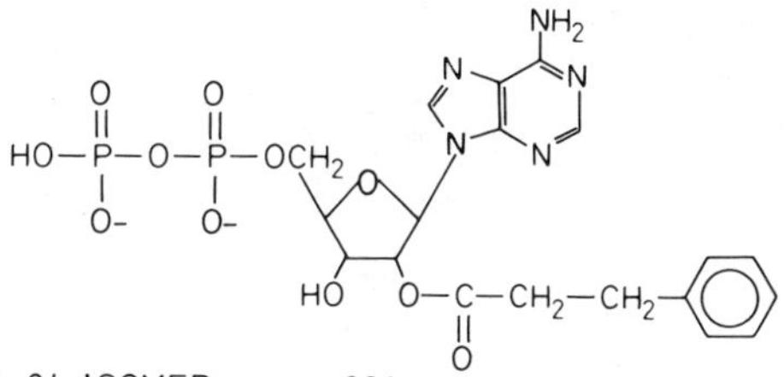

+3′-ISOMER 231

by enzymic hydrolysis with α-chymotrypsin (129). Reaction of triethyl orthoisovalerate with nucleoside 5′-diphosphates leads to 2′,3′-*O*-isovaleroyloxymethylene intermediates (**237–241**), which upon acid hydrolysis give an isomeric mixture of 2′(3′)-*O*-isovaleroylnucleoside 5′-diphosphates (**242–246**), likewise suitable for single-step addition by means of polynu-

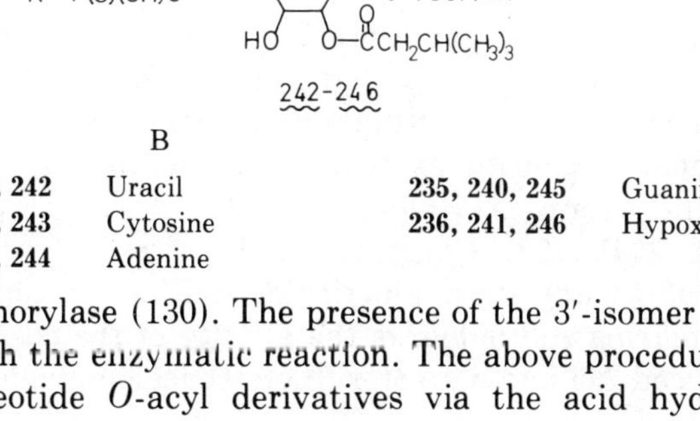

232–236 $\xrightarrow[H^+]{(CH_3)_2CH\text{-}CH_2C(OC_2H_5)_3}$ 237–241 $\xrightarrow{H^+/H_2O}$ 242–246 + 3′-ISOMER

R = $P(O)(OH)O^-$

	B		B
232, 237, 242	Uracil	**235, 240, 245**	Guanine
233, 238, 243	Cytosine	**236, 241, 246**	Hypoxanthine
234, 239, 244	Adenine		

cleotide phosphorylase (130). The presence of the 3′-isomer does not seem to interfere with the enzymatic reaction. The above procedure for the synthesis of nucleotide *O*-acyl derivatives via the acid hydrolysis of the respective ortho esters was first introduced by Zemlicka and Chladek for the synthesis of 2′,(3′)-*O*-glycyladenosine 5′-diphosphate (**250**) and 5′-triphosphate **251** respectively (131).

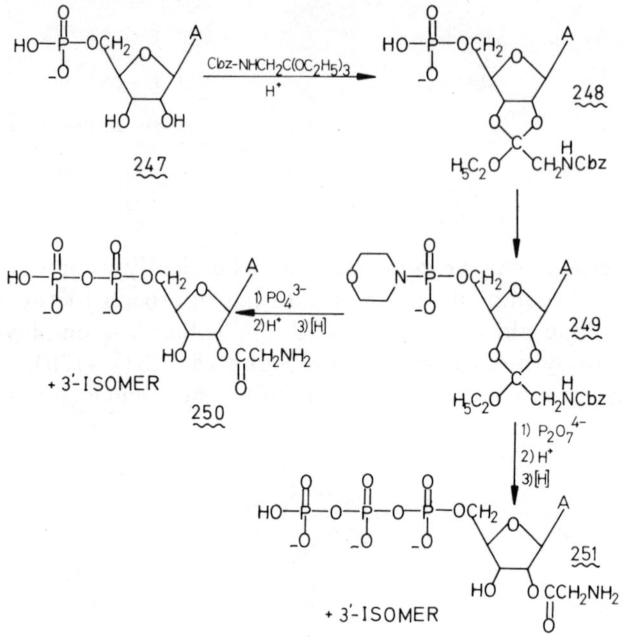

Pyrimidine nucleosides of 4-thio-D-ribofuranose were first described by Whistler and Block (132). 9-(4-Thio-β-D-ribofuranosyl)adenine (**252**) was

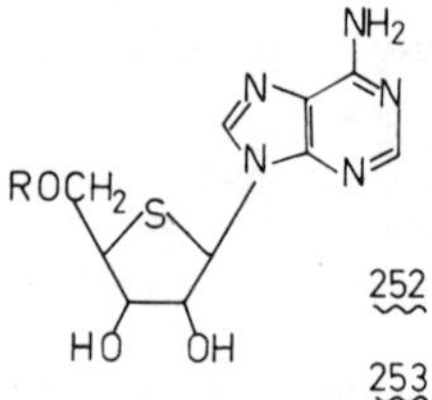

synthesized by condensation of 2,3,5-tri-*O*-acetyl-4-thio-D-ribofuranosyl-chloride with chloromercuri-6-benzamidopurine (132). Phosphorylation of **252** to 9-(4-thio-β-D-ribofuranosyl)adenine 5′-phosphate (**253**) involved standard procedures (133). Nicotinamide mononucleotide and **248** reacted in the presence of dicyclohexylcarbodiimide to give P^1-5′-*O*-(1-β-D-ribofuranosylnicotinamide) P^2-5′-*O*-(9-(4-thio-β-D-ribofuranosyl)adenine) pyrophosphate (**254**) (133). Compound **254** was reduced by a number of dehydrogenases. Steady-state kinetic as well as static measurements showed that the ring sulfur lowers the affinity of the coenzyme for these enzymes. Reduced **254** had a greater fluorescence emission quantum yield

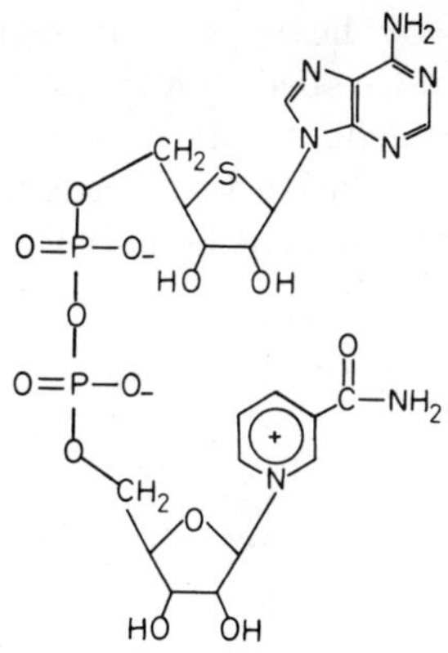

254

compared to NADH. The ORD spectra of **252** and **253** showed larger negative Cotton effects than adenosine or adenosine 5′-phosphate. This could be an indication for a larger population of molecules in the anti conformation because of the repulsion between the nonbonding electrons of N(3) and the ring sulfur atom.

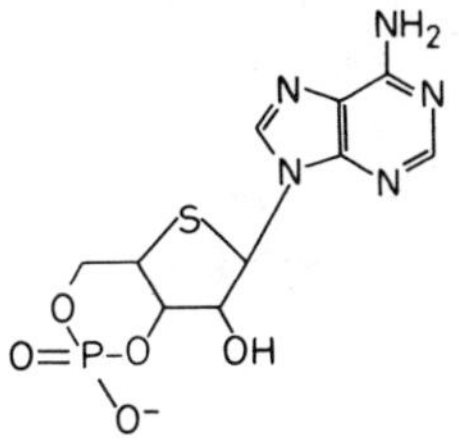

255

Cyclization of **253** by means of dicyclohexylcarbodiimide afforded 9-(4-thio-β-D-ribofuranosyl)adenine 3′,5′-cyclic phosphate (**255**) (134). Compound **255** was a substrate for phosphodiesterase from bovine heart with kinetic parameters comparable to those of cAMP (134).

2,4,6-Trinitrobenzene-1-sulfonate reacts with adenosine 5′-triphosphate to 2′,(3′)-*O*-(2,4,6-trinitrophenyl) adenosine 5′-triphosphate (**256**). Com-

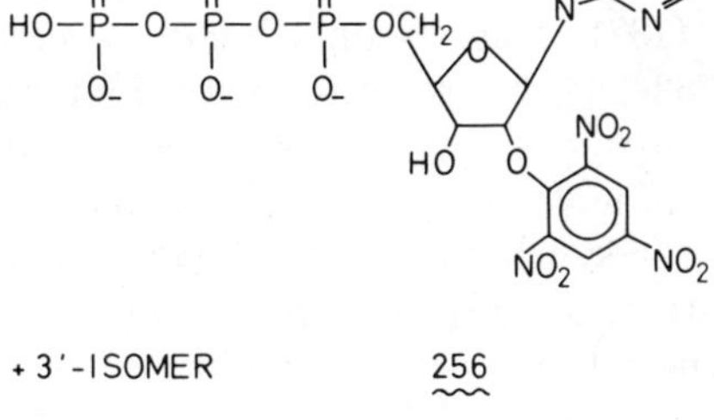

+ 3′-ISOMER **256**

pound **256** is hydrolyzed by heavy meromyosin. Binding to the latter causes an absorption difference spectrum in the visible region (135).

A question of permanent general interest concerns the orientation of the adenine ring in nucleotides relative to the sugar part and the phosphate moiety in solution. A second covalent bond between the adenine and sugar

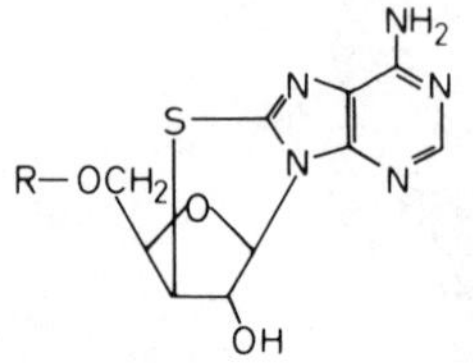

257, R = H

258, R = P(O)(OH)O⁻

257, 258

moiety in addition to the glycosidic bond would fix the orientation of the heterocyclic substituent. In 8,3′-thioanhydroadenosine (**257**) this is achieved by a thioether bridge between adenine and ribose. Phosphorylation of **257** by phorphorus oxychloride gives 8,3′-thioanydroadenosine 5′-phosphate (**258**) (136). It is interesting that phosphorylation of **257** does

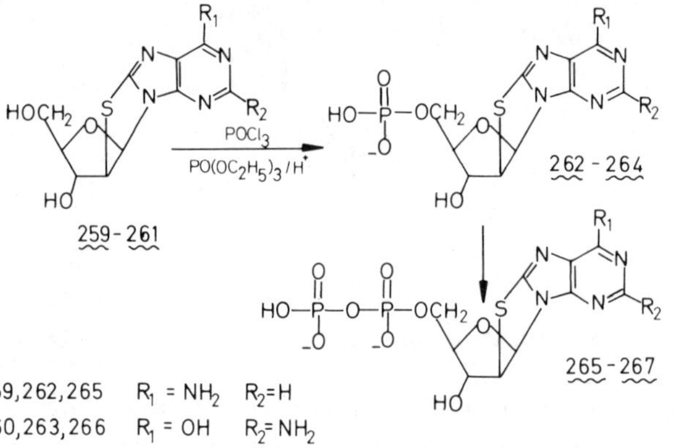

$POCl_3$

$PO(OC_2H_5)_3/H^+$

259-261

262-264

265-267

259,262,265	R_1 = NH_2	R_2=H
260,263,266	R_1 = OH	R_2= NH_2
261,264,267	R_1 = OH	R_2= H

not influence the CD spectrum of **257** in contrast to 8-bromoadenosine. In this case phosphorylation drastically effects the CD spectrum, indicating that 8-bromoadenosine is not rigidly fixed in a certain conformation. A series of analogous 8,2′-thioanhydropurine nucleotides were prepared by Ogilvie and Slotin (137). The 5′-diphosphates were obtained from the morpholidates of **262–264** by reaction with inorganic phosphate. The synthetic scheme for the synthesis of 8,2′-anhydro-8-oxy-9-(β-D-arabinofuranosyl)adenine 3′-phosphate (**272**) is outlined in detail, because it exemplifies a useful synthetic approach by which purine anhydronucleotides can be obtained (138). The formation of the side product **271** may be caused by the relative drastic conditions necessary to remove the

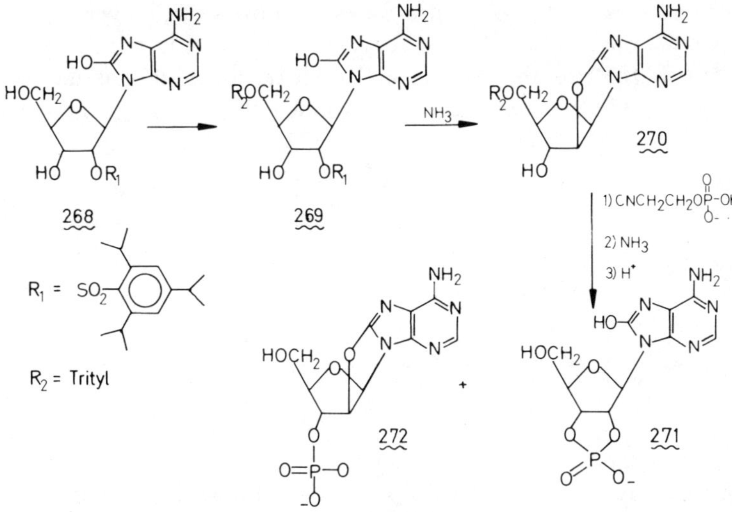

2-cyanoethyl group. Hampton and Susaki (139) synthesized 8,2′-anhydro-8-oxy-9-(β-D-arabinofuranosyl)adenine 5′-phosphate (**273**) and 8,3′-anhydro-8-oxy-9-(β-D-xylofuranosyl)adenine 5′-phosphate (**274**) to investigate their substrate properties for AMP aminohydrolase, AMP kinase, and snake venom 5′-nucleotidase. Compounds **273** and **274** were found to

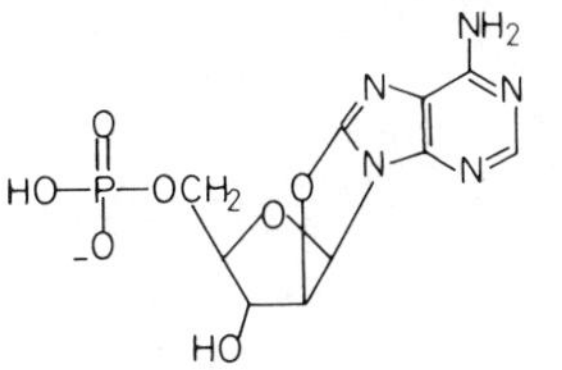

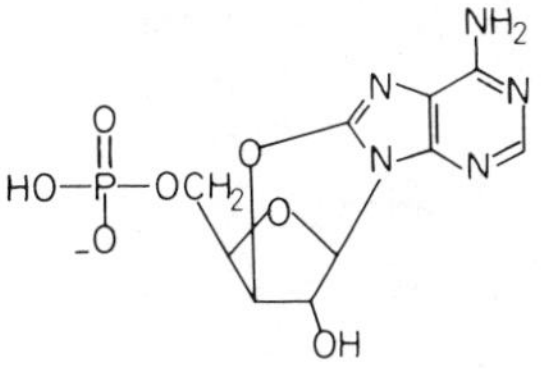

be extremely poor substrates for the above enzymes, although the affinities of the analogs for AMP aminohydrolase as well as AMP kinase were higher than that of AMP itself. These findings provide evidence for the conformation about the glycosidic bond in enzyme bound AMP in which the H(8) of the adenine is oriented in the vicinity of C(4′).

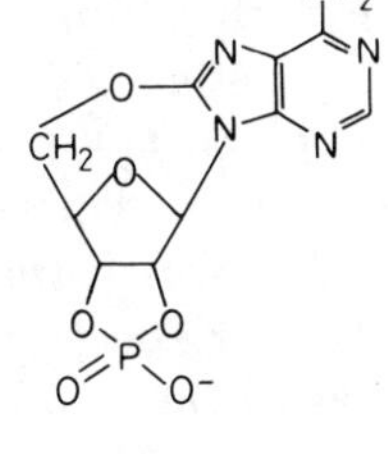

The synthesis of 8,5′-anhydro-8-oxy-adenosine 2′,3′-cyclic phosphate (**275**) by phosphorylation of the respective nucleoside with 2-cyanoethyl phosphate in the presence of dicyclohexylcarbodiimide was described by Ikehara et al. (140). Compound **275** was a substrate for ribonuclease M from *Aspergillus saitoi* (140).

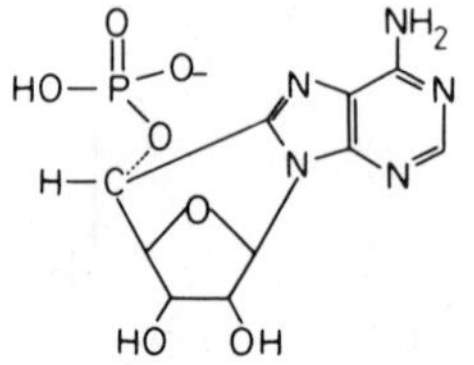

276 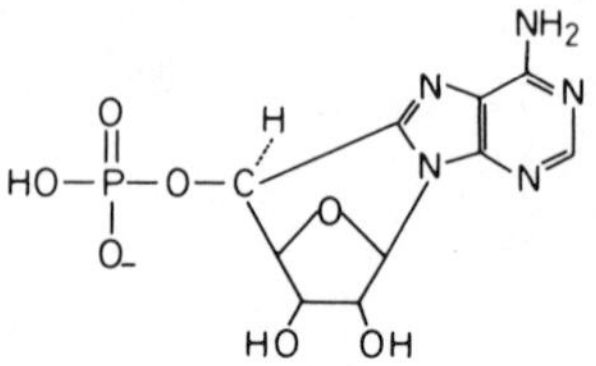 277

8,5′-Cycloadenosine, which possesses a new epimeric center at C(5′), was obtained by Harper and Hampton (141). Phosphorylation of the mixture of epimers led to **276** and **277**, which could not be separated (142). The substrate properties of the epimeric mixture **276** and **277** for 5′-nucleotidase, AMP aminohydrolase, and rabbit muscle AMP kinase were

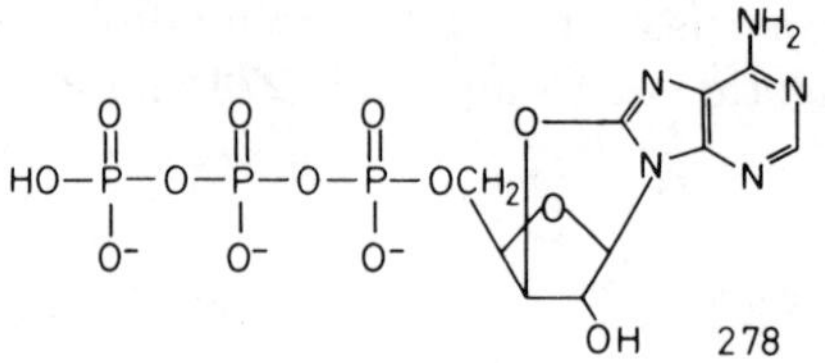

278

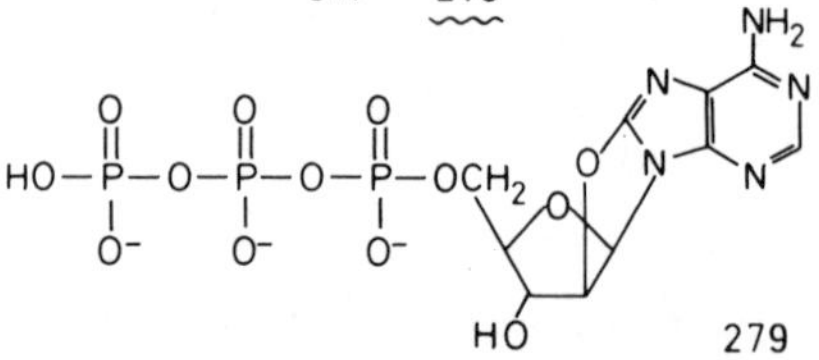

279

investigated (142). Compounds **276** and **277** were inhibitors for the two latter enzymes. 5′-Nucleotidase hydrolyzed about one-third of the epimeric mixture, indicating that only one epimer is a substrate and furthermore that the ratio of epimers in the mixture is approximately 1:3 (142). Hampton et al., proposed that the epimer that is hydrolyzed has the gauche-trans conformation about the C(5′)—C(4′) bond and is identical with **276**. Raleigh and Blackburn (143) achieved the stereoselective photochemical synthesis of 8,5′-cycloadenosine and phosphorylated it to the 5′-phosphate. The product, to which the authors assigned configuration **277** on the basis of its ^{1}H nmr spectrum, was not hydrolyzed by 5′-nucleotidase. The conformation of the C(5′)—C(4′) bond in **277** is trans-gauche. It thus appears that nucleotides that are substrates of 5′-

nucleotidase from snake venom require gauche-trans or gauche-gauche conformation about the C(5′)—C(4′) bonds (143).

Boos et al., employed the 5′-triphosphates of 8,3′-anhydro-8-oxy-9-(β-D-xylofuranosyl)adenine (**278**) and 8,2′-anhydro-8-oxy-9-(β-D-arabinofuranosyl)adenine (**279**) in investigations of structure-function relationships of nucleotides for the nucleotide carrier in rat liver mitochondria (144). Neither **278** nor **279** showed carrier mediated exchange with the endogenous nucleotide pool across the inner mitochondrial membrane. Compound **279** binds, however, to the carrier protein and inhibits translocation (144). It is concluded that the transfer step requires nucleotides with a nonrigidly fixed anti conformation of the glycosidic bond (144).

5.4. ALIPHATIC ANALOGS OF NUCLEOTIDES

This section deals with synthetic attempts to substitute heterocyclic bases by hydroxyalkanes with the aim of obtaining open-chain sugar analogs of nucleosides or nucleotides. Chemical manipulations of the sugar moiety of nucleotides leading to ring-opened derivatives is also considered.

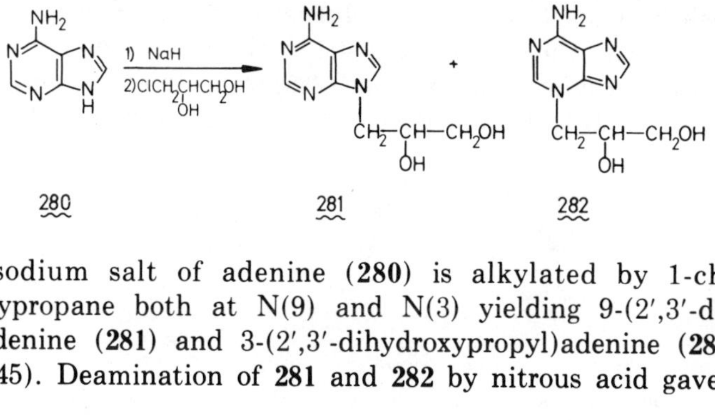

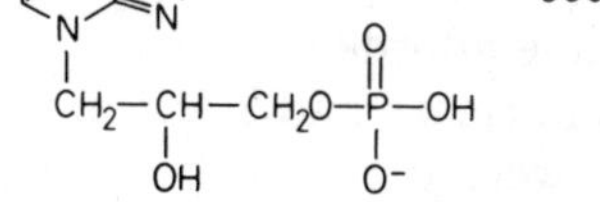

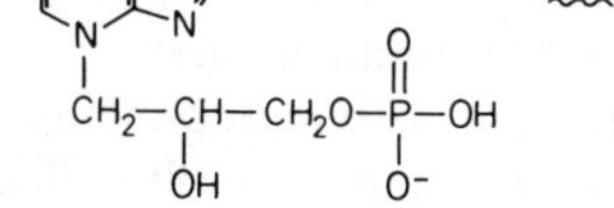

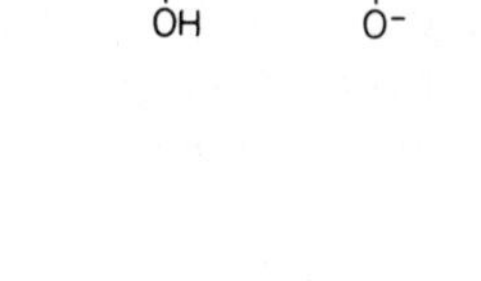

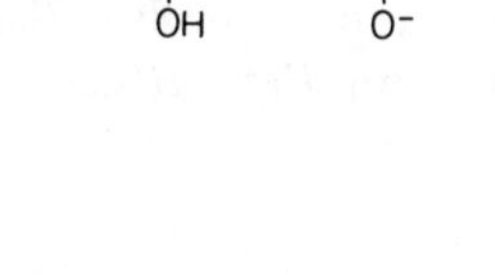

The sodium salt of adenine (**280**) is alkylated by 1-chloro-2,3-dihydroxypropane both at N(9) and N(3) yielding 9-(2′,3′-dihydroxypropyl)adenine (**281**) and 3-(2′,3′-dihydroxypropyl)adenine (**282** respectively (145). Deamination of **281** and **282** by nitrous acid gave the corresponding hypoxanthine derivatives (145). Phosphorylation by phosphorus oxychloride led to the 3′-phosphates **283–286** exclusively (145). This selective phosphorylation of the primary 3′-OH group is only observed, if the reaction mixture is subjected to an acid hydrolysis step. It is suggested that this phosphorylation gives a 2′,3′-cyclic phosphoester intermediate, which is selectively hydrolyzed in acid to the 3′-phosphate (145). The derivatives **283–286**, which possess a chiral center at C(2′), are racemic mixtures of the R and S form. Holy and Ivanova (146) prepared *R*,*S*-1-(2′,3′-dihydroxypropyl) derivatives of thymine (**291**) and uracil (**292**) by alkylation of pyrimidine bases with allyl bromide and subsequent hydroxylation of the allylic double bond. The synthesis of the S isomers of **291** and **292** was

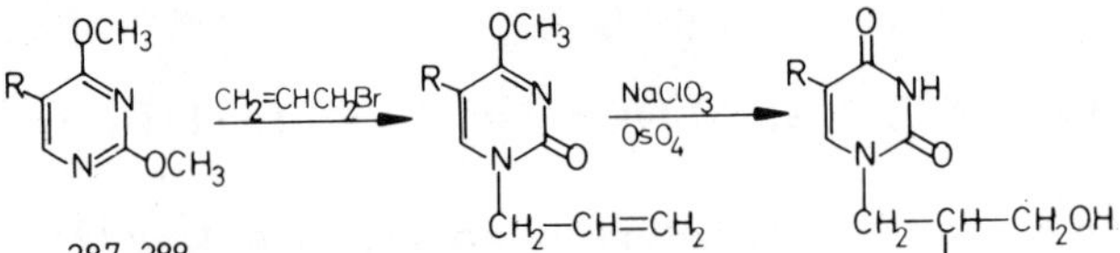

287, 289, 291 R = CH_3
288, 290, 292 R = H

accomplished by reaction of 3-O-p-toluenesulfonyl-1,2-*O*-isopropylidene-D-glycerol (**293**) with the sodium salts of 4-methoxypyrimidine-2-one (**294**) or 4-methoxy-5-methylpyrimidine-2-one (**295**) respectively. Removal of blocking groups by acid hydrolysis gave *S*-1-(2′,3′-dihydroxypropyl)uracil (**296**)

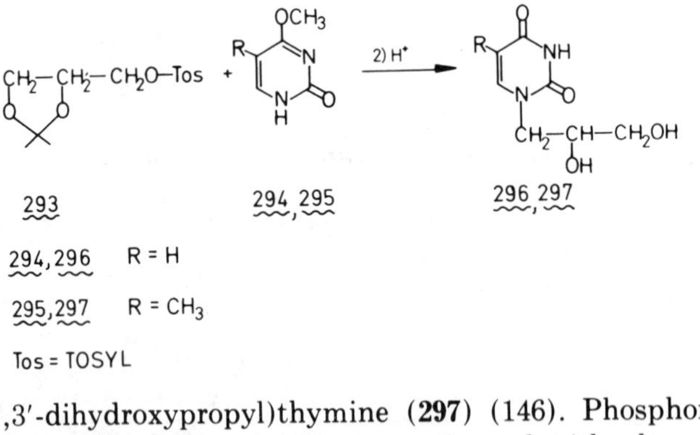

294, 296 R = H
295, 297 R = CH_3
Tos = TOSYL

and *S*-1-(2′,3′-dihydroxypropyl)thymine (**297**) (146). Phosphorylation of compounds **291**, **292**, **296**, and **297** was performed with phosphorus oxychloride. Workup of the respective reaction mixtures under acidic conditions gave exclusively the 3′-phosphates **298–301**. Cyclization of the 3′-phosphates **298–301** by dicyclohexylcarbodiimide afforded the 2′,3′-cyclic phosphates **302–305**. Although pancreatic ribonuclease and T_2 ribonuclease did not hydrolyze compounds **302–305**, they were found to be substrates for ribonucleases from *Penicillium brevicompactum*, *P. chrysogenum*, *P. claviforme* and *Aspergillus clavatus* (146). Only the S isomers of compounds

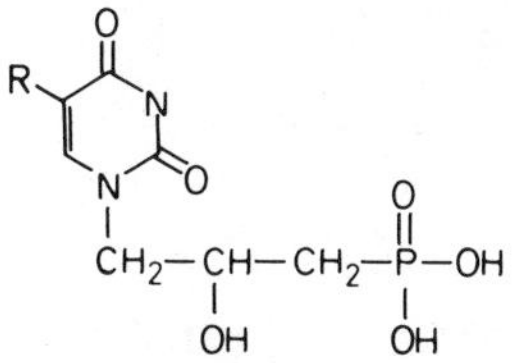

298-301

298, R = H R,S

299, R = H S

300, R = CH_3 R,S

301, R = CH_3 S

	R	Configuration		R	Configuration
298	H	R,S	**300**	CH_3	R,S
299	H	S	**301**	CH_3	S

302–305 turned out to be substrates. This allowed the resolution of racemic mixtures by means of hydrolysis with ribonuclease. The separated R and S forms of one compound exhibited CD spectra with Cotton effects of similar amplitudes but opposite sign (146).

Chemical synthesis of *R,S*-1-(2′,3′-dihydroxypropyl)thymine 2′-phosphate (**309**) required protection of the 3′-OH group of **306** by reaction with

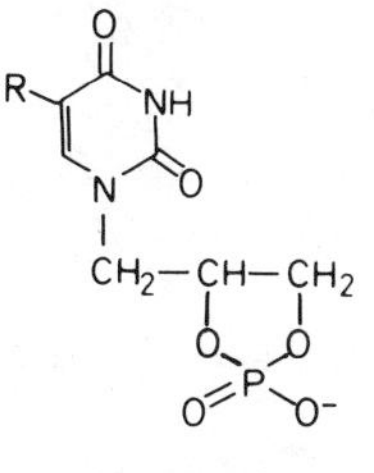

302, R = H R,S

303, R = H S

304, R = CH_3 R,S

305, R = CH_3 S

302-305

	R	Configuration		R	Configuration
302	H	R,S	**304**	CH_3	R,S
303	H	S	**305**	CH_3	S

dimethoxytritylchloride. Phosphorylation of *R,S*-1-(2′,3′-dihydroxy-3′-dimethoxytrityl)thymine (**307**) with 2-cyanoethyl phosphate and dicyclohexylcarbodiimide furnished **308**. Removal of blocking groups from **308** led to the 2′-phosphate **309**. Reaction of 3-O-p-toluenesulfonyl-1,2-isopropylidene-D-glycerol with the sodium salts of adenine, uracil, and

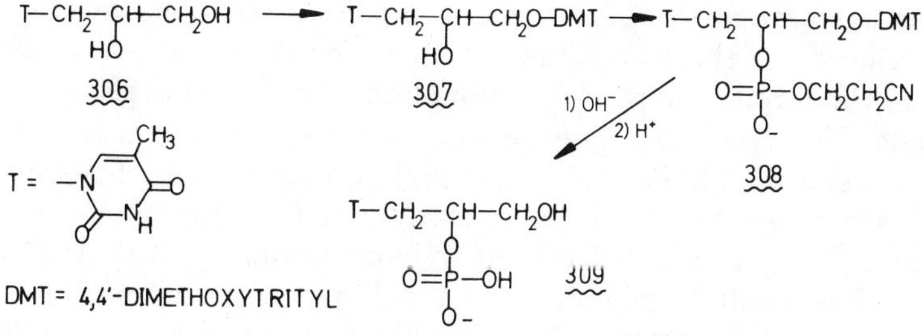

B—CH$_2$—CH(OH)—CH$_2$OH

310 - 314

	B		B
310	Adenine	**313**	Cytosine
311	Uracil	**314**	Hypoxanthine
312	Thymine		

thymine gave the S-2′,3′-dihydroxypropyl derivatives **310–312** after removal of the isopropylidene groups by acid hydrolysis (147). Alkylation of adenine occurred at N(9) exclusively, whereas in the case of uracil and thymine reaction at both nitrogen atoms N(3) and N(1) took place. Compound **313** was prepared by transformation of **311** via the 4-thiouracil derivative and its amination. Deamination of **310** with nitrous acid gave **314**.

Tos—OCH$_2$ … OCH$_3$ —ADENINE→ 316 —1) H$^+$ 2) NaJO$_4$ 3) NaBH$_4$→ 317 —H$^+$→ 318

315 316 317 318

Tos = TOSYL

Holy employed an original procedure for the synthesis of R-9(2′,3′-dihydroxypropyl)adenine (**318**) starting from methyl-5-O-p-toluenesulfonyl-2,3-O-isopropylidene-D-riboside (**315**) (147). Inversion of configuration in pyrimidine derivatives can be achieved by formation and subsequent cleavage of 2′,2-anhydro compound **321**. This reaction is in close similarity to the transformation of ribonucleosides into arabinonucleosides via formation of a 2′,2-anhydronucleoside and its subsequent hydrolysis. The reaction sequence from **319** to **322** leads to inversion of configuration at C(2′). The 3,4-dihydroxybutyl derivatives of thymine and adenine are accessible by a slight modification of the above described procedures (147). R,S-1-(3,4-Dihydroxybutyl)thymine (**323**) results from reaction of homoallyl bromide with 2,4-dimethoxythymine followed by hydroxylation of the homoallylic double bond. R,S-9-(3,4-dihydroxy-

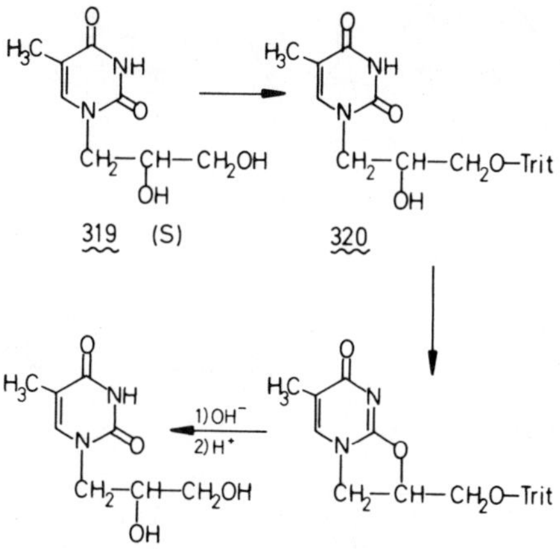

319 (S) 320 321 322 (R)

Trit = $(C_6H_5)_3C$

butyl)adenine (**324**) can be obtained by reaction of 1,2-*O*-isopropylidene-4-*p*-toluenesulfonylbutanetriol with adenine. Phosphorylation of compounds **323** and **324** by phosphorus oxychloride took place specifically at the 4′-OH groups to give *R*,*S*-1-(3′,4′-dihydroxybutyl)thymine 4′-phosphate (**325**) as well as *R*,*S*-9-(3′,4′-dihydroxybutyl)adenine 4′-phosphate (**326**). Neither

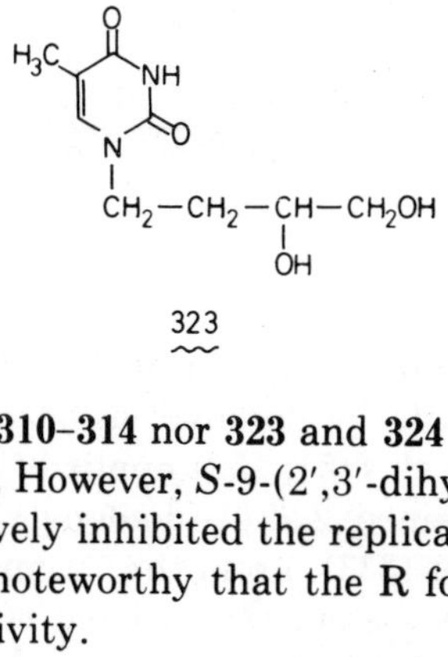

323 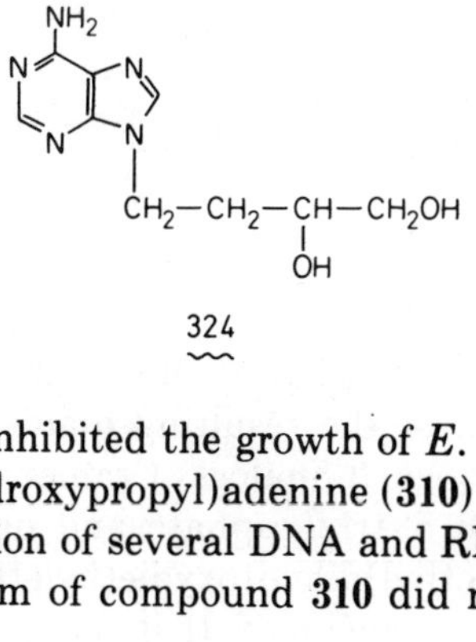 324

compounds **310–314** nor **323** and **324** inhibited the growth of *E. coli* to any extent (147). However, *S*-9-(2′,3′-dihydroxypropyl)adenine (**310**) in cell cultures effectively inhibited the replication of several DNA and RNA viruses (148). It is noteworthy that the R form of compound **310** did not exhibit antiviral activity.

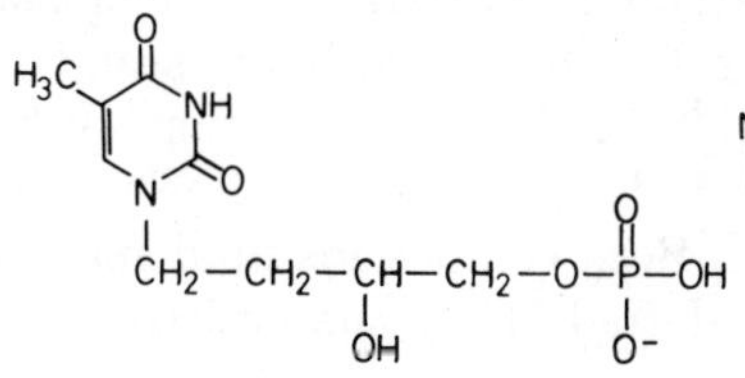

325 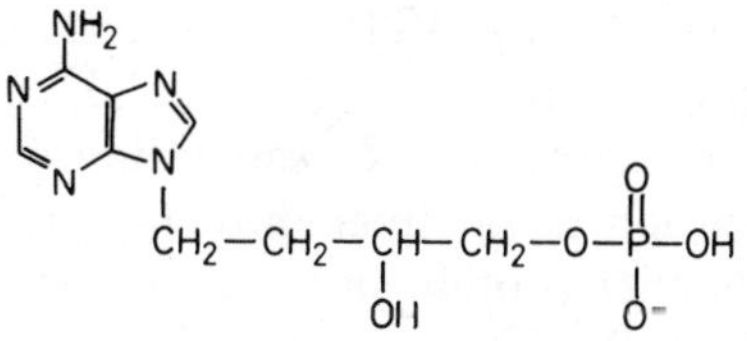 326

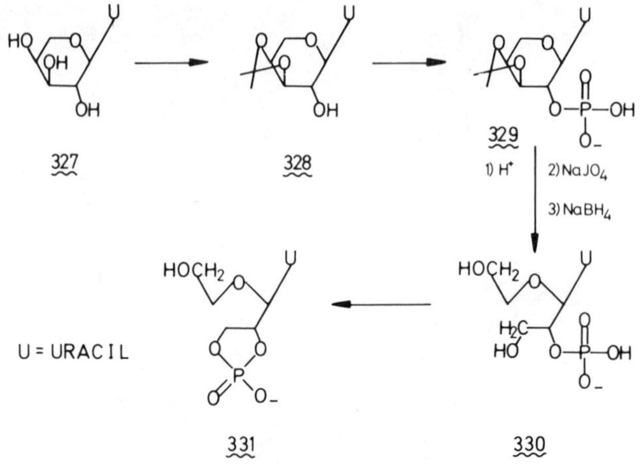

The synthesis of a conformational analog of uridine 2′,3′-cyclic phosphate was described by Mikhailov (149). Starting from 1-(α-L-arabinopyranosyl)uracil (**327**) Mikhailov accomplished the synthesis of 2′(*R*),3′-dihydroxy-1′(*R*)-β-hydroxyethyl-1′-(uracil-1-yl)propane 2′,3′-cyclic phosphate (**331**). *R*,*S*-9-(2′,3′-Dihydroxypropyl)adenine 3′-triphosphate (**332**) was described by Prasolov et al. (150) and found to be a competitive inhibitor of moderate affinity for tryptophanyl tRNA synthetase.

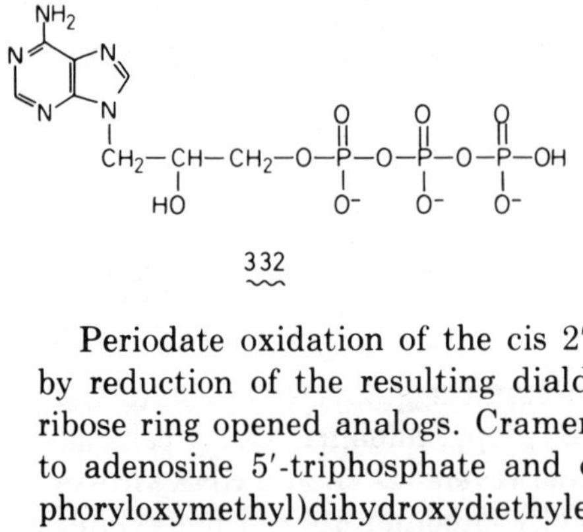

Periodate oxidation of the cis 2′,3′-OH groups in nucleotides followed by reduction of the resulting dialdehyde by borohydride leads to stable ribose ring opened analogs. Cramer et al. (151, 152) applied this reaction to adenosine 5′-triphosphate and obtained 2,2′-(1-(9-adenyl)-1′-(triphosphoryloxymethyl)dihydroxydiethylether (**333**). From the ^{1}H nmr spectrum these authors concluded that **333** possesses neither a fixed conformation about the glycosidic bond nor a preferred orientation of the exocyclic groups (152). Cramer et al. (151) investigated the extent to which **333** was able to substitute ATP as a substrate for tRNA synthetases and tRNA nucleotidyl transferase. Boos et al. (153) studied the functions of **333** as well as those of the corresponding 5′-diphosphate in adenine nucleotide translocation in rat liver mitochondria. Both compounds bind specifically to the carrier protein but are not transferred. They competitively inhibit the uptake of adenine nucleotides.

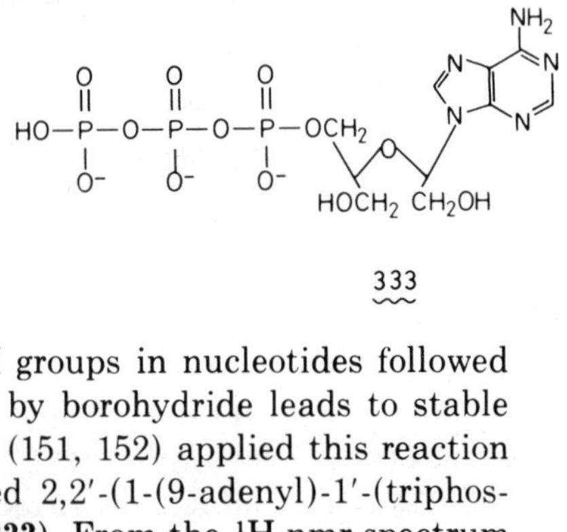

5.5 MISCELLANEOUS

Nucleotides with six-membered sugar rings have not gained much interest because of the inability of the respective nucleoside analogs to participate in cellular metabolism. 9-(β-D-Glucopyranosyl)adenine phosphates were obtained by hydrolysis of 6-benzamido-9-(2′,3′,4′-tri-*O*-acetyl-6-diphenyl-

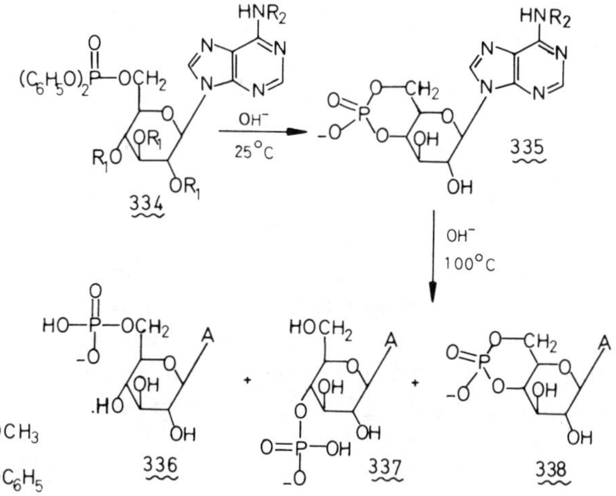

$R_1 = COCH_3$

$R_2 = COC_6H_5$

R = ADENINE

phosphoryl-β-D-glucopyranosyl)purine (**334**), which in turn was prepared by condensation of protected diphenylphosphorylglucose with 6-benzamidopurine. Mild alkaline hydrolysis gave 6-benzamido-9-(β-D-glucopyranosyl)purine 4′,6′-cyclic phosphate (**335**) as an intermediate. Hydrolysis of **335** under more forcing conditions gave 9-(β-D-glucopyranosyl)purine 6′-phosphate (**336**), 9-(β-D-glucopyranosyl)purine 4′-phos-

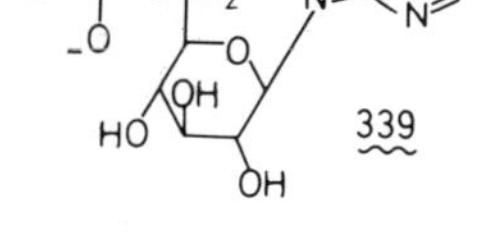

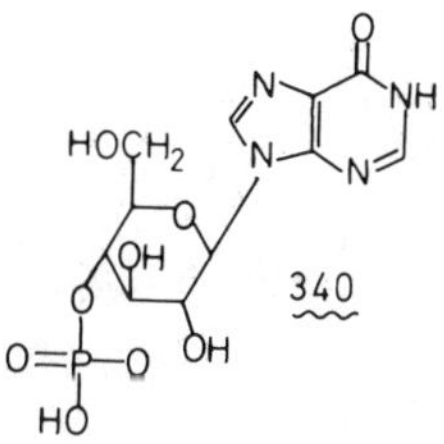

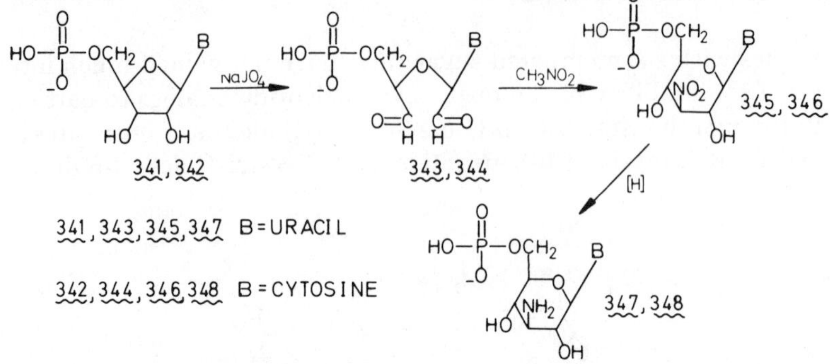

NaJO4
CH3NO2
[H]
341, 342
343, 344
345, 346
347, 348
341, 343, 345, 347 B = URACIL
342, 344, 346, 348 B = CYTOSINE

phate (**337**), and 9-(β-D-glucopyranosyl)purine 4′,6′-cyclic phosphate (**338**) respectively (157). Compounds **336** and **337** were deaminated to give 9-(β-D-glucopyranosyl)hypoxanthine 6′-phosphate (**339**) and 9-(β-D-glucopyranosyl)hypoxanthine 4′-phosphate (**340**) respectively (154).

Baer and Fischer discovered that dialdehydes derived from glucosides by treatment with periodate condensed with nitromethane to formation of 3-nitro-3-deoxypyranosides (155). This method was further developed by Fox et al. (156) and Lichtenthaler (157). Takei and Kuwada (158) applied this reaction to uridine 5′-phosphate (**341**) and cytidine 5′-phosphate (**342**). The gluco configuration of 1-(3′-amino-3′-deoxy-β-D-glucopyranosyl)uracil 6′-phosphate (**347**) and 1-(3′-amino-3′-deoxy-β-D-glucopyranosyl)cytosine 6′-phosphate (**348**) was affirmed by ^{1}H-nmr spectroscopy (158). No reports have been made on biochemical properties of the above mentioned compounds, nor have synthetic efforts been undertaken to synthesize 3′-amino-3′-deoxy-β-D-glucopyranosyl analogs of the natural nucleoside 5′-triphosphates.

Holy reported the preparation of cyclic phosphates derived from deoxy and ribopyranosylthymine employing transesterification with triethylphosphite followed by oxidative cyclization of the respective phosphites with hexachloroacetone (158). 1-(2′-Deoxy-β-D-ribopyranosyl)thymine

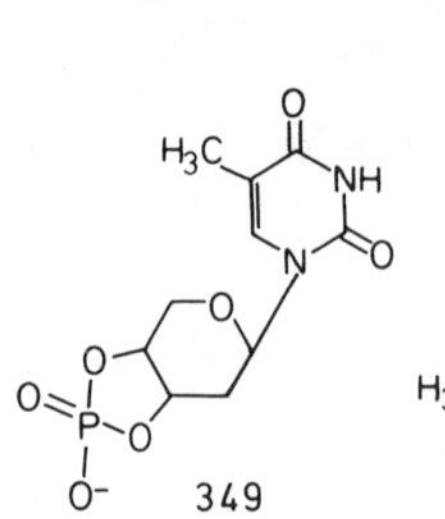

349

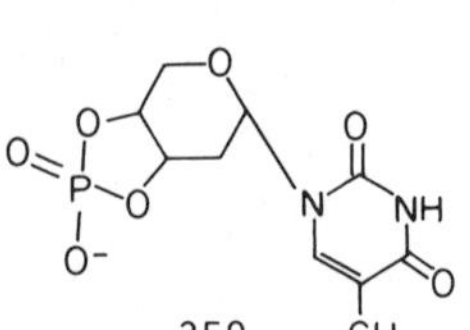

350

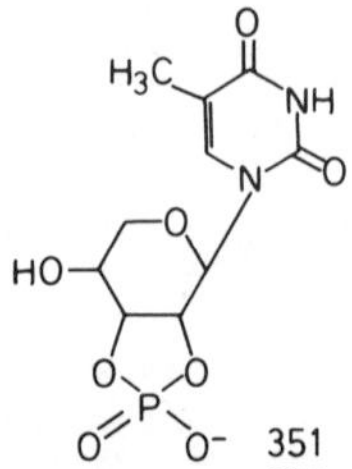

351

3′,4′-cyclic phosphate (**349**), 1-(2′-deoxy-α-D-ribopyranosyl)thymine 3′,4′-cyclic phosphate (**350**) as well as 1-(β-D-ribopyranosyl)thymine 2′,3′-cyclic phosphate (**351**) were not substrates for pancreatic ribonuclease A (159). It was not determined whether the above compounds are inhibitors of ribonuclease A. Such information would be of interest.

The phosphorylation of 9-α-D-mannofuranosyladenine was described by Taylor et al. (160). Treatment of the unprotected nucleoside analog with phosphorus oxychloride gave unexpectedly 9-(α-D-mannofuranosyl)-adenine 5′-phosphate (**352**) as the only product. It is not quite clear why phosphorylation of 9-(2′,3′-*O*-isopropylidene-α-D-mannofuranosyl)adenine by phosphorus oxychloride furnished a mixture of 9-(α-D-mannofuranosyl)adenine 6′-phosphate (**353**) and **352** after removal of blocking

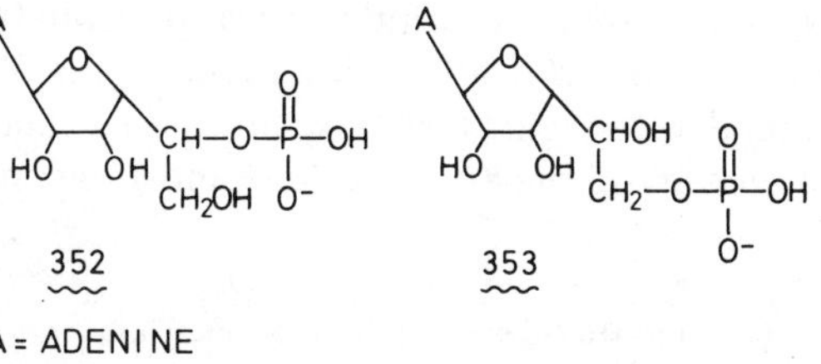

A = ADENINE

groups. The isopropylidene group seems to have an influence on the formation for a cyclic phosphodiester intermediate or its hydrolysis by acid. The structural resemblance between compounds **352** as well as **353** and adenosine 5′-phosphate is rather vague. Nevertheless, **353** proved to be potent uncompetitive inhibitor of AMP kinase (160).

Analogs of adenosine 5′-phosphate and 5′-triphosphate bearing substituents at C(5′) were described by Kappler and Hampton (161). Reaction of 2′,3′-*O*-isopropylideneadenosine 5′-aldehyde (**354**) with nitromethane afforded the epimers 9-(6-deoxy-6-nitro-2′,3′-*O*-isopropylidene-α-L-talofuranosyl)adenine (**355**) and 9-(6-deoxy-6-nitro-2′,3′-isopropylidene-β-

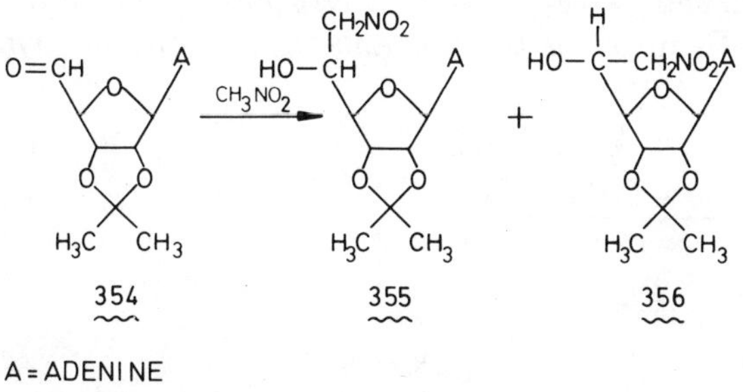

A = ADENINE

D-allofuranosyl)adenine (**356**). Catalytic hydrogenation of **355** and **356** gave the corresponding talo-C(5′)-aminomethyl nucleoside **357** and allo-C(5′)-aminomethyl nucleoside **358**. After acetylation of the respective primary amino groups, phosphorylation was performed with 2-cyanoethyl phosphate in the presence of dicyclohexylcarbodiimide. Direct phos-

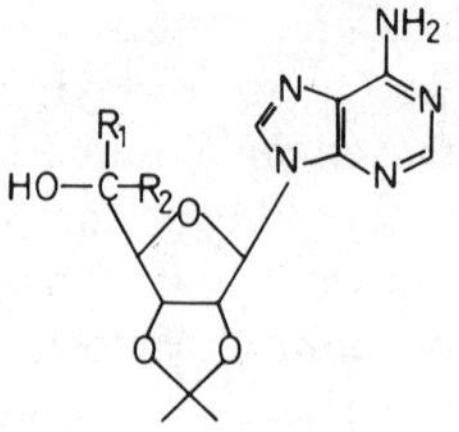

357, 358

357, $R_1 = CH_2NH_2$, $R_2 = H$

358, $R_1 = H$, $R_2 = CH_2NH_2$

phorylation of **357** and **358** did not appear to be feasible. 9-(6-Acetamidomethyl-6-deoxy-α-L-talofuranosyl)adenine 5′-phosphate (**359**) and 9-(6-acetamidomethyl-6-deoxy-β-D-allofuranosyl)adenine 5′-phosphate (**360**) were obtained in moderate yields (161). A suitable method for the preparation of 9-(6-aminomethyl-6-deoxy-α-L-talofuranosyl)adenine 5′-phosphate

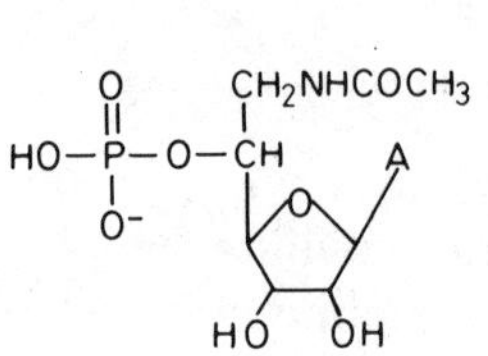

359 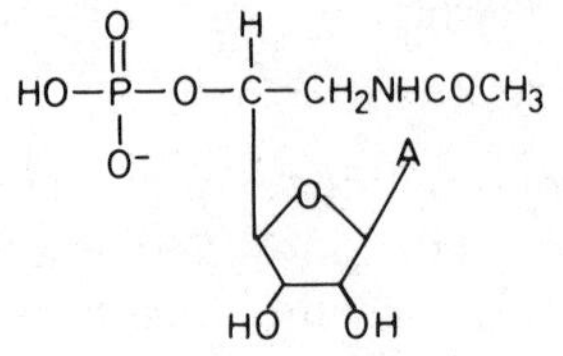 360

A = ADENINE

(**363**) and 9-(6-aminomethyl-6-deoxy-β-D-allofuranosyl)adenine 5′-phosphate (**364**) involved protection of the aminomethyl group by reaction with phenyl chloroformate (161). Phosphorylation of compounds **361** and **362** was accomplished with 2-cyanoethyl phosphate and dicyclohexylcarbodiimide. Removal of the 2-cyanoethyl group and hydrolysis of the

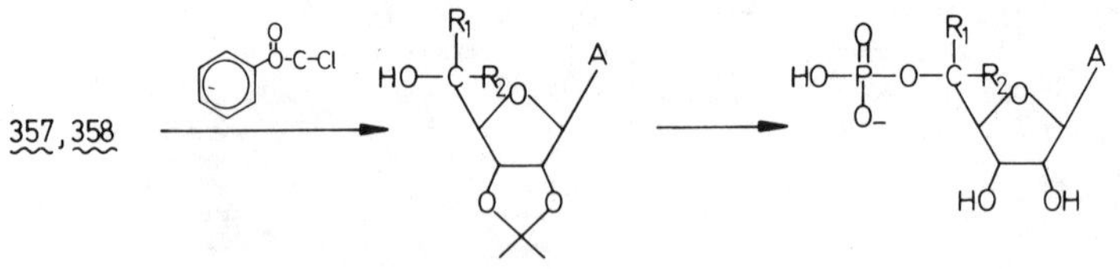

357, 358 → 361, 362 → 363, 364

361, $R_1 = CH_2NHCOOC_6H_5$, $R_2 = H$

362, $R_1 = H$, $R_2 = CH_2NHCOOC_6H_5$

363, $R_1 = CH_2NH_2$, $R_2 = H$

364, $R_1 = H$, $R_2 = CH_2NH_2$ A = ADENINE

365,366

365, $R_1 = CH_2NHCH_2CH_2CN$

366, $R_2 = H$

A = ADENINE

phenylurethane occurred simultaneously upon treatment with alkali. Byproducts were isolated from the respective reaction mixtures to which the structures **365** and **366** were assigned. The authors demonstrated that compounds **365** and **366** are formed by alkylation with cyanoethylene, which is liberated via β-elimination from the 2-cyanoethylphosphate esters

367,368

367, $R_1 = CH_2NHCOCH_3$, $R_2 = H$

368, $R_1 = H$, $R_2 = CH_2NHCOCH_3$

during removal of the blocking groups by alkaline treatment (161). Conversion of *C*(5′)-acetaminomethyl derivatives **359** and **360** to the 5′-triphosphates **367** and **368** involved anionic displacement of diphenylphosphate from the corresponding P^1-diphenyl P^2-nucleoside 5′-pyrophosphate by inorganic pyrophosphate (161).

REFERENCES

1. G. D. Morton, J. E. Lancaster, G. E. Van Lear, W. Fulmor, and W. E. Meyer, *J. Am. Chem. Soc.* **91**, 1535–1537 (1969).
2. W. Schroeder and H. Hoeksema, *J. Am. Chem. Soc.* **81**, 1767–1768 (1959).
3. H. Hoeksema, G. Slomp, and E. E. van Tamelen, *Tetrahedron Lett.* 1787–1795 (1964).
4. W. W. Lee, A. Benitez, L. Goodman, and R. R. Baker, *J. Am. Chem. Soc.* **82,** 2648–2649 (1960).
5. E. J. Reist, A. Benitez, L. Goodman, B. R. Baker and W. W. Lee, *J. Org. Chem.* **27,** 3274–3279 (1962).
6. G. A. LePage, in *Antineoplastic and Immunosuppressive Agents*, Vol. 2, p. 426, A. C. Sartorelli and D. G. Jones, Eds., Springer, Berlin, 1975.
7. J. J. Furth and S. S. Cohen, *Cancer Res.* **27,** 1528–1533 (1967).
8. S. S. Cohen, in *Progress in Nucleic Acid Research and Molecular Biology*, Vol. 5, pp. 1–88, Academic Press, New York, 1966.
9. T. H. Haskell, A. Arbor, and D. R. Watson, U.S. Patent 861485 (1969).

10. H. Follmann and H. P. C. Hogenkamp, *Biochemistry* **10,** 186–192 (1971).
11. M. Hubert-Habart and L. Goodman, *Chem. Commun.* **1969,** 740–741.
12. W. J. Wechter, *J. Org. Chem.* **34,** 244–247 (1969).
13. W. W. Lee, L. V. Fisher, and L. Goodman, *J. Heterocycl. Chem.* **8,** 179–180 (1971).
14. G. R. Revankar, J. H. Huffman, L. B. Allen, R. W. Sidwell, R. K. Robins, and R. L. Tolman, *J. Med. Chem.* **18,** 721–726 (1975).
15. T. A. Khwaja, R. Harris, and R. K. Robins, *Tetrahedron Lett.* **46,** 4681–4684 (1972).
16. A. M. Mian, R. Harris, R. W. Sidwell, R. K. Robins, and T. A. Khwaja, *J. Med. Chem.* **17,** 259–263 (1974).
17. M. Kaneko, M. Kimura, B. Shimizu, J. Yano, and M. Ikehara, *Chem. Pharm. Bull.* (*Tokyo*) **25,** 1892–1898 (1977).
18. D. Johnson, M. MacCoss, and S. Narindrasorasak, *Biochem. Biophys. Res. Commun.* **71,** 144–149 (1976).
19. W. Bergmann and D. C. Burke, *J. Org. Chem.* **20,** 1501–1507 (1955).
20. L. I. Pizer and S. S. Cohen, *J. Biol. Chem.* **235,** 2387–2392 (1960).
21. P. T. Cardeilhac and S. S. Cohen, *Cancer. Res.* **24,** 1595–1603 (1964).
22. J. Smrt, *Coll. Czech. Chem. Commun.* **32,** 3958–3965 (1967).
23. W. J. Wechter, *J. Med. Chem.* **10,** 762–773 (1967).
24. W. J. Wechter, *J. Org. Chem.* **34,** 244–247 (1969).
25. R. A. Long, G. L. Szekeres, T. A. Khwaja, R. W. Sidwell, L. N. Simon, and R. K. Robins, *J. Med. Chem.* **15,** 1215–1218 (1972).
26. W. Kreis and W. J. Wechter, *Proc. Am. Ass. Cancer Res.* **13,** 62 (1972).
27. A. Holy and F. Sorm, *Coll. Czech. Chem. Commun.* **34,** 1929–1953 (1969).
28. J. Nagyvary and C. M. Tapiero, *Tetrahedron Lett.* **40,** 3481–3484 (1969).
29. J. Nagyvary, *J. Am. Chem. Soc.* **91,** 5409–5410 (1969).
30. M. Ikehara and S. Uesugi, *Chem. Pharm. Bull.* (*Tokyo*) **21,** 264–269 (1973).
31. J. Nagyvary and R. G. Provenzale, *Biochemistry* **8,** 4769–4775 (1969).
32. R. G. Provenzale and J. Nagyvary, *Biochemistry* **9,** 1744–1752 (1970).
33. G. A. LePage, in *Antineoplastic and Immunosuppressive Agents*, Vol. 2, pp. 430–431, A. C. Sartorelli and D. G. Johns, Eds., Springer, Berlin, 1975.
34. J. L. York and G. A. LePage, *Can. J. Biochem.* **44,** 19–36 (1966).
35. P. Roy-Burman, Analogues of Nucleic Acid Components, in *Recent Results in Cancer Research*, pp. 32–33, P. Rentchnick Ed., Springer, Berlin, 1970.
36. J. J. Furth and S. S. Cohen, *Cancer. Res.* **28,** 2061–2067 (1968).
37. P. J. Ortiz, *Biochem. Biophys. Res. Commun.* **46,** 1728–1733 (1972).
38. W. Plunkett and S. S. Cohen, *Cancer. Res.* **35,** 415–422 (1975).
39. K. M. Rose and S. T. Jacob, *Biochem. Biophys. Res. Commun.* **81,** 1418–1424 (1978).
40. G. Bunick, and D. Voet, *Acta crystallogr.* **B30,** 1651–1660 (1974).
41. W. A. Creasey, in *Antineoplastic and Immunosuppressive Agents*, Vol. 2, p. 232, A. C. Sartorelli and D. G. Johns, Eds., Springer, Berlin, 1975.
42. R. L. Momparler, *Biochem. Biophys. Res. Commun.* **34,** 465–471 (1969).
43. N. B. Furlong and C. Gresham, *Nature* (*New Biology*) **233,** 212–214 (1971).
44. E. Wist, H. Krokan, and H. Prydz, *Biochemistry* **15,** 3647–3652 (1976).

45. R. L. Momparler, T. P. Brent, A. Labitan, and V. Krygier, *Mol. Pharmacol.* **7,** 413–419 (1971).
46. M. Staub, H. R. Warner, and P. Reichard, *Biochem. Biophys. Res. Commun.* **46,** 1824–1829 (1972).
47. R. L. Momparler, *Mol. Pharmacol.* **8,** 362–370 (1972).
48. S. Suzaki, A. Yamazaki, A. Kamimura, K. Mitsugi and I. Kumashiro, *Chem. Pharm. Bull. (Tokyo)* **18,** 172–176 (1970).
49. D. B. Ellis and G. A. LePage, *Mol. Pharmacol.* **1,** 231–238 (1965).
50. M. Asai, H. Hieda, and B. Shimizu, *Chem. Pharm. Bull. (Tokyo)* **15,** 1863–1870 (1967).
51. B. Shimizu, A. Saito, T. Nishimura, and M. Miyaki, *Chem. Pharm. Bull. (Tokyo)* **15,** 2011–2014 (1967).
52. N. J. Leonard and R. A. Laursen, *Biochemistry* **4,** 354–365 (1965).
53. A. Holy and F. Sorm, *Coll. Czech. Chem. Commun.* **34,** 3523–3532 (1969).
54. A. Holy and F. Sorm, *Coll. Czech. Chem. Commun.* **34,** 3383–3401 (1969).
55. A. Holy and F. Sorm, *Coll. Czech. Chem. Commun.* **36,** 3282–3299 (1971).
56. J. Simuth and A. Holy, *Nucleic Acids Res.*, Special Publication No. 1, 165–168 (1975).
57. T. Nishimura, B. Shimizu, and M. Futai, *Biochim. Biophys. Acta* **129,** 654–657 (1966).
58. A. Holy, *Coll. Czech. Chem. Commun.* **38,** 100–114 (1973).
59. H. R. Bentley, K. G. Cunningham, and F. S. Spring, *J. Chem. Soc.* **1951,** 2301–2305.
60. E. A. Kaczka, E. L. Dulaney, C. O. Gitterman, H. B. Woodruff, and K. Folkers, *Biochem. Biophys. Res. Commun.* **14,** 452–455 (1964).
61. H. Klenow, *Biochim. Biophys. Acta* **76,** 354–365 (1963).
62. H. T. Shigeura and C. N. Gordon, *J. Biol. Chem.* **240,** 806–810 (1965).
63. M. Sprinzl, K. H. Scheit, H. Sternbach, F. v. d. Haar, and F. Cramer, *Biochem. Biophys. Res. Commun.* **51,** 881–887 (1973).
64. R. I. Gumport, E. B. Edelheit, T. Uematsu and R. J. Suhadolnik, *Biochemistry* **15,** 2804–2809 (1976).
65. H. C. Tole, J. F. Jolly, and J. A. Boezi, *J. Biol. Chem.* **250,** 1723–1733 (1975).
66. H. T. Shigeura and G. E. Boxer, *Biochem. Biophys. Res. Commun.* **17,** 758–763 (1964).
67. M. Sprinzl and F. Cramer, *Nature (New Biology)* **245,** 3–5 (1973).
68. R. C. Adelman, F. J. Ballard, and S. Weinhouse, *J. Biol. Chem.* **242,** 3360–3365 (1967).
69. B. M. Chassy and R. J. Suhadolnik, *J. Biol. Chem.* **243,** 3538–3541 (1968).
70. I. Özer and K. H. Scheit, *Eur. J. Biochem.* **85,** 173–180 (1978).
71. J. J. K. Novak and F. Sorm, *Coll. Czech. Chem. Commun.* **38,** 1173–1178 (1973).
72. M. Cashel, E. Hamel, P. Shapshak, and M. Bouquet, in *Control of Ribosome Synthesis*, p. 279, O. Maaloe and N. J. Kjeldgaard, Eds., Munksgaard, Copenhagen, 1976.
73. E. Hamel, *Biochim. Biophys. Acta* **414,** 326–340 (1975).
74. E. Hamel, *Eur. J. Biochem.* **63,** 431–440 (1976).
75. N. N. Gerber and H. A. Lechevalier, *J. Org. Chem.* **27,** 1731–1732 (1962).

76. B. R. Baker, R. E. Schaub, and H. M. Kissman, *J. Am. Chem. Soc.* **77,** 5911–5915 (1955).
77. L. H. Pugh and N. N. Gerber, *Cancer Res.* **23,** 640–647 (1963).
78. L. H. Pugh, H. A. Lechevalier, and M. Solotorovsky, *Antibiot. and Chemother.* **12,** 310–317 (1962).
79. J. T. Truman and H. Klenow, *Mol. Pharmacol.* **4,** 77–86 (1968).
80. H. T. Shigeura, G. E. Boxer, M. L. Meloni, and S. D. Sampson, *Biochemistry* **5,** 994–1004 (1966).
81. V. W. Armstrong, and F. Eckstein, *Eur. J. Biochem.* **70,** 33–38 (1976).
82. T. H. Fraser and A. Rich, *Proc. Nat. Acad. Sci. (US)* **70,** 2671–2675 (1973).
83. K. E. Pfitzner and J. G. Moffatt, *J. Org. Chem.* **29,** 1508–1511 (1964).
84. M. J. Robins, J. R. McCarthy, Jr., and R. K. Robins, *Biochemistry* **5,** 224–231 (1966).
85. L. Toji and S. S. Cohen, *Proc. Nat. Acad. Sci. (US)* **63,** 871–877 (1969).
86. A. F. Russel, and J. G. Moffatt, *Biochemistry* **8,** 4889–4896 (1966).
87. J. P. H. Verheyden and J. G. Moffatt, *J. Am. Chem. Soc.* **86,** 2093–2095 (1964).
88. M. R. Atkinson, M. P. Deutscher, A. Kornberg, A. F. Russell, and J. G. Moffatt, *Biochemistry* **8,** 4897–4904 (1969).
89. L. Toji and S. S. Cohen, *J. Bact.* **103,** 323–328 (1970).
90. P. Langen, G. Etzold, R. Hintsche, and G. Kowollik, *Acta Biol. Med. Ger.* **23,** 759–766 (1969).
91. T. A. Khwaja and C. Heidelberger, *J. Med. Chem.* **10,** 1066–1070 (1967).
92. G. R. Owen, J. P. Verheyden, and J. G. Moffatt, *J. Org. Chem.* **41,** 3010–3017 (1976).
93. G. P. Lampson, A. A. Tytell, A. K. Field, M. M. Nemes, and M. R. Hilleman, *Proc. Nat. Acad. Sci. (US)* **58,** 782–789 (1967).
94. J. Hobbs, H. Sternbach, M. Sprinzl, and F. Eckstein, *Biochemistry* **11,** 4336–4344 (1972).
95. B. Janik, M. P. Kotick, T. H. Kreiser, L. F. Reverman, R. G. Sommer, and D. P. Wilson, *Biochem. Biophys. Res. Commun.* **46,** 1153–1160 (1972).
96. M. Ikehara, T. Fukui, and N. Kakiuchi, *Nucleic Acids Res.* **5,** 1877–1888 (1978).
97. J. Hobbs and F. Eckstein, *Nucleic Acids Res.* **2,** 1987–1994 (1975).
98. J. C. Catlin and W. Guschlbauer, *Biopolymers* **14,** 51–72 (1975).
99. R. P. Glinski, M. Sami Khan, R. L. Kalamas, C. L. Stevens, and M. B. Sporn, *Chem. Commun.* **1970,** 915–916.
100. D. Wagner, J. P. H. Verheyden, and J. G. Moffatt, *J. Org. Chem.* **37,** 1876–1878 (1971).
101. J. P. H. Verheyden, D. Wagner, and J. G. Moffatt, *J. Org. Chem.* **36,** 250–254 (1971).
102. P. F. Torrence, J. A. Waters, and B. Witkop, *J. Am. Chem. Soc.* **94,** 3638–3639 (1972).
103. P. F. Torrence, A. M. Bobst, J. A. Waters, and B. Witkop, *Biochemistry* **12,** 3962–3972 (1973).
104. J. Hobbs, H. Sternbach, and F. Eckstein, *Biochem. Biophys. Res. Commun.* **46,** 1509–1515 (1972).

105. M. Ikehara, T. Fukui, and N. Kakiuchi, *Nucleic Acids Res.* **3,** 2089-2099 (1976).
106. J. Hobbs, H. Sternbach, M. Sprinzl, and F. Eckstein, *Biochemistry* **12,** 5138-5145 (1973).
107. R. H. Hall, *The Modified Nucleosides in Nucleic Acids*, p. 167, New York, Columbia University Press, 1971.
108. T. A. Khwaja and R. K. Robins, *J. Am. Chem. Soc.* **88,** 3640-3643 (1966).
109. A. D. Broom and R. K. Robins, *J. Am. Chem. Soc.* **87,** 1145-1146 (1965).
110. J. B. Gin and C. A. Dekker, *Biochemistry* **7,** 1413-1420 (1968).
111. D. M. G. Martin and C. B. Reese, *J. Chem. Soc.* 1731-1738 (1968).
112. Y. Furukawa, K. Kobayashi, Y. Kanai, and M. Honjo, *Chem. Pharm. Bull. (Tokyo)* **13,** 1273-1278 (1965).
113. F. Rottman and K. Heinlein, *Biochemistry* **7,** 2634-2641 (1968).
114. F. Rottman and K. L. Johnson, *Biochemistry* **8,** 4354-4361 (1969).
115. J. Simuth, P. Strehlke, U. Niedballa, H. Vorbrüggen, and K. H. Scheit, *Biochim. Biophys. Acta* **228,** 654-663 (1971).
116. B. Zmudzka, C. Janion, and D. Shugar, *Biochem. Biophys. Res. Commun.* **37,** 895-901 (1969).
117. B. E. Dunlap, K. H. Friderici, and F. Rottman, *Biochemistry*, **10,** 2581-2587 (1971).
118. F. Rottman, K. Friederici, P. Comstock, and M. K. Khan, *Biochemistry* **13,** 2762-2771 (1974).
119. B. C. Pal, and D. G. Schmidt, *Prep. Biochem.* **3,** 563-567 (1973).
120. I. Tazawa, S. Tazawa, J. L. Alderfer, and P. O. P. Ts'o, *Biochemistry* **11,** 4931-4937 (1972).
121. M. Khurshid, A. Khan, and F. Rottman, *FEBS-Lett.* **28,** 25-28 (1972).
122. J. T. Kusmierek, M. Kielanoeska, and D. Shugar, *Biochem. Biophys. Res. Commun.* **53,** 406-412 (1973).
123. G. F. Gerard, F. Rottman, and J. A. Boezi, *Biochem. Biophys. Res. Commun.* **46,** 1095-1101 (1972).
124. G. F. Gerard, F. Rottman, and J. A. Boezi, *Biochemistry* **10,** 1974-1981 (1971).
125. V. A. Aivasashvili, R. Sh. Bibilashvili, and V. L. Florentjev, *Mol. Biol. (USSR)* **11,** 661-669 (1977).
126. J. K. Mackey and P. T. Gilham, *Nature (London)* **233,** 551-553 (1971).
127. G. N. Bennett and P. T. Gilham, *Biochemistry* **14,** 3152-3158 (1975).
128. M. Ikehara, S. Tanaka, T. Fukui, and E. Ohtsuka, *Nucleic Acids Res.* **3,** 3203-3211 (1976).
129. Y. Kikuchi, K. Hirai, and K. Sakaguchi, *J. Biochem.* **77,** 469-472 (1975).
130. G. C. Walker and O. C. Uhlenbeck, *Biochemistry* **14,** 817-824 (1975).
131. J. Zemlicka and S. Chladek, *Coll. Czech. Chem. Commun.* **33,** 3293-3303 (1968).
132. R. L. Whistler and A. Block, *Abstr. 7th Int. Congr. Biochem., Tokyo, Japan*, 661, 1967.
133. D. J. Hoffman and R. L. Whistler, *Biochemistry* **9,** 2367-2372 (1970).
134. A. K. M. Anisuzzaman, W. C. Lake, and R. L. Whistler, *Biochemistry* **12,** 2041-2045 (1973).

135. T. Hiratsuka and K. Uchida, *Biochim. Biophys. Acta* **320,** 635–647 (1973).
136. M. Ikehara, S. Uesugi, and K. Yoshida, *Biochemistry* **11,** 830–836 (1972).
137. K. K. Ogilvie and L. A. Slotin, *Can. J. Chem.* **51,** 2397–2405 (1973).
138. M. Ikehara, T. Nagura, and E. Ohtsuka, *Chem. Pharm. Bull. (Tokyo)* **22,** 2578–2586 (1974).
139. A. Hampton and T. Sasaki, *Biochemistry* **12,** 2188–2191 (1973).
140. M. Ikehara, T. Nagura, and E. Ohtsuka, *Chem. Pharm. Bull. (Tokyo)* **22,** 123–127 (1974).
141. P. J. Harper and A. Hampton, *J. Org. Chem.* **37,** 795–797 (1972).
142. A. Hampton, P. J. Harper, and T. Sasaki, *Biochemistry* **11,** 4965–4969 (1972).
143. J. A. Raleigh, and B. J. Blackburn, *Biochem. Biophys. Res. Commun.* **83,** 1061–1066 (1978).
144. K. S. Boos, E. Schlimme, and M. Ikehara, *Zeitschrift für Naturforsch.* **33c,** 552–556 (1978).
145. T. Seita, K. Yamauchi, M. Kinoshita, and M. Imoto, *Bull. Chem. Soc. Japan* **45,** 926–928 (1972).
146. A. Holy and G. S. Ivanova, *Nucleic Acids Res.* **1,** 19–34 (1974).
147. A. Holy, *Coll. Czech. Chem. Commun.* **40,** 187–214 (1975).
148. E. DeClercq, J. Descamps. P. DeSomer, and A. Holy, *Science* **200,** 563–564 (1978).
149. S. N. Mikhailov, *Biorg. Khim. (USSR)* **4,** 639–644 (1978).
150. V. S. Prasolov, A. M. Kritsyn, S. N. Mikhailov, and V. L. Florentev, *Doklad. Akad. Nauk SSSR* **221,** 1226–1228 (1975).
151. F. Cramer, F. v. d. Haar, and E. Schlimme, *FEBS-Lett.* **2,** 136–139 (1968).
152. F. v. d. Haar, E. Schlimme, M. Gomez-Guillen, and F. Cramer, *Eur. J. Biochem.* **24,** 296–302 (1971).
153. K. S. Boos, E. Schlimme, D. Bojanovski, and W. Lamprecht, *Eur. J. Biochem.* **60,** 451–458 (1975).
154. A. Nohara, K. Imai, and M. Honjo, *Chem. Pharm. Bull. (Tokyo)* **14,** 491–495 (1966).
155. H. H. Baer and H. O. L. Fischer, *Proc. Natl. Acad. Sci. (US)* **44,** 991–993 (1958).
156. K. A. Watanabe and J. J. Fox, *Chem. Pharm. Bull. (Tokyo)* **12,** 975–976 (1964).
157. F. W. Lichtenthaler, H. P. Albrecht, and G. Olfermann, *Angew. Chem.* **77,** 131(1965).
158. S. Takei and Y. Kuwada, *Chem. Pharm. Bull. (Tokyo)* **16,** 944–948 (1968).
159. A. Holy, *Coll. Czech. Chem. Commun.* **34,** 3510–3522 (1969).
160. M. J. Taylor, B. D. Kohn, W. G. Taylor, and P. Cohn, *Carbohydr. Res.* **30,** 133–142 (1973).
161. F. Kappler and A. Hampton, *J. Org. Chem.* **40,** 1378–1385 (1975).

SIX

Methods of Phosphorylation

It cannot be the objective of this chapter to provide a complete survey of the methods employed in the synthesis of esters, amides, and anhydrides of phosphoric acid. This is beyond the scope of this book. Fortunately, Khorana (1) and Michelson (2) have provided a competent and comprehensive treatment of this subject from its early beginning until 1962. These accounts of an approximately 50-year long struggle to find solutions to the problems of the phosphorylation of hydroxyl groups and phosphate anions represent required reading, not only because of the historical background they provide. In this period of nucleotide chemistry, the foundations and principles of phosphorylation were laid on which our present work is based. From this area methods of phosphorylation originated that with or without modification are still in use. Of interest is an early attempt of Clark et al. (3) to classify the then known phosphorylating reagents and to derive a theoretical basis of structure-reactivity principles for those agents. A good and concise review of phosphorylating reagents, especially those employed in the synthesis of phosphate esters, which covers the literature up to 1976, was presented by Slotin (4). In this chapter the phosphorylation procedures developed since 1962 are reviewed. The aim is to discuss respective methods from a practical point of view, namely that of an experimentalist who is in need of the most suitable phosphorylation procedure for the synthesis of a particular nucleotide.

In connection with the synthesis of nucleotides the problem of phosphorylation is manifested in two different forms: (1) phosphorylation of OH groups in nucleosides, and (2) formation of P—O—P bonds in nucleoside polyphosphates. At a first glance the two types of phosphorylation seem to have so much in common that a division appears arbitrary. Indeed, in early studies, a particular phosphorylating reagent was employed both in the phosphorylation of sugar hydroxyl groups as well as in the synthesis of P—O—P bonds, although in the latter case almost always with unsatisfactory results. The difference between the two synthetic problems is well illustrated by the different phosphorylation procedures that are now successfully applied to the phosphorylation of OH groups in nucleosides and the formation of P—O—P bonds.

This chapter is therefore divided into two major parts. The first is concerned with the synthesis of phosphate esters and the second deals with procedures for the formation of the P—O—P bonds of nucleoside polyphosphates.

6.1 SYNTHESIS OF PHOSPHATE ESTERS

The synthesis of phosphate esters can obviously be achieved in two ways: (1) alkylation of phosphate or phosphate esters by an alkyl halide (Scheme 6.1), and (2) reaction of an alcohol with an activated phosphate (Scheme 6.2). The limitations of both reactions are immediately evident. Phos-

$$HO{-}P(=O)(O^-){-}O^- + Hal{-}CH_2{-}R \longrightarrow HO{-}P(=O)(O^-){-}O{-}CH_2{-}R + Hal^-$$

Scheme 6.1

$$HO{-}P(=O)(O^-){-}X^- + HO{-}CH_2{-}R \longrightarrow HO{-}P(=O)(O^-){-}O{-}CH_2{-}R + X^-$$

Scheme 6.2

phoric acid is tribasic and consequently alkylation of the respective anions should lead to a mixture of mono-, di-, and triesters. Moreover, application of this reaction to the synthesis of nucleoside phosphates requires the availability of halogeno derivatives of nucleosides. Because of these obstacles, this reaction type is unsuitable with one exception—the

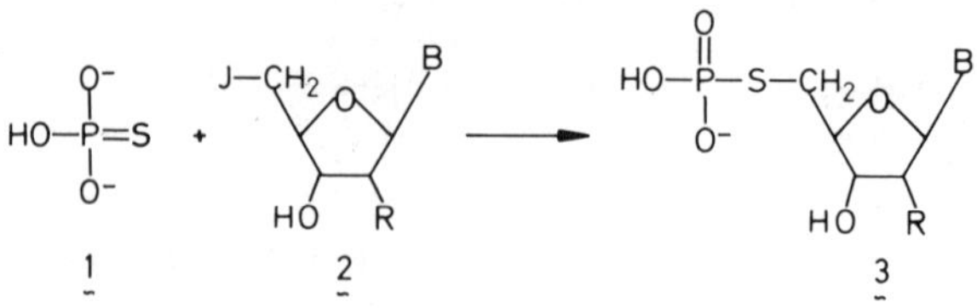

$$\underset{1}{HO{-}P(=S)(O^-){-}O^-} + \underset{2}{J{-}CH_2{-}\text{nucleoside}} \longrightarrow \underset{3}{HO{-}P(=O)(O^-){-}S{-}CH_2{-}\text{nucleoside}}$$

B = HYPOXANTHINE, ADENINE, URACIL, THYMINE

R = H, OH

preparation of nucleoside phosphate esters with P—S—C bonds. In this case the specific alkylation of thiophosphate by halogeno nucleosides is exploited for the synthesis of nucleoside-*S*-phosphate esters. Alkylation of thiophosphate (**1**) by 5′-iodo 5′-deoxynucleosides (**2**) results in the formation of nucleoside 5′-*S*-phosphates (**3**) (5–7).

In Scheme 6.2 a phosphoryl group is transferred to a nucleophilic center. This requires activation of the phosphate by a substituent *X* which possesses the general property of being easily eliminated as X^-. Since free OH groups in a compound of type **4** would effectively compete with other nucleophilic groups in substitution reactions, the consequences would be the undesirable instability of **4** due to numerous sidereactions. Further-

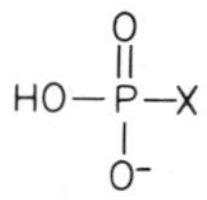

$$HO-P(=O)(O^-)-X$$

4

more, the anionic form of **4** is much more nucleophilic than the OH groups to be phosphorylated. To circumvent these difficulties, the OH groups in **4** have been protected by esterification. Suitable substituents *X* in activated **4** are halogen atoms such as chlorine and bromine. Diesterphosphorchloridates and monoesterphosphordichloridates therefore constitute an important class of stable phosphorylating reagents. Anhydrides of phosphoric acid, inorganic pyrophosphate being the simplest case, are not reactive enough to phosphorylate OH groups of general low nucleophilicity. Certain tetraesterpyrophosphates, however, have been shown to be highly reactive phosphorylating agents. Anhydrides of phosphoric acid esters became useful as phosphorylating compounds when a variety of reagents were discovered that reacted with monoesterphosphates to form primary intermediates which rapidly condense to highly reactive polyphosphates. Another approach employed in the phosphorylation of hydroxyl groups is transesterification with reactive phosphate esters.

6.1.1 Phosphoesterchloridates

The first class of phosphorylating agents which are discussed are ester phosphorochloridates. In general terms they can be regarded as mixed anhydrides. In Table 6.1 those diesterphosphorochloridates are listed that are currently employed in phosphorylation reactions. The introduction of

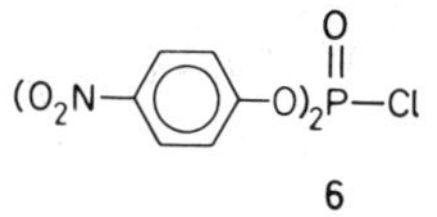

$$(O_2N-C_6H_4-O)_2P(=O)-Cl$$

6

$$(C_6H_5-O)_2P(=O)-Cl$$

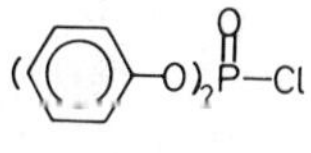

5

Table 6.1 Diesterphosphorochloridates

Compound	Reference
Diphenyl phosphorochloridate (**5**)	8, 9
Bis-(4-nitrophenyl)-phosphorochloridate (**6**)	11, 12
Bis-(2-cyanoethyl)-phosphorochloridate (**7**)	13
Bis-(2,2,2-trichloroethyl) phosphorochloridate (**8**)	14
2-Chloro-2-oxo-P^{V}-1,3,2- benzodioxaphosphole (**9**)	15
Bis-(2-tert.-butylphenyl) phosphorochloridate (**10**)	16

diphenyl phosphorochloridate (**5**) meant a major breakthrough and established a new concept in phosphorylation (8, 9). Compound **5** fulfills many requirements of a useful phosphorylating reagent: it is (1) easily accessible, (2) of sufficient stability, and (3) highly reactive under appropriate conditions. Phosphorylation of OH groups with **5** leads to fairly stable neutral triester intermediates, which are easy to purify. One phenyl group is readily removed by mild alkaline hydrolysis. Removal of both phenyl ester moieties from the phosphorylated product requires alkaline hydrolysis under forced condition or hydrogenation over a platinum catalyst. The high stability of the second phenyl group is a serious shortcoming of this phosphorylating agent. Nevertheless, **5** is still used successfully in special cases. Two procedures have been developed that help to avoid the drastic conditions necessary for the removal of the phenyl ester groups. The neutral diphenyl triester obtained after phosphorylation was transesterified by means of sodium benzoxide to the respective dibenzyl ester, which can be cleaved by catalytic hydrogeneration much more easily than phenyl esters (10). The second method invokes mild alkaline hydrolysis of one phenyl group from a nucleoside 5′-phosphorodiphenyl ester and enzymatic hydrolysis of the remaining 5′-phosphate phenyl ester (10). The latter approach proved to be quite useful on a small scale. Bis-(4-nitrophenyl)-phosphorochloridate (**6**) became important as a reagent for the synthesis of nucleoside 5′-phosphate 4-nitrophenyl esters; when Borden and Smith (17) discovered that the latter cyclized in the presence of strong base to form nucleoside 3′,5′-cyclic phosphates. Bis-(2-

$$(CN{-}CH_2CH_2O)_2\overset{\overset{\displaystyle O}{\|}}{P}{-}Cl$$

7

cyanoethyl)phosphorochloridate (**7**) has not gained practical importance as a phosphorylating agent, since its preparation due to instability turned out to be unpleasant (13). This is regrettable in view of the mild experimental conditions required for the removal of the 2-cyanoethyl groups. Bis-(2,2,2-trichloroethyl)-phosphorochloridate (**8**) was described by Eck-

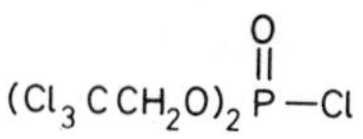

$$(Cl_3CCH_2O)_2P(=O)-Cl$$

8

stein and Scheit (14) as a crystalline, fairly stable and easily accessible compound. Reaction of **8** with unprotected nucleosides leads selectively to 5′-phosphoro bis-(2,2,2-trichloroethyl) triesters, which can be isolated and crystallized. Removal of the 2,2,2-trichloroethyl groups can be achieved by treatment with Cu-Zn in dimethylformamide, zinc dust in acetic acid, or zinc dust in boiling pyridine (14, 18, 19). Khwaja et al. investigated the phosphorylation of sugar hydroxyl groups with the highly reactive 2-

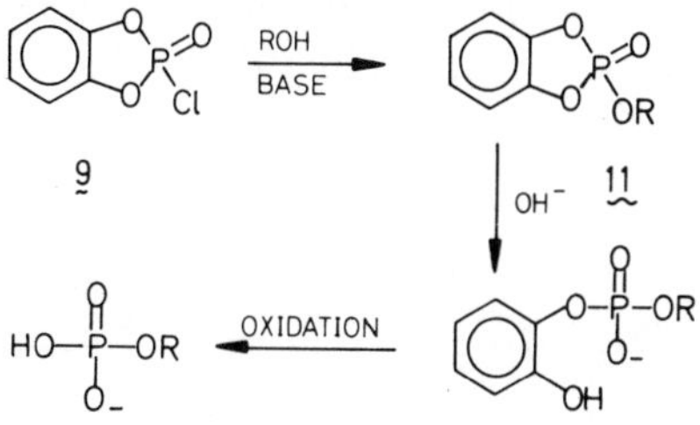

ROH / BASE — **9** → **11**; OH^- → **12**; OXIDATION → $HO-P(=O)(O^-)-OR$

chloro-2-oxo-P^V-1,3,2-benzodioxaphosphole (**9**) (15). The intermediate triester **11** is quantitatively converted to the *O*-hydroxyphenylphosphate ester **12**. Formation of the desired monoester-phosphate occurs upon oxidation of **12** by bromine water, periodic acid in aqueous solution, or lead (IV) acetate in organic solvent. The necessity of removing the protecting

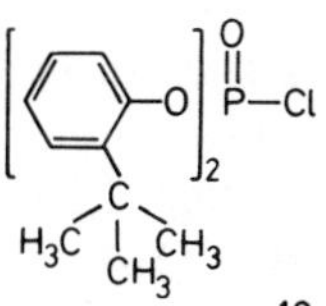

10

ester function by an oxidative process might limit the applicability of this reagent. [Bis-(2-*t*-butylphenyl)]phosphorochloridate (**10**) phosphorylates unprotected nucleosides specifically in the 5′-position (16). The advantage of the bis-(2-*t*-butylphenyl) triesters formed lies in their stability toward acidic as well as alkaline hydrolysis. Hydrogenation in the presence of Adams catalyst removes the protecting ester groups.

Monoester phosphorodichloridates are listed in Table 6.2. Monoester phosphorodichloridates no longer play a significant role in the preparation of nucleoside monophosphates. Phosphorylation with phosphorodichlori dates under controlled conditions allows the sequential utilization of the two activating chlorine atoms, thus offering the opportunity of synthesiz-

Table 6.2 Monoesterphosphorochloridates

Compound	Reference
4-Nitrophenyl phosphorodichloridate	20
2-Chloromethyl-4-nitrophenyl phosphorodichloridate (**13**)	21
N-Acetyl-5-iodo-3-indolyl phosphorodichloridate (**18**)	22
Methyl phosphorodichloridate (**16**)	23

ing unsymmetrical phosphodiesters. This property of phosphorodichloridates was used to advantage in the so-called "triester approach" of oligonucleoside synthesis and is reviewed elsewhere (24). An interesting type of monoester phosphorodichloridate is represented by 2-chloromethyl-4-nitro-phenyl phosphorodichloridate (**13**), in which after phos-

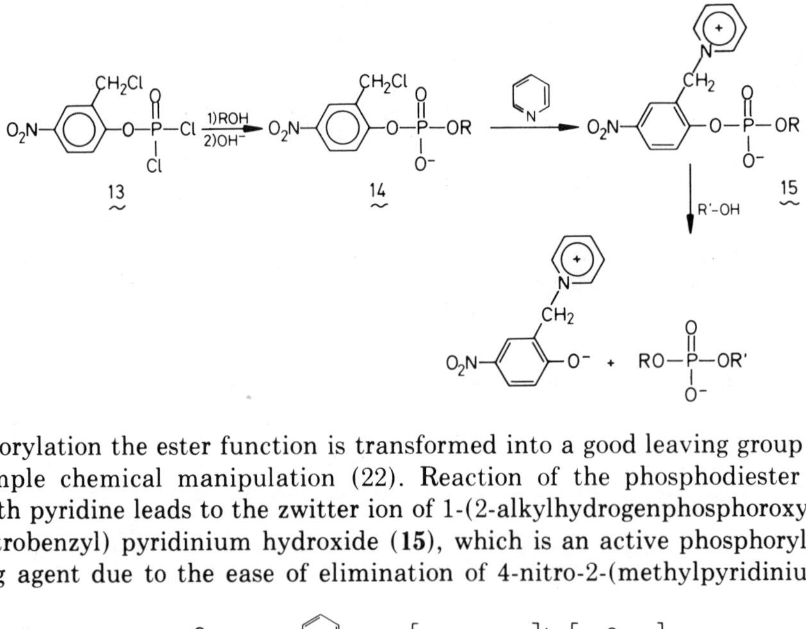

phorylation the ester function is transformed into a good leaving group by simple chemical manipulation (22). Reaction of the phosphodiester **14** with pyridine leads to the zwitter ion of 1-(2-alkylhydrogenphosphoroxy-5-nitrobenzyl) pyridinium hydroxide (**15**), which is an active phosphorylating agent due to the ease of elimination of 4-nitro-2-(methylpyridinium)

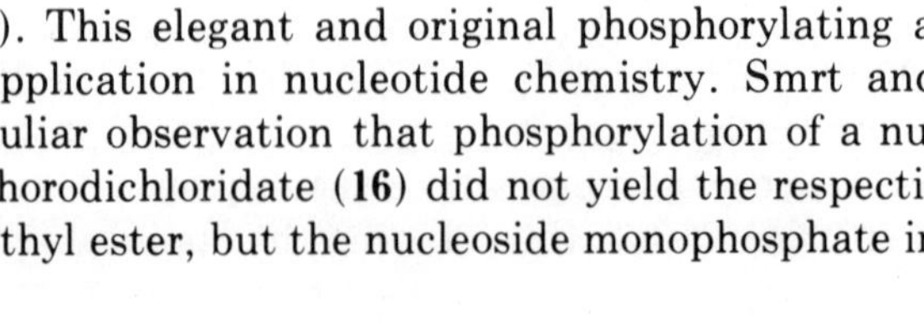

phenolate (25). This elegant and original phosphorylating agent has not found wide application in nucleotide chemistry. Smrt and Catlin (23) made the peculiar observation that phosphorylation of a nucleoside with methyl phosphorodichloridate (**16**) did not yield the respective nucleoside phosphate methyl ester, but the nucleoside monophosphate instead. These authors noticed a reaction between pyridine and **16** with formation of crystalline adduct to which they assigned structure **17** on the basis of chemical and spectroscopic evidence. Compound **17** was found to phosphorylate nucleosides to nucleoside monophosphates. Phosphorylation of nucleosides with ethyl phosphorodichloridate gave a mixture of the nucleoside monophosphate and nucleotide ethyl ester, in which the latter predominated (23).

N-Acetyl-5-iodo-3-indolyl phosphorodichloridate (**18**) is employed as a reagent for the preparation of nucleoside phosphate-3-indolyl esters (**19**), which are required as chromogenic substrates for phosphodiesterases (22).

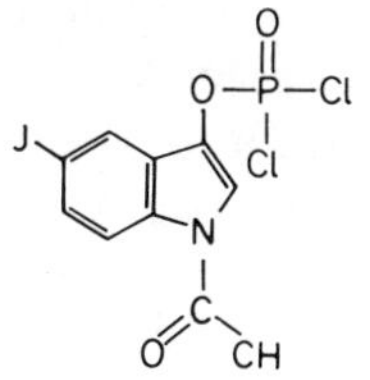

18 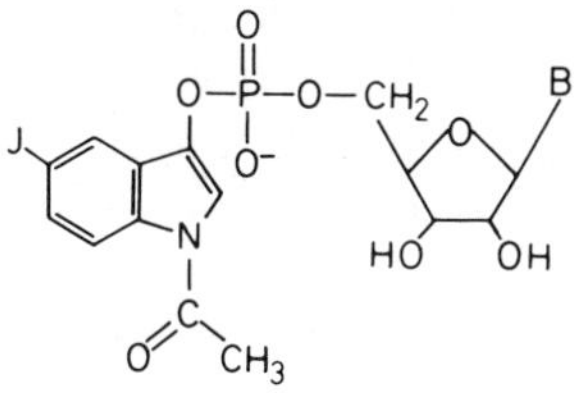 **19**

Phosphorus oxychloride was introduced as a phosphorylating agent by Neuberg and Pollack (26). Fischer improved on this phosphorylating procedure by carrying out the reaction in pyridine at low temperature (27). However, the application of this reagent to the phosphorylation of nucleosides generally gave low yields as well as being unspecific. Yoshikawa et al. reinvestigated the experimental conditions for satisfactory phosphorylation of nucleosides with phosphorus oxychloride (28–31). They observed that by employing trialkylphosphate as solvent in the presence of traces of moisture 5′-phosphorylated unprotected ribonucleosides were obtained in good yields. This discovery has made phosphorus oxychloride one of the most common phosphorylating agents of the past decade. It has been suggested that phosphorus oxychloride interacts with the trialkylphosphate, either trimethylphosphate or triethylphosphate, to form the true phosphorylating species **20**. An alternative

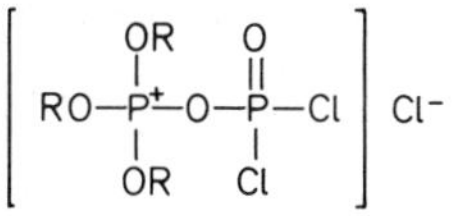

$$\left[(RO)_3P^{+}\text{—O—}P(O)Cl_2 \right] Cl^{-}$$

20

procedure for the use of phosphorus oxychloride in the selective phosphorylation of the 5′-OH groups was reported by Sowa and Ouchi (32), who employed mixtures of pyridine, water, and acetonitrile as solvent. These authors claim that under the conditions reported, an adduct is

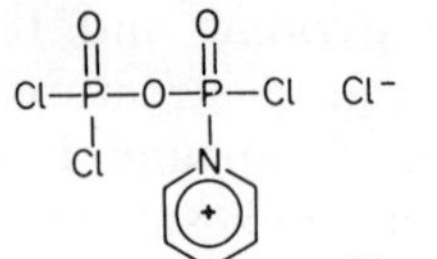

21

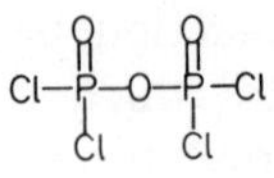

22

formed between pyrophosphoryl-chloride and pyridiniumchloride to which the structure trichloropyrophosphopyridinium chloride (**21**) was assigned. The phosphorylation of nucleosides by phosphorus oxychloride in trialkylphosphate was reexamined by Dawson et al. (33). The authors convincingly demonstrated by anion exchange chromatography and ^{31}P nmr spectroscopy of reaction mixtures that phosphorylation was not as totally selective toward the 5′-OH groups as claimed by Yoshikawa et al. (28). Modifications of the method improved the yield of 5′-nucleotide but were unable to avoid the concomitant formation of 2′(3′)-monophosphates and 5′,3′(2′)-triphosphates. According to Dawson et al. (33), the monophosphorylated nucleosides can contain between 3 to 15 percent of the respective 2′(3′)-monophosphates.

When pyrophosphoryl chloride (**22**) is allowed to react with nucleosides in *m*-cresol or *o*-chlorophenol as solvent, selective phosphorylation of the 5′-OH groups occurs (34). The specificity of the phosphorylation strongly depends on the use of the above solvents, which may therefore participate in the reaction, the mechanism of which is not yet known.

6.1.2 Phosphoric Acid Anhydrides

Pyrophosphoric acid (35) and linear polyphosphoric acid (36–38) can be used in the preparation of nucleoside monophosphates only under forcing conditions. The sodium salt of trimetaphosphate (**23**) reacts with ethylene glycol at elevated temperature (39). The formation of 2-hydroxyethyl phosphate (**25**) instead of the linear tri-polyphosphate ester (**24**) was explained by the following mechanism (**23–25**) (39). Reaction of **25** with

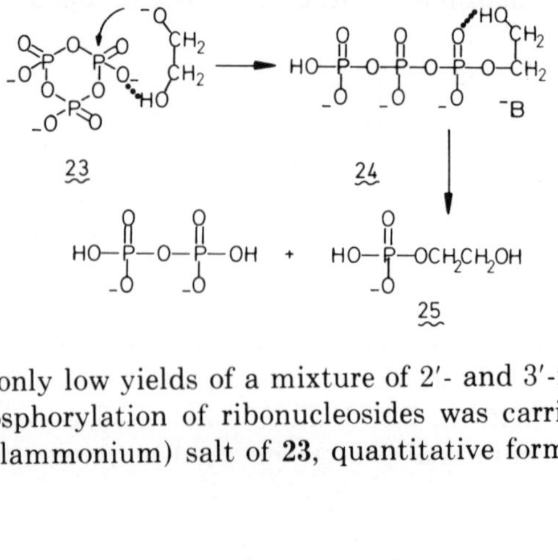

23

24

25

adenosine gave only low yields of a mixture of 2′- and 3′-monophosphates (40). When phosphorylation of ribonucleosides was carried out with the tris-(tetramethylammonium) salt of **23**, quantitative formation of ribonucleoside 2′(3′)-monophosphates was observed. No phosphorylation of the 5′-OH groups occurred. This method seems suitable for the preparation of ribonucleoside 2′(3′)-phosphates, provided the nucleoside withstands strongly basic reaction conditions (41). It is worthwhile mentioning that 5′-amino-5′-deoxynucleosides react with trimetaphosphate to form linear

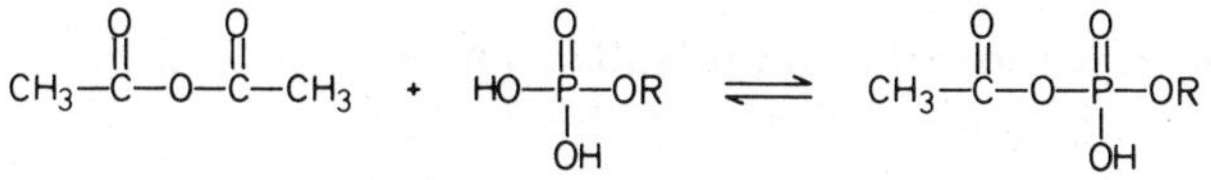

$$CH_3-\overset{O}{\overset{\|}{C}}-O-\overset{O}{\overset{\|}{C}}-CH_3 + HO-\overset{O}{\overset{\|}{\underset{OH}{\underset{|}{P}}}}-OR \rightleftharpoons CH_3-\overset{O}{\overset{\|}{C}}-O-\overset{O}{\overset{\|}{\underset{OH}{\underset{|}{P}}}}-OR$$

26

nucleoside 5′-*N*-triphosphates in good yields (42, 43). Mixed anhydrides (**26**) between monoesterphosphates and carboxylic acids can be easily prepared by an exchange reaction (44, 45). However, chemical studies showed that nucleophilic attack occurs more rapidly at the carbonyl group, hence application of compounds such as **26** to the phosphorylation of OH groups is not possible (45).

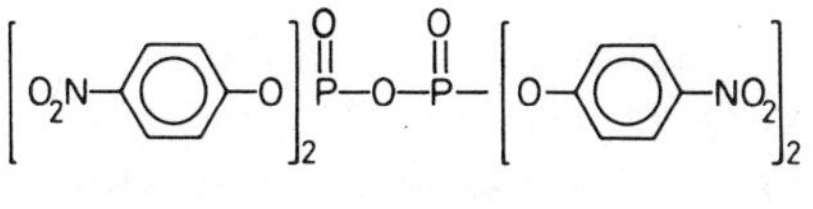

$$[O_2N-C_6H_4-O]_2\overset{O}{\overset{\|}{P}}-O-\overset{O}{\overset{\|}{P}}[O-C_6H_4-NO_2]_2$$

27

Bis-(4-nitrophenyl)phosphate is converted to tetra-(4-nitrophenyl) pyrophosphate (**27**) by the action of di-*p*-tolylcarbodiimide (1, 46). Compound **27** is a powerful phosphorylating agent, which reacts with OH groups of suitably blocked nucleosides to give nucleoside phosphate bis-(4-nitrophenyl) esters (46). Monoesterphosphates can be activated with dicyclohexylcarbodiimide to an intermediate, which reacts with primary and secondary OH groups at room temperature to form unsymmetric diesterphosphates (1). Experimental evidence suggests that the reactive inter-

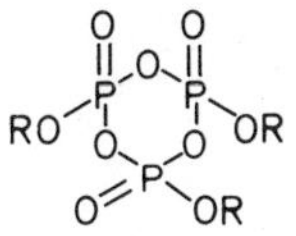

28

mediate may be a trimetaphosphate ester (47, 48) (**28**). A number of monoesterphosphates have been introduced as phosphorylating agents in the presence of dicyclohexylcarbodiimide, in which the alkyl groups serve as protecting groups and must therefore be removed after phosphorylation.

The most versatile proved to be 2-cyanoethylphosphate (49) (**29**) and 2,2,2-trichloroethylphosphate (**30**) (50). Both monoesterphosphates are similar in that the alkyl groups can be removed by β-elimination. β-Elimination of the 2-cyanoethyl group leads to formation of 2-cyanoethylene. It should be borne in mind that this highly reactive species could be involved in side reactions with the product, such as alkylation (51).

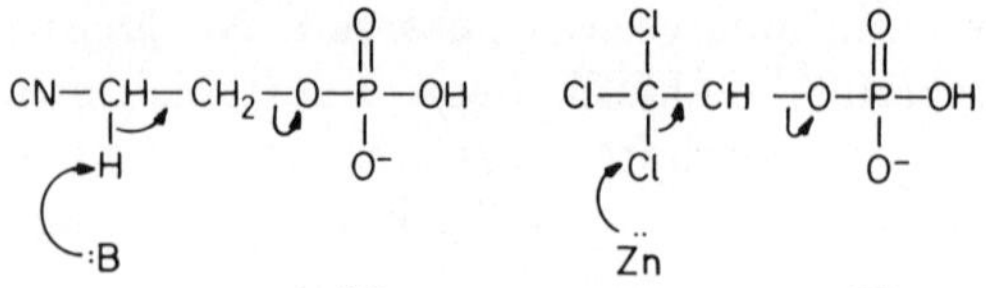

As shown by Khorana (52), activation of monoesterphosphates to the reactive intermediate **28** can be performed equally well with aryl sulfonyl chlorides. Numerous attempts have been reported to use new reagents for the activation of monoesterphosphates, but none have proved to be as efficient as dicyclohexylcarbodimide or certain arylsulfonyl chlorides such as 2,4,6-triisopropylbenzenesulfonyl chloride.

6.1.3 Phosphorylation by Transesterification

2-Methylthio-2-oxo-P^V-4H-1,3,2-benzodioxaphosphorin (**31**) reacts with alcohol in the presence of cyclohexylamine, the dioxaphosphorin ring being opened to give an alkyl-O-salicyl-S-methylphosphorothioate (**32**). It is suggested that a benzyl cation is formed which reacts with cyclohexylamine. The stable O-alkyl, S-methylphosphorothioate (**33**) can be isolated and oxidation with iodine yields the desired monoesterphosphate (53, 54). Reaction of unprotected ribonucleosides with **31** afforded the 2′,3′-cyclic phosphates as major products (55). Employing ribonucleosides protected at the 2′,3′-cis diol nucleoside 5′-phosphorothioate S-methylesters (**34**) were the exclusive products after removal of the protecting group. Oxidation of **34** by iodine in aqueous solution gave nucleoside 5′-phosphates (55), whereas oxidation in anhydrous pyridine led to nucleoside 3′,5′-cyclic phosphates (55). This is a rather capricious phosphorylation procedure to the cyclic phosphate. Whether it stands up to the

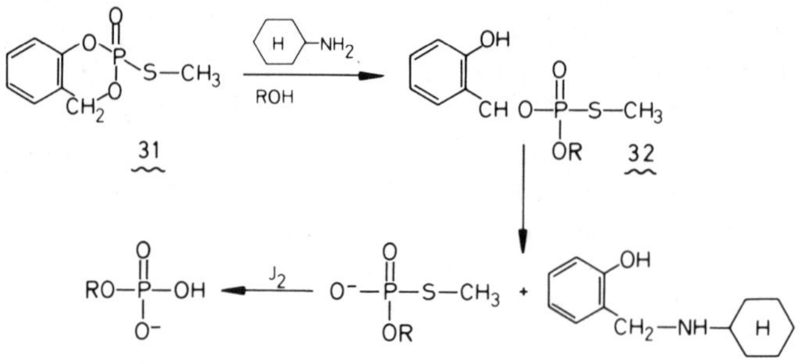

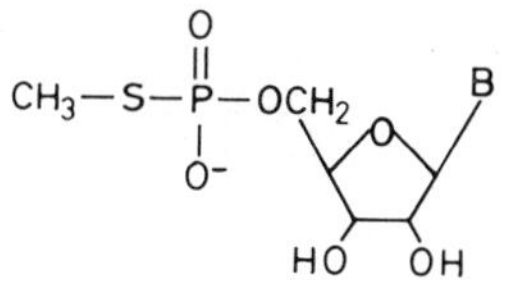

34

straightforward phosphorylation of a nucleoside by bis(4-nitrophenyl) phosphorochloridate and cyclization of the bis-(4-nitrophenyl) ester of the nucleoside according to Borden and Smith (17) is doubtful. Highly reactive five-membered cyclic phosphoester derivates have been described by Ramirez et al. (56, 57). 4,5-Dimethyl-2-(1-imidazolyl)-2-oxo-P^V-1,2,3-

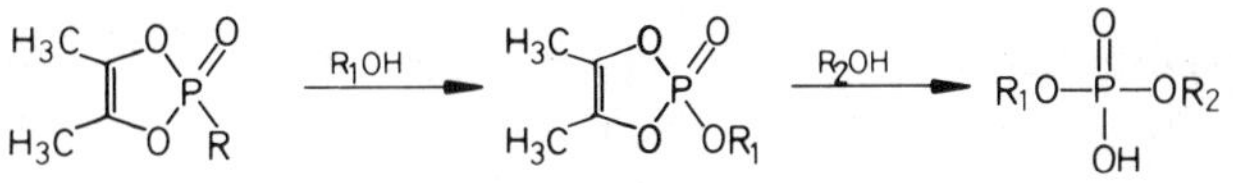

35,36 37

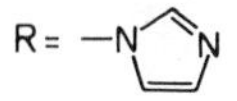

35 R=

36 R=Cl

dioxaphosphate (**35**) and 2-chloro-4,5-dimethyl-2-oxo-P^V-1,3,2-dioxaphosphole (**36**) react with primary or secondary alcohols. The five-membered cyclic triester (**37**) can then undergo further reaction with an alcohol to give unsymmetrical phosphodiesters. Compounds **35** and **36** should be useful in the synthesis of nucleoside monophosphates, as well as nucleotide

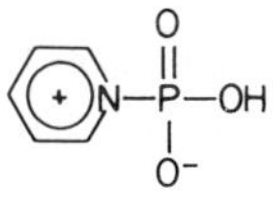

38

esters. Their high reactivity, however, requires suitably protected nucleosides for the initial transesterification step.

4-Nitrophenylphosphate gives very efficient transesterification with 2′,3′-*O*-isopropylidene nucleosides at elevated temperature (58), provided the reaction is carried out in dry pyridine which seems to catalyze this reaction (59). Hata and Chong suggest that this phosphorylation proceeds via a reactive *N*-phosphopyridinium intermediate (**38**) (59).

Why the authors exclude the formation of reactive polyphosphate intermediates is not obvious. The reaction products obtained depend critically on the conditions employed. An excess of pyridine leads to formation of nucleoside polyphosphates in high yield.

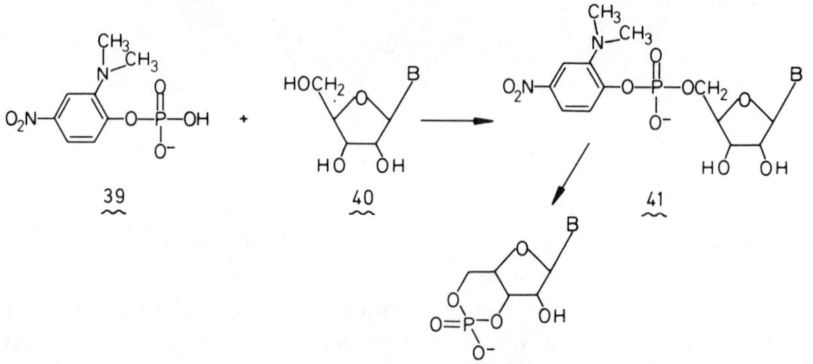

2-(N,N-Dimethylamino)-4-nitrophenyl phosphate (**39**) likewise phosphorylates alcohols by transesterification (60). The transesterification is acid catalysed and gives monoesterphosphates in good yields. An elegant reaction is the specific condensation of **39** with ribonucleoside (**40**) in the presence of dicyclohexylcarbodiimide to give the respective 5′-2-(N-N-dimethylamino)-4-nitrophenyl phosphate (**41**), which then cyclizes in refluxing pyridine in the presence of acetic acid and triethylamine to the corresponding 3′,5′-cyclic phosphate (61–63) (**42**). Activated diesterphosphates with the 8-quinolyl moiety as leaving group (**43**) have been employed in the synthesis of unsymmetric phosphate diesters (64). α-Pyridylphosphate esters (**44**) were introduced much earlier for the same purpose (65). Neither compounds of type **43** nor **44** were of any use in the phosphorylation of nucleosides. 4-Nitrophenyl, 5-(2-pyridyl)phosphorothioate (**45**) transfers the 4-nitro-phenylphosphate residue in a transesterification reaction to the OH groups of nucleosides (66). Tris-(8-hydroxychinolyl) phosphate (**46**) is reactive enough to be used in the phosphorylation of nucleosides (67–69). Hydrolysis of the nucleotide bis(8-hydroxychinolyl) phosphates is achieved with aqueous copper (II) chloride.

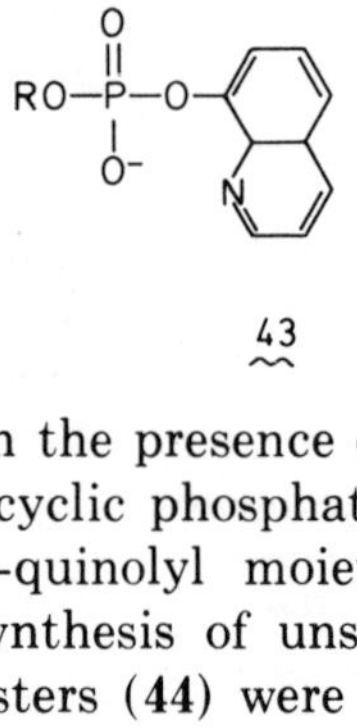

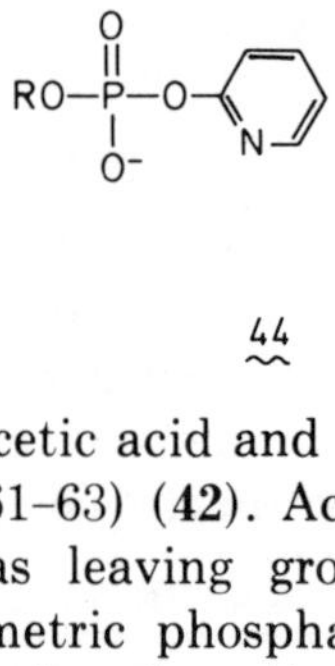

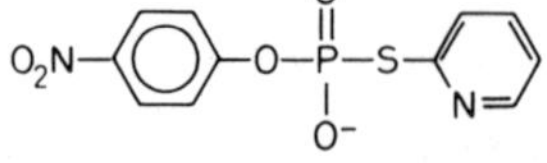

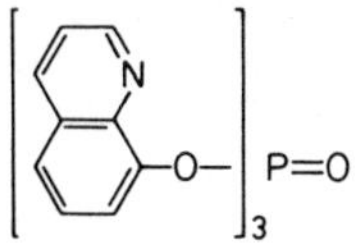

46

6.1.4 Miscellaneous

Phosphorous acid, which is equivalent to a monoester phosphate with respect to its dissociation constants, reacts readily with nucleotides in the presence of an activating reagent such as di-*p*-tolylcarbodiimide or an arylsulfonylchloride to nucleoside phosphites (70). The interest in nucleoside phosphites results from the fact that they can be oxidized to the respective nucleoside phosphates and therefore represent potential precursors of the latter. Oxidation of purine nucleoside phosphites was achieved with permanganate in aqueous solution (70). Application of this oxidation procedure to pyrimidine derivatives is, however, not feasible. Holy and Sorm (71) prepared nucleoside phosphites by acid or base catalysed transesterification with triphenylphosphite in good yields. The high reactivity of triphenylphosphite demands protected nucleosides if specific phosphitylation is desired (72) (see Chapter 4, Section 4). The less reactive triethylphophite (**47**) reacts specifically with the cis diol grouping

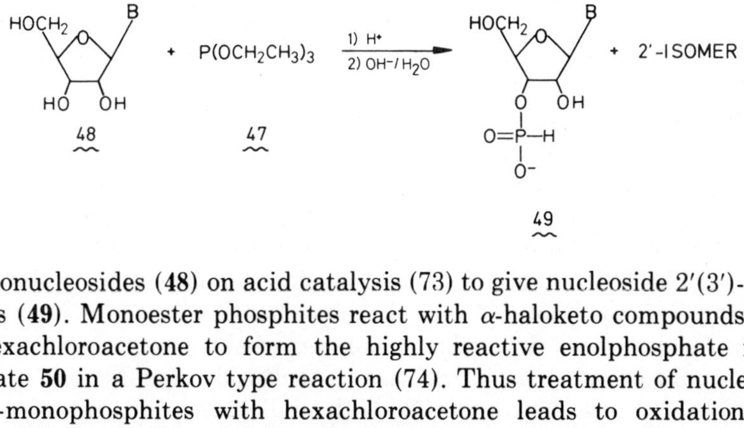

HOCH2, B, HO, OH + $P(OCH_2CH_3)_3$ —1) H^+, 2) OH^-/H_2O→ HOCH2, B, O, OH, O=P—H, O⁻ + 2'-ISOMER

48 47 49

in ribonucleosides (**48**) on acid catalysis (73) to give nucleoside 2′(3′)-phosphites (**49**). Monoester phosphites react with α-haloketo compounds such as hexachloroacetone to form the highly reactive enolphosphate intermediate **50** in a Perkov type reaction (74). Thus treatment of nucleoside 2′(3′)-monophosphites with hexachloroacetone leads to oxidation and intramolecular phosphorylation resulting in formation of 2′,3′-cyclic phosphates in high yields (74). Acid catalyzed reaction of ribonucleosides with triethylphosphite and subsequent oxidative cyclization by means of hexachloroacetone represents an elegant and efficient route to nucleoside 2′,3′-cyclic phosphates. Treatment of nucleoside 5′-phosphites with hexachloroacetone leads to the simultaneous formation of nucleoside 3′,5′-cyclic phosphate and P^1,P^2-di(nucleoside-5′)pyrophosphate from the intermediate enolphosphate (74). An elegant procedure for the oxidation of nulceoside phosphites was described by Hata and Sekine (75). The usefulness of this method is exemplified in the case of thymidine 5′-phosphite (**51**). The initial step involves silylation of **51** by reaction with trimethylsilylchloride. The bis-trimethylsilylphosphite **52** reacts at room temperature with 2,2′-dipyridyl disulfide to give an intermediate with the

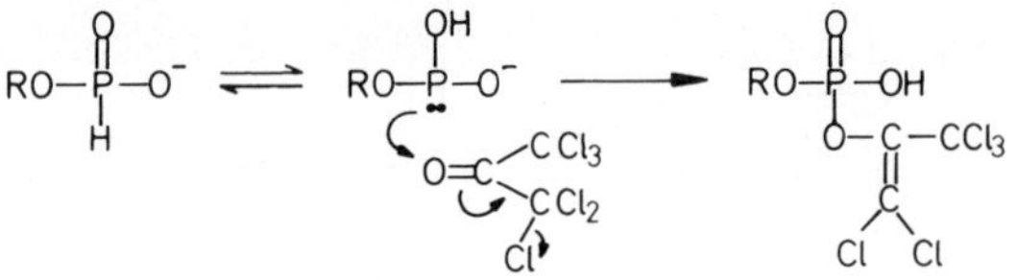

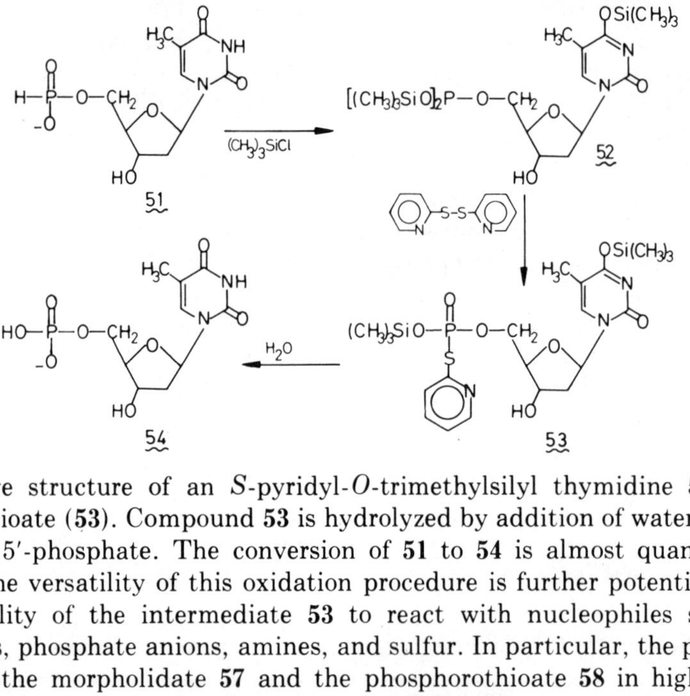

tentative structure of an *S*-pyridyl-*O*-trimethylsilyl thymidine 5′-phosphorothioate (**53**). Compound **53** is hydrolyzed by addition of water to thymidine 5′-phosphate. The conversion of **51** to **54** is almost quantitative (75). The versatility of this oxidation procedure is further potentiated by the ability of the intermediate **53** to react with nucleophiles such as alcohols, phosphate anions, amines, and sulfur. In particular, the preparation of the morpholidate **57** and the phosphorothioate **58** in high yields starting from the easy accessible **50** is a remarkable synthetic alternative, which should prove useful in the case of nucleoside analogs. The reaction of 5′-azido-5′-deoxy nucleosides (**59**) with trialkyl phosphite was employed by Freist et al. (76) in the preparation of thymidine 5′-*N*-phosphate phenylester (**61**).

An interesting phosphorylation procedure, which leads from nucleosides directly to 3′,5′-cyclic phosphates, was introduced by Marumoto et al. (77). Trichloromethylphosphonodichloridate (**62**) reacts specifically with the 5′-OH groups of unprotected nucleosides. The resulting nucleoside 5′-trichlorophosphonate (**64**) cyclizes upon treatment with potassium *t*-

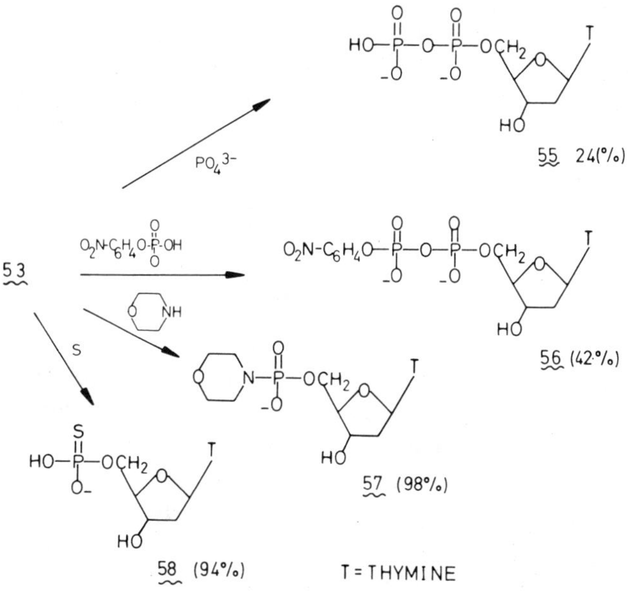

PO_4^{3-}

55 24(%)

56 (42%)

57 (98%)

58 (94%)

T = THYMINE

butoxide to the 3′,5′-cyclic phosphate. It is assumed that this cyclization proceeds through a nucleoside 5′-metaphosphate intermediate.

A new and efficient way to activate monoesterphosphates involving an oxidation-reduction cycle was discovered by Mukaiyama and Hashimoto (66, 78–80). Triphenylphosphine reacts with 2,2′-dipyridyldisulfide to form a triphenylphosphonium salt **65** which is attacked by monoesterphosphate

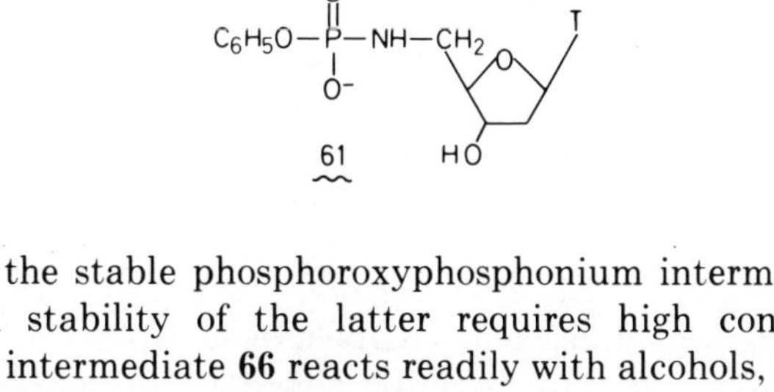

$(C_6H_5O)_2P$

59

60

OH^-

61

anions to give the stable phosphoroxyphosphonium intermediate **66**. The formation and stability of the latter requires high concentrations of reactants. The intermediate **66** reacts readily with alcohols, phosphates, or amines to give diesterphosphates, pyrophosphates, and phosphoroamidates respectively. The condensation of monoesterphosphate and alcohol involves the oxidation of triphenylphosphine to triphenylphosphineoxide and reduction of 2,2′-dipyridyldisulfide to pyrid-2-thione. This phosphorylation procedure was applied to the phosphorylation of protected nucleosides with 2-cyanoethyl phosphate, formation of nucleotide morpholidates, as well as the intramolecular cyclization of nucleoside monophosphates to the corresponding cyclic phosphates.

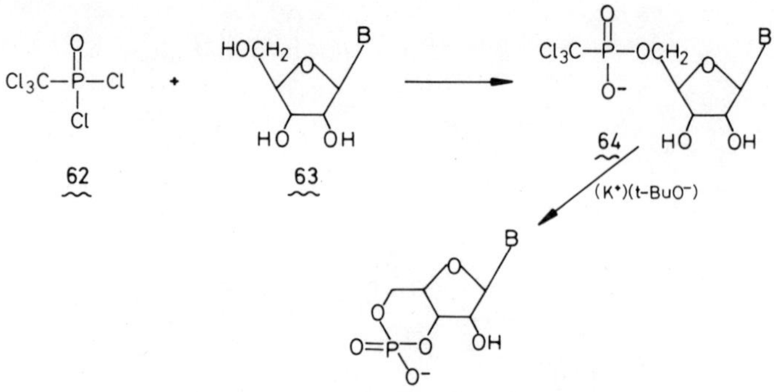

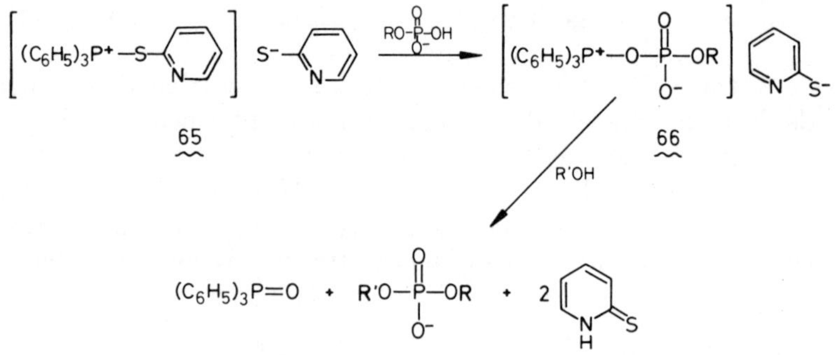

Takaku et al. (81) observed that under the conditions of the above described phosphorylation procedure 5′-*S*-pyridylthio-5′-deoxyribonucleosides were formed as side products. In a variation of Mukaiyama's procedure these authors employed a combination of triphenylphosphite and 2,2′-dipyridyl diselenide as activating reagent for monoesterphosphate. It is suggestive to assume compounds of type **67** as the phosphorylating intermediates. Phosphorylation of protected nucleosides with 2-cyanoethylphosphate by this procedure occurred in high yields. It is also worth mentioning that the above method of activating monoesterphosphates was successfully applied to oligonucleotide synthesis (81).

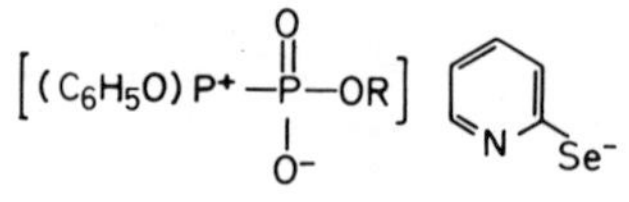

6.2 SYNTHESIS OF NUCLEOSIDE POLYPHOSPHATES

The first synthesis of adenosine 5′-diphosphate and 5′-triphosphate by Todd and coworkers was an application of the classical chemical procedure for the synthesis of anhydrides, namely the reaction of the salt of an acid with an acid chloride (2, 82, 83). The battle, which those ingenious organic chemists won against surmounting difficulties, hindered by the lack of modern chemical and analytical techniques, deserves respectful admiration. The following procedures for the synthesis of nucleoside polyphosphates are presently used: (1) displacement reactions with

C_6H_5O C_6H_5O P—O—P—OCH_2 O⁻ B HO OH

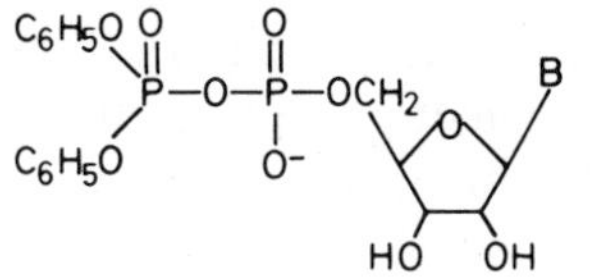

68

triester pyrophosphates, (2) reaction of nucleoside phosphoramidates with phosphate or pyrophosphate anions, and (3) reaction with reactive phosphodiesters.

6.2.1 Displacement Reactions with Triester Pyrophosphates

Reaction of a nucleoside 5′-phosphate with diphenyl phosphorochloridate leads to formation of the relatively stable P^1-diphenyl P^2-(nucleoside 5′-) pyrophosphate (**68**) (84). Nucleophilic attack at the P—O—P bond in **68** occurs such that the more stable anion, namely the anion of the stronger

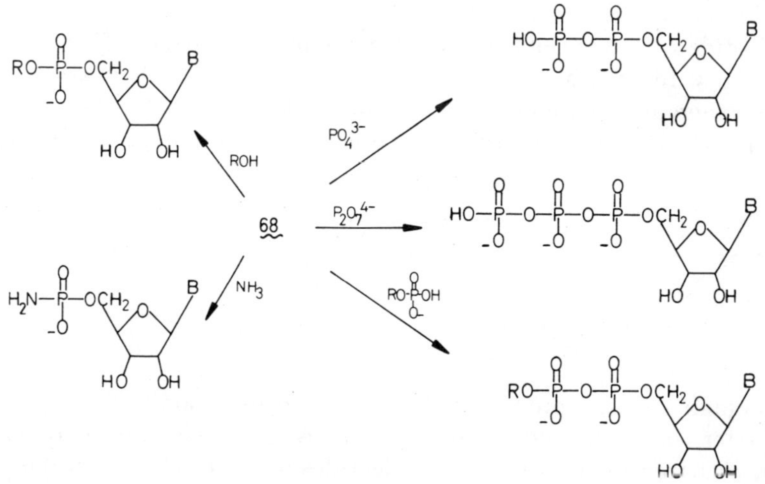

RO—P—OCH_2 B HO OH — ROH — 68 — PO_4^{3-} → HO—P—O—P—OCH_2 B HO OH; $P_2O_7^{4-}$ → HO—P—O—P—O—P—OCH_2 B HO OH; RO-P-OH → RO—P—O—P—OCH_2 B HO OH; NH_3 → H_2N—P—OCH_2 B HO OH

Scheme 6.3

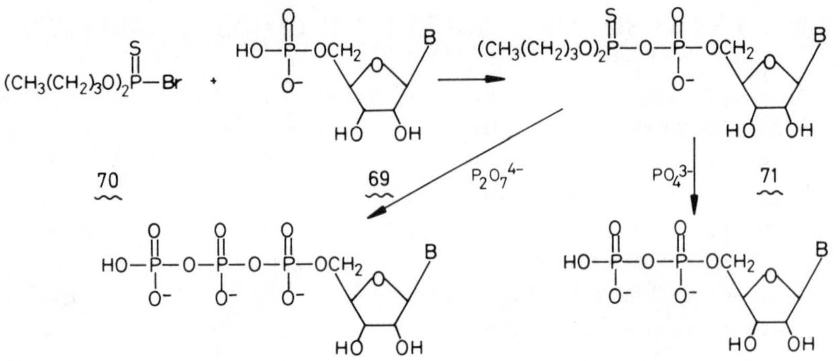

$(CH_3(CH_2)_3O)_2P(S)$—Br + HO—P(O)(O⁻)—OCH₂-nucleoside (HO, OH) → $(CH_3(CH_2)_3O)_2P(S)$—O—P(O)(O⁻)—OCH₂-nucleoside

70 **69** $P_2O_7^{4-}$ PO_4^{3-} **71**

HO—P(O)(O⁻)—O—P(O)(O⁻)—O—P(O)(O⁻)—OCH₂-nucleoside (HO, OH)   HO—P(O)(O⁻)—O—P(O)(O⁻)—OCH₂-nucleoside (HO, OH)

acid diphenyl phosphoric acid, is formed. Triesterpyrophosphates of type **68** react with a broad range of phosphate anions and other nucleophiles as outlined in Scheme 6.3. This method, which was introduced by Michelson over 15 years ago, had been used ever since and because of its simplicity and reproducibility it continues to remain one of the principal methods for the conversion of nucleotides into nucleoside polyphosphates. Hata et al. (85) obtained stable nucleoside 5′-phosphoric di-*n*-butylphosphinothioic anhydrides by reaction of nucleoside 5′-phosphates (**69**) with di-*n*-butylphosphinethioyl bromide (**70**). The anhydrides **71** are formed practically quantitatively and can be isolated. Compounds of type **71** appeared to be stable to water. The anhydrides reacted in anhydrous pyridine in the presence of silver acetate with either mono(tri-*n*-butylammonium) phosphate or bis(tri-*n*-butylammonium) pyrophosphate to give in good yields the 5′-diphosphates or 5′-triphosphates respectively.

6.2.2 Nucleoside Phosphoramidates

Nucleoside 5′-phosphoramidates (**72**) were prepared by activation of a nucleotide with dicyclohexylcarbodiimide in the presence of ammonia (86, 87). They reacted smoothly at room temperature with inorganic phosphate, inorganic pyrophosphate, or monoester phosphates with formation of a P—O—P bond. Poor solubility of the phosphoramidates (**72**) in suitable organic solvents prompted the search for phosphoramidate derivatives of better solubility and reasonable reactivity. Nucleotide morpholidates (**73**), prepared by activation of nucleotides with dicyclohexylcarbodiimide in the presence of morpholine, turned out to be the best compromise

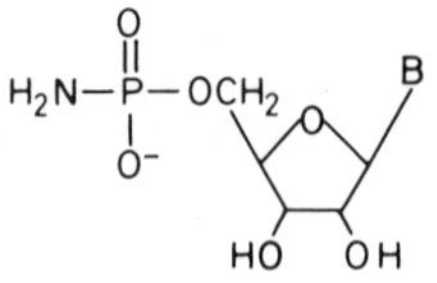

H_2N—P(O)(O⁻)—OCH₂-nucleoside (HO, OH)

72

between ease of preparation, solubility, and reactivity (88, 89). Under proper experimental conditions they are quantitatively formed and can be stored with some precautions in solid form for a long time. Originally, the reaction of compounds **73** with phosphate or pyrosphosphate was carried out in pyridine until Moffatt (90) noted that nucleoside 5′-triphosphate

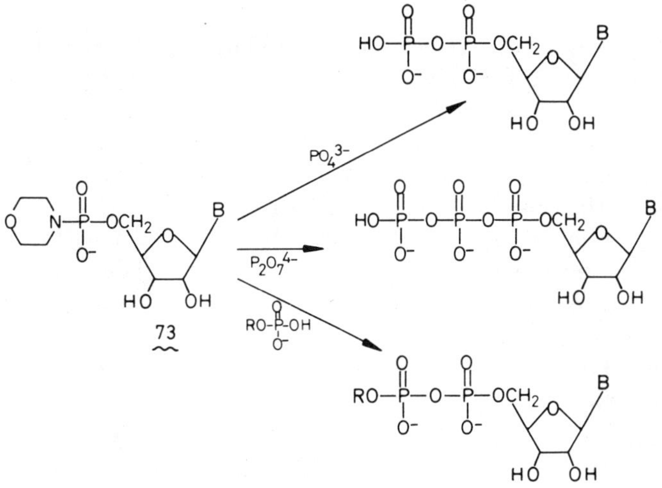

Scheme 6.4

decomposed under the chosen conditions. He suggested instead the use of solvents such as dimethylformamide or dimethylsulfoxide in the reactions outlined in Scheme 6.4. The conversion of a nucleoside into its respective nucleoside 5′-polyphosphates requires two steps, phosphorylation and conversion of the nucleotide into the morpholidate. The reagent 2,2,2-tribromoethyl phosphomorpholinochloridate **74**, introduced by van Boom

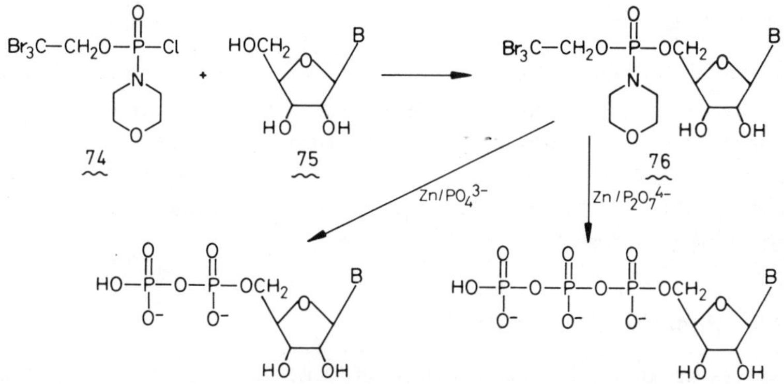

et al. (91) allows the preparation of a nucleotide morpholidate in one step. Phosphorylation of unprotected ribonucleosides with **74** occurs specifically in the 5′-position. The neutral intermediate **76**, which withstands mild

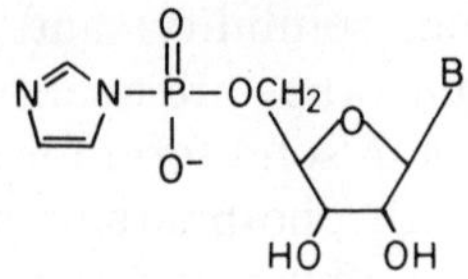

77

alkaline or acidic conditions, can be isolated and reacted with tri-*n*-butylammonium salts of phosphate or pyrophosphate in the presence of zinc dust to give the respective nucleoside 5′-polyphosphate.

Cramer et al. (92–94) discovered that nucleotide imidazolidates (**77**) have chemical properties similar to nucleotide morpholidates. One advantage is the possibility of converting nucleotides into imidazolidates in situ by reaction with 1,1′-carbonyldiimidazole. This reaction can be conveniently carried out on a small scale (95). The choice of a nucleotide imidazolidate in an attempted synthesis of a nucleoside polyphosphate may be indicated if nucleotide analogs do not withstand the more vigorous experimental conditions necessary for the preparation of nucleotide morpholidates.

78 79 80

1) PO_4^{3-}
2) H^+/H_2O

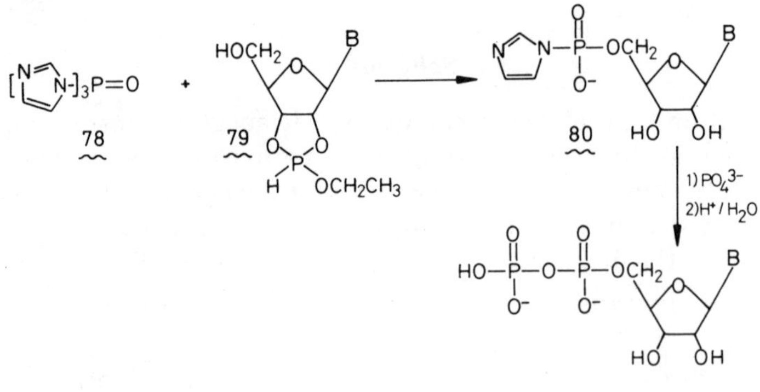

Trisimidazolyl phosphinoxide (**78**), easily prepared from phosphorus oxychloride and imidazole, phosphorylates protected nucleosides under mild conditions to nucleoside phosphoimidazolidates (**80**). The latter are stable enough above pH 7 to be isolated by chromatographic procedures. Compound **80** can then be employed in reactions with phosphate, pyrophosphate or monoesterphosphate (96).

6.2.3 Reactive Phosphate Esters

Reactive esters of nucleotides, which could be potentially employed in the synthesis of nucleoside polyphosphates, have never gained importance. Their preparation is laborious, involves low yields, and the properties in general do not match those of the known activated forms of nucleotides used in nucleotide polyphosphate synthesis. If one reactive phosphate ester, namely 2-cyanoethyl, 2-pyridylphosphate (**81**) is mentioned here, then it is only for the sake of demonstrating a general synthetic concept. 2-Pyridyl, alkyl phosphodiesters were shown to react with monoesterphosphates to form unsymmetric diester pyrophosphates (96). Analogously, the 2-pyridyl ester (**81**) reacted with nucleoside 5′-phosphates (**82**) to give P^1-2-(cyanoethyl) P^2-(nucleoside 5′-) pyrophosphate (**83**). Mild alkaline hydrolysis converted **83** into the corresponding nucleoside 5′-diphosphate (97). Because compound **81** (98) can be prepared as the stable, crystalline cyclohexylammonium salt, it represents a reagent for the conversion of a given nucleotide to the respective nucleoside 5′-diphosphate.

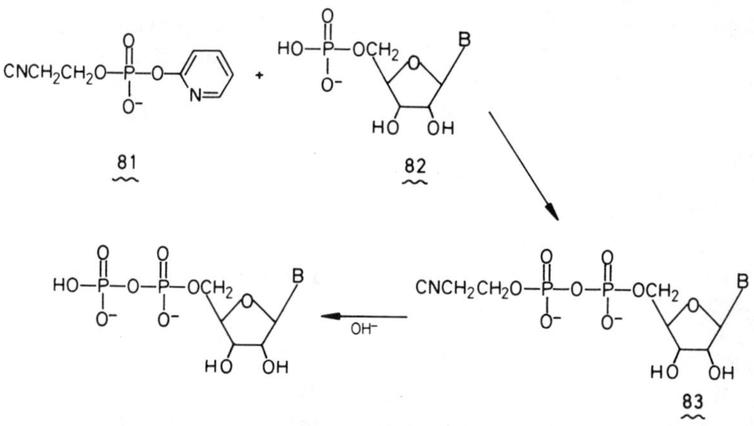

REFERENCES

1. H. G. Khorana, *Some Recent Developments in the Chemistry of Phosphate Esters of Biological Interest*, Wiley, New York, 1961.
2. A. M. Michelson, *The Chemistry of Nucleosides and Nucleotides*, Academic Press, New York, 1963.
3. V. M. Clark, D. W. Hutchinson, A. J. Kirby, and S. G. Warren, *Angew. Chemie* **76,** 704–712 (1964).
4. L. A. Slotin, *Synthesis*, 737–752 (1977).
5. A. Hampton, L. W. Brox, and M. Bayer, *Biochemistry* **8,** 2303–2311 (1969).
6. A. Stütz and K. H. Scheit, *Eur. J. Biochem.* **50,** 343–349 (1975).
7. K. H. Scheit and A. Stütz, *J. Carbohydr. Nucleosides Nucleotides* **1,** 485–490 (1974).
8. K. Zeile and H. Meyer, *Zeitschrift Physiol. Chemie* **265,** 131 (1938).
9. P. Brigl, and H. Müller, *Berichte* **72,** 2121–2130 (1939).
10. G. H. Jones and J. G. Moffatt, *J. Am. Chem. Soc.* **90,** 5337–5338 (1968).
11. K. Schimmelschmidt and H. Kappenberger, *Chem. Abstr.* **55,** 12355d (1957).
12. A. Murayama, B. Jastorff, F. Cramer, and H. Hettler, *J. Org. Chem.* **36,** 3029–3033 (1971).
13. H. Witzel, H. Mirbach, and K. Dimroth, *Angew, Chem.* **72,** 751 (1960).

14. F. Eckstein and K. H. Scheit, *Angew. Chem.* **79,** 317–318 (1967).
15. T. A. Khwaja, C. B. Reese, and J. M. B. Stewart, *J. Chem. Soc.* **1970,** 2092–2100.
16. J. Hes and M. P. Mertes, *J. Org. Chem.* **39,** 3767–3769 (1974).
17. K. F. Borden and M. Smith, *J. Org. Chem.* **31,** 3241–3246 (1966).
18. A. Franke, K. H. Scheit, and F. Eckstein, *Chem. Ber.* **101,** 2998–3001 (1968).
19. K. H. Scheit, *Biochim. Biophys. Acta* **157,** 632–633 (1968).
20. A. F. Turner and H. G. Khorana, *J. Am. Chem. Soc.* **81,** 4651–4656 (1959).
21. T. Hata, Y. Mushika, and T. Mukaiyama, *J. Am. Chem. Soc.* **91,** 4532–4535 (1969).
22. K. C. Tsou, K. W. Lo, S. L. Ledis, and E. E. Miller, *J. Med. Chem.* **15,** 1221–1225 (1972).
23. J. Smrt and J. Catlin, *Tetrahedron Lett.* 5081–5084 (1970).
24. R. I. Zhdanov and S. M. Zhenodarova, *Synthesis* 222–245 (1975).
25. Y. Mushika, T. Hata, and T. Mukaiyama, *Bull. Chem. Soc. (Japan)* **44,** 232–235 (1971).
26. C. Neuberg and H. Pollak, *Biochem. Z.* **23,** 515 (1910).
27. E. Fischer, *Berichte* **47,** 3193–3205 (1914).
28, M. Yoshikawa, T. Kato, and T. Takenishi, *Tetrahedron Lett.* 5065–5068 (1967).
29. M. Yoshikawa and T. Kato, *Bull. Chem. Soc. (Japan)* **40,** 2849–2853 (1967).
30. K. Kusashio and M. Yoshikawa, *Bull. Chem. Soc. (Japan)* **41,** 142–149 (1968).
31. M. Yoshikawa, T. Kato, and T. Takenishi, *Bull. Chem. Soc. (Japan)* **42,** 3505–3508 (1969).
32. T. Sowa and S. Ouchi, *Bull. Chem. Soc. (Japan)* **48,** 2084–2090 (1975).
33. W. H. Dawson, R. L. Cargill, and R. Dunlap, *J. Carbohydr. Nucleosides Nucleotides* **4,** 363–375 (1977).
34. K. I. Imai, *J. Org. Chem.* **34,** 1547–1550 (1969).
35. E. W. Haeffner, *Biochim. Biophys. Acta* **212,** 182–184 (1970).
36. T. Ueda and I. Kawai, *Chem. Pharm. Bull. (Tokyo)* **18,** 2303–2308 (1970).
37. M. Honjo, Y. Furukawa, and K. Kobayashi, *Chem. Pharm. Bull. (Tokyo)* **14,** 1061–1065 (1966).
38. Y. Sanno and A. Nohara, *Chem. Pharm. Bull. (Tokyo)* **16,** 2056–2059 (1968).
39. W. Feldmann, *Chem. Ber.* **100,** 3850–3860 (1967).
40. D. J. Cosgrove, *Soil. Biol. Chem.* **4,** 387 (1972).
41. R. Saffhill, *J. Org. Chem.* **35,** 2881–2883 (1970).
42. D. B. Trowbridge, D. M. Yamamoto, and G. L. Kenyon, *J. Am. Chem. Soc.* **94,** 3816–3824 (1972).
43. R. L. Letsinger, J. S. Wilkes, and L. B. Dumas, *J. Am. Chem. Soc.* **94,** 289–293 (1972).
44. A. W. D. Avison, *J. Chem. Soc.* **1955,** 732–738.
45. H. G. Khorana and J. P. Vizsolyi, *J. Am. Chem. Soc.* **81,** 4660–4664 (1959).
46. J. G. Moffatt and H. G. Khorana, *J. Am. Chem. Soc.* **79,** 3741–3746 (1957).
47. G. Weimann and H. G. Khorana, *J. Am. Chem. Soc.* **84,** 4329–4341 (1962).
48. G. M. Blackburn, M. J. Brown, M. R. Harris, and D. Shire, *J. Chem. Soc.* **1969,** 676–683.

49. G. Tener, *J. Am. Chem. Soc.* **83,** 159–168 (1961).
50. F. Eckstein, *Angew. Chem.* **77,** 912–913 (1965).
51. F. Kappler and A. Hampton, *J. Org. Chem.* **40,** 1378–1385 (1975).
52. R. Lohrmann and H. G. Khorana, *J. Am. Chem. Soc.* **88,** 829–833 (1966).
53. M. Eto, H. Ohkawa, K. Kobayashi, and T. Hosai, *Agric. Biol. Chem.* **32,** 1056 (1968).
54. H. Ohkawa, and M. Eto, *Agric. Biol. Chem.* **33,** 443 (1969).
55. M. Eto, H. Ohkawa, K. Kobayashi, and T. Hosai, *Agric. Biol. Chem.* **38,** 2081 (1974).
56. F. Ramirez, H. Okazaki, and J. F. Marecek, Synthesis 637–638 (1975).
57. F. Ramirez, J. F. Marecek, and H. Okazaki, *J. Am. Chem. Soc.* **97,** 7181–7182 (1975).
58. T. Hata and K. Chong, *Bull. Chem. Soc.* (*Japan*) **45,** 654–655 (1972).
59. K. J. Chong and T. Hata, *Bull. Chem. Soc.* (*Japan*) **44,** 2741–2744 (1971).
60. Y. Taguchi and Y. Mushika, *J. Org. Chem.* **40,** 2310–2313 (1975).
61. Y. Taguchi and Y. Mushika, *Chem. Pharm. Bull.* (*Tokyo*) **23,** 1586–1588 (1975).
62. Y. Taguchi and Y. Mushika, *Tetrahedron Lett.* **24,** 1913–1916 (1975).
63. Y. Taguchi and Y. Mushika, *Bull. Chem. Soc.* (*Japan*) **48,** 1528–1532 (1975).
64. H. Takaku and Y. Shimada, *Chem. Pharm. Bull.* (*Tokyo*) **21,** 445–447 (1973).
65. W. Kampe, *Chem. Ber.* **98,** 1038–1044 (1965).
66. T. Mukaiyama and M. Hashimoto, *Bull. Chem. Soc.* (*Japan*) **44,** 196–199 (1971).
67. H. Takaku and Y. Shimada, *Tetrahedron Lett.* 1279–1282 (1974).
68. H. Takaku, Y. Shimada, and K. Anai, *Bull. Chem. Soc.* (*Japan*) **47,** 779–780 (1974).
69. H. Takaku and Y. Shimada, *Chem. Pharm. Bull.* (*Tokyo*) **22,** 1743–1747 (1974).
70. J. A. Schofield and A. Todd, *J. Chem. Soc.* **1961,** 2316–2320.
71. A. Holy and F. Sorm, *Coll. Czech. Chem. Commun.* **31,** 1544–1561 (1966).
72. A. Holy, *Coll. Czech. Chem. Commun.* **32,** 3064–3067 (1967).
73. A. Holy, and F. Sorm, *Coll. Czech. Chem. Commun.* **31,** 1562–1568 (1966).
74. A. Holy, J. Smrt, and F. Sorm, *Coll. Czech. Chem. Commun.* **30,** 3309–3319 (1965).
75. T. Hata and M. Sekine, *Tetrahedron Lett.* 3943–3946 (1974).
76. W. Freist, K. Schattka, F. Cramer, and B. Jastorff, *Chem. Ber.* **105,** 991–999 (1972).
77. R. Marumoto, T. Nishimura, and M. Honjo, *Chem. Pharm. Bull.* (*Tokyo*) **23,** 2295–2300 (1975).
78. T. Mukaiyama and M. Hashimoto, *Bull. Chem. Soc.* (*Japan*) **44,** 2284 (1971).
79. T. Mukaiyama and M. Hashimoto, *Tetrahedron Lett.* 2425–2428 (1971).
80. T. Mukaiyama and M. Hashimoto, *J. Am. Chem. Soc.* **94,** 8528–8532 (1972).
81. H. Takaku, Y. Shimada, Y. Nakajima, and T. Hata, *Nucleic Acids Res.* **3,** 1233–1247 (1976).
82. J. Baddiley, and A. R. Todd, *J. Chem. Soc.* **1947,** 648–651.
83. J. Baddiley, and A. M. Michelson, and A. R. Todd, *J. Chem. Soc.* **1949,** 582–586.

84. A. M. Michelson, *Biochim. Biophys. Acta* **91,** 1–13 (1964).
85. T. Hata, K. Furusawa, and M. Sekine, *J. Chem. Soc. Chem. Commun.* **1975,** 196–197.
86. R. W. Chambers and J. G. Moffatt, *J. Am. Chem. Soc.* **80,** 3752–3756 (1958).
87. R. W. Chambers, P. Shapiro, and V. Kurkov, *J. Am. Chem. Soc.* **82,** 970–975 (1961).
88. J. G. Moffatt and H. G. Khorana, *J. Am. Chem. Soc.* **83,** 649–658 (1961).
89. S. Roseman, J. J. Distler, J. G. Moffatt, and H. G. Khorana, *J. Am. Chem. Soc.* **83,** 659–663 (1961).
90. J. G. Moffatt, *Can. J. Chem.* **42,** 599–604 (1964).
91. J. H. van Boom, R. Crea, W. C. Luyten, and A. B. Vink, *Tetrahedron Lett.* 2779–2782 (1975).
92. F. Cramer, H. Schaller, and H. A. Staab, *Chem. Ber.* **94,** 1612–1621 (1961).
93. H. Schaller, H. A. Staab, and F. Cramer, *Chem. Ber.* **94,** 1621–1633 (1961).
94. F. Cramer and H. Neunhoeffer, *Chem. Ber.* **95,** 1664–1669 (1962).
95. D. E. Hoard and D. G. Ott, *J. Am. Chem. Soc.* **87,** 1785–1788 (1965).
96. K. H. Scheit, *Chem. Ber.* **101,** 2998–3001 (1968).
97. K. H. Scheit and W. Kampe, *Chem. Ber.* **98,** 1045–1048 (1965).
98. W. Kampe, *Chem. Ber.* **98,** 1038–1044 (1965).

SEVEN

Reactive Derivatives of Nucleotides

The recently published Volume 46 of *Methods in Enzymology*, "Affinity Labeling," edited by W. B. Jacoby and M. Wilchek (1), deals with several aspects of the application of "active-site-directed" reagents in the modification of proteins. It covers both the general theoretical basis of the concept affinity labeling, and also the experimental approach as demonstrated for specific protein systems. Examples of chemically reactive nucleotide analogs employed as "active-site-directed" reagents are also included in this work. However, the material presented on this subject necessarily represents only a selected part of the available experimental data and it is therefore impossible to obtain sufficient information from this source about the synthesis and chemistry of reactive nucleotide analogs in general. This chapter reviews the chemical procedures employed in the preparation of reactive nucleotide derivatives, the chemical reactions which those species can undergo, and their application. In the following, the reasons that motivate the chemical synthesis of "active-site-directed" chemical reagents are briefly discussed. Many important biological processes involve the interactions of small molecules with proteins. These interactions can be classified as (1) the interaction of substrates with enzymes and their subsequent chemical transformation, (2) the interactions of cofactors with enzymes, and (3) the binding of effector molecules to enzymes. All of these small molecules share a common feature, namely the ability to specifically and reversibly interact with the biological macromolecule. The often very specific interactions raise the problem of the topology or structural organization of what is called the "active site" of an enzyme (or the ligand binding site of a protein in general). Spectroscopic methods such as fluorescence, nmr, and esr when applied to ligand-protein complexes can provide informations on this protein. Often, however, the size of the complex demands sensitivities of these techniques which cannot be attained. X-Ray structural analysis of a protein-ligand complex gives the most thorough information about the topology of the active site, but its application requires contentment of many experimental parameters. A simple and straightforward approach toward obtaining knowledge about the topology of a ligand binding site of a protein is the synthesis of a ligand analog bearing a chemically reactive function. If in the ideal case the specificity of interaction with the protein is preserved despite the chemical modification, the ligand analog will bind and react with the functional groups of amino acids located at or near to the active site. These reactive ligand analogs have been named "active-site-directed" irreversible inhibitors (2, 3). The properties of these reagents should meet the following criteria.

1. The analog must possess the structural elements of the natural ligand, ensuring a specific interaction with the protein. This provides higher local concentrations of the reagent at the active site thus favoring specific labeling.
2. The ligand analog bearing a permanently reactive function should be of medium chemical reactivity. Too high a reactivity would lead to the indiscriminative labeling of functional groups on amino acids not located at or close to the active site. The effect of the difference in local concentrations would therefore be canceled.

It is obvious that irreversible, "active-site-directed" inhibitors are also of interest from a chemotherapeutical point of view. Chemically reactive ligands, which possess alkylating or acylating functions, require the presence of nucleophilic groups of amino acids located near the active site for successful labeling. In enzymes this is indeed often the case. However, proteins that bind, but do not chemically transform ligands, need not necessarily have nucleophilic functions at the ligand binding site. In these cases "active-site-directed" reagents are required that are able to attack C—H bonds. Such species, capable of C—H insertion are carbenes and nitrenes. Reagents of this type cannot be applied as such, but have to be generated *in situ* from a stable precursor. In protein chemistry only one process of carbene or nitrene generation from a precursor is applicable, that of photolysis. Carbenes can be generated by photolysis from diazoalkanes, diazirines and α-ketodiazo compounds (4, 5). The pioneering experiments with "active-site-directed" reagents capable of generating carbenes were reported by Westheimer and coworkers (6, 7). Nitrenes are photolytically produced from alkyl and aryl azides, acyl azides, and isocyanates. In general, nitrenes are more selective than carbenes, due to their longer lifetimes compared to carbenes. The absorption maxima of arylazides allow photolysis at wavelengths above 350 nm, whereas alkyl azides with absorption maxima around 290 nm are in general not suitable as photogenerated affinity-labeling agents because of photoinactivation of the protein. The possible chemical reactions of a nitrene are outlined in Scheme 7.1.

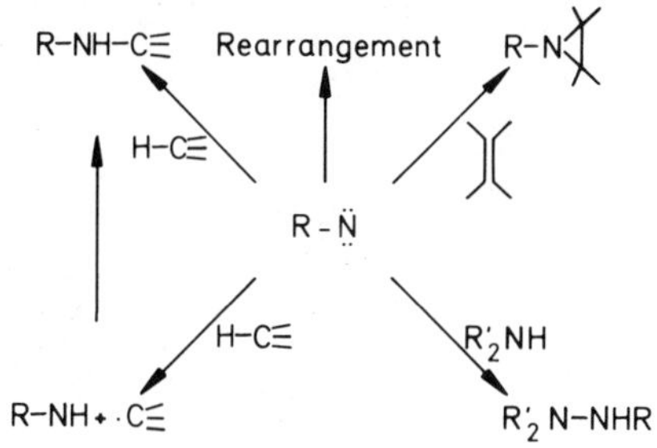

R–NH–C≡ Rearrangement R–N
H–C≡
R–N̈
H–C≡
R'_2NH
R–NH· ·C≡
R'_2N–NHR

Scheme 7.1

The first report of the use of an arylnitrene as an affinity-labeling reagent was made by Knowles (8). Reviews concerning the application of photogenerated reagents in affinity labeling of proteins were given by Knowles (8, 9).

A chemical reactive analog of a ligand with reasonable affinity for its macromolecular receptor molecules possesses an additional valuable property. The reactive group can be exploited to covalently bind the respective ligand to an insoluble matrix for the purpose of affinity chromatography. The application of chemically reactive nucleotide analogs in the preparation of affinity absorbents is therefore included in this chapter as well. Methods of preparation and the properties of insoluble matrixes suitable for reaction with reactive nucleotides are not discussed. A detailed description of this subject is given in *Methods in Enzymology*, Volume 34.

As can be seen from the foregoing discussion, nucleotide analogs carrying a chemical reactive function can be grouped into two major categories: (1) chemically reactive derivatives of nucleotides and (2) photogenerated reactive derivatives of nucleotides. Because of the structural elements of a nucleotide, the two main categories are subdivided in accordance with the structural element of the nucleotide analog bearing the functional group, for example, heterocyclic base, sugar or phosphate moiety.

7.1 REACTIVE DERIVATIVES OF NUCLEOTIDES

7.1.1 Phosphate-Modified Nucleotides

Cuatrecasas et al. (10, 11) reported the first synthesis of an alkylating nucleotide derivative to be used as an "active-site-directed" reagent for an enzyme. 3′-(4-Nitrophenylphosphoryl) 2′-deoxy thymidine 5′-phosphate (**1**) was found to be a potent competitive inhibitor of *Staphylococcal* nuclease. Catalytic hydrogenation of **1** gave 3-(4-aminophenylphosphoryl)2′-deoxy-thymidine 5′-phosphate (**2**), which could be specifically

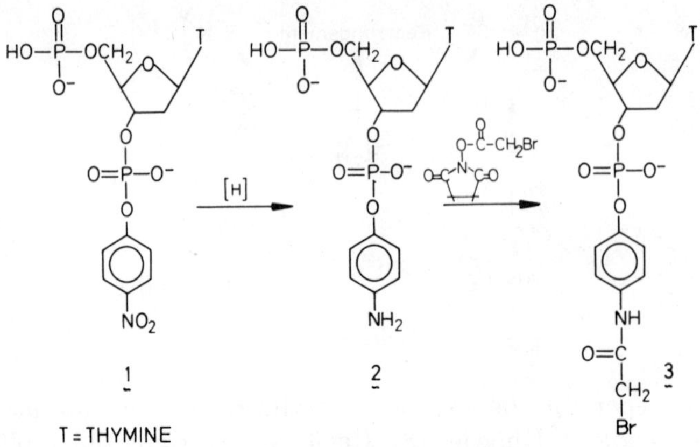

T = THYMINE

acylated by bromoacetyl-*N*-hydroxysuccinimide ester to 3′-(4-bromoacetamidophenylphosphoryl) 2′-deoxythymidine 5′-phosphate (**3**). 5′-*O*-(4-Nitrophenylphosphoryl) 2′-deoxythymidine, a poor substrate of *Staphylococcal* nuclease was converted by analogous procedures to 5′-*O*-(4-bromoacetamidophosphoryl)2′-deoxythymidine (**4**). Reaction with bromoacetyl-

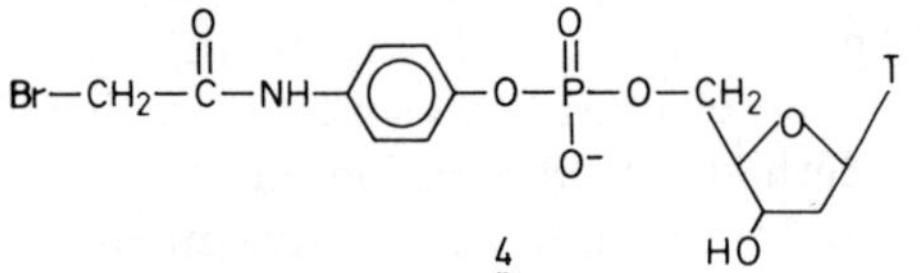

T=THYMINE

N-hydroxysuccinimide ester in partially aqueous solution occurs specifically at the 4-amino group. It is noteworthy that under the reaction conditions employed no alkylation of the thymine moiety took place. Compounds **3** and **4** irreversibly inactivated *Staphylococcal* nuclease. Inhibition was due to covalent substitution of lysine and tyrosine residues. The

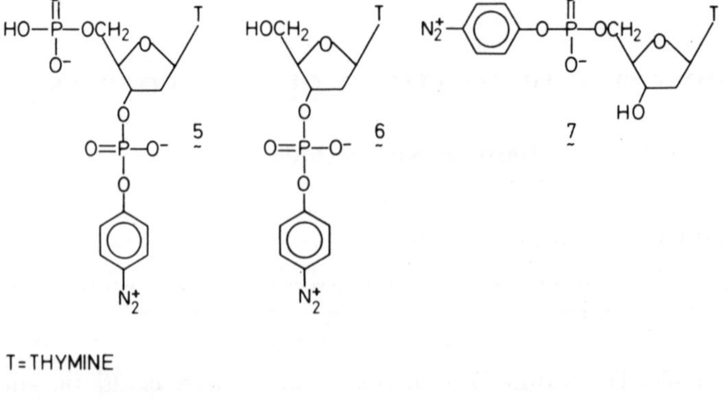

T=THYMINE

specificity was ascertained by protection of the enzyme from modification in the presence of the strong competitive inhibitor 2′-deoxythymidine 3′,5′-diphosphate.

Cuatrecasas (12) obtained the diazonium derivatives **5**, **6**, and **7** by diazotization of 3′-*O*-(4-aminophenylphosphoryl) 2′-deoxythymidine 5′-phosphate (**3**), 3′-*O*-(4-aminophenylphosphoryl) 2′-deoxythymidine, and

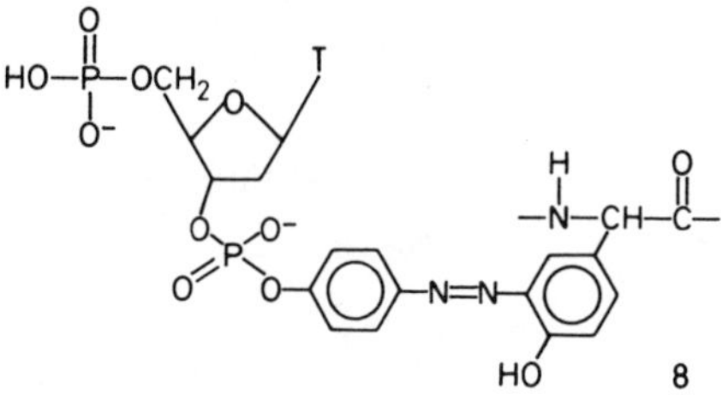

T = THYMINE

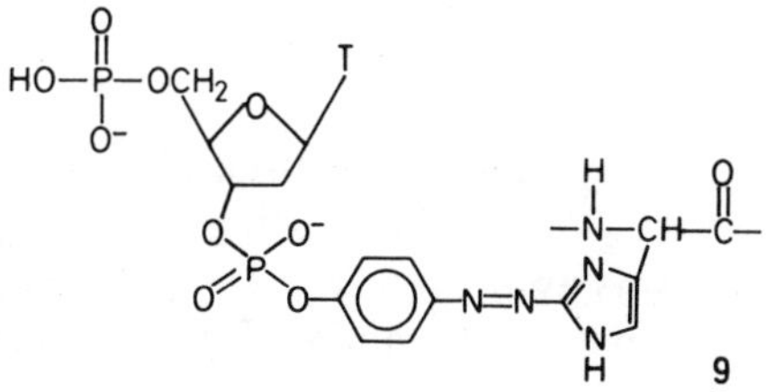

5′-*O*-(4-aminophenylphosphoryl)2′-deoxythymidine respectively. Compounds **5** and **6** were specific, irreversible inhibitors of *Staphylococcal* nuclease. Specificity was demonstrated by successful competition experiments. Chemical modification with these reagents took place on tyrosine and histidine residues. Reaction of the diazonium derivative **5** with tyrosine or histidine residues should lead to modified aminoacids of tentative structures **8** and **9**, which were assigned on the basis of comparative studies with model compounds. In the above cases chemically reactive derivatives of an effective inhibitor of the enzyme were employed in affinity-labeling experiments. Chemical modification of an inhibitor is often superior to that of a substrate, because the enzymic transformation of a reactive substrate into its corresponding product, which has a lower affinity to the enzyme, is in general much faster than the slow chemical substitution by the nucleophilic functions on the enzyme.

Glinski and Sporn (13) synthesized a series of haloacetamidophenyl phosphate esters (**10**–**12**) of 2′-deoxythymidine as potential "active-site-directed" irreversible inhibitors of an exoribonuclease from cell nuclei (14). 2′-Deoxythymidine-5′,3′-di-(3-bromoacetamidophenyl)phosphate, which belongs to type **12**, was found to be an irreversible inhibitor of RNA-dependent DNA polymerase from Rauscher Murine Leukemia virus (15).

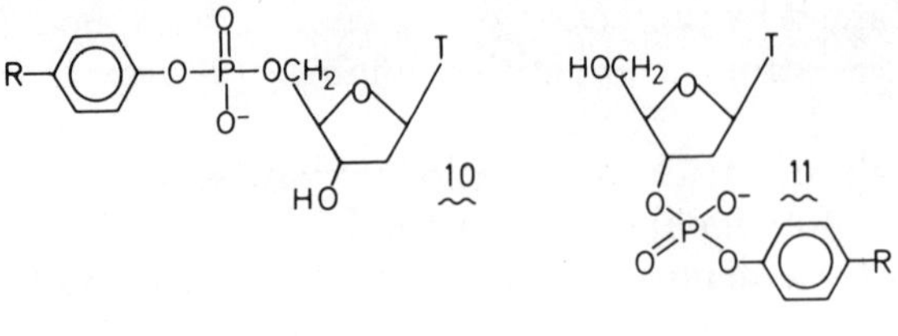

10 **11**

R=$NHCOCH_2Cl$, $NHCOCH_2Br$, $NHCOCH_2J$

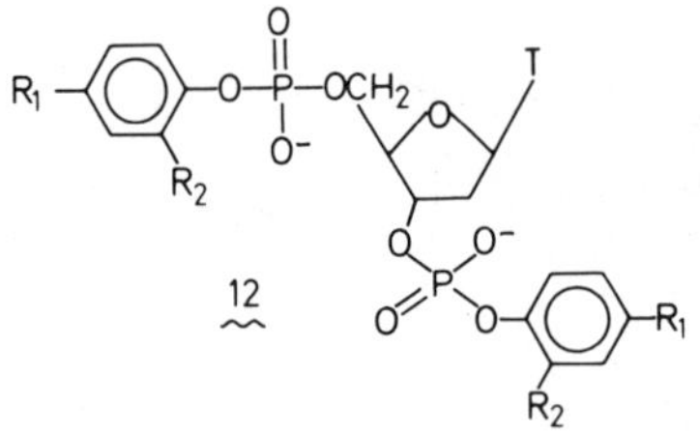

12

R_1 = $NHCOCH_2Cl$, $NHCOCH_2Br$, $NHCOCH_2J$
R_2 = H
R_1 = H
R_2 = $NHCOCH_2Br$
T = THYMINE

The reagent was competitive with template and noncompetitive with respect to substrates. It is interesting to note that an "active-site-directed" inhibitor was employed in view of its potential application as an antiviral agent. However, mammalian DNA polymerase was also inhibited by this reagent (15).

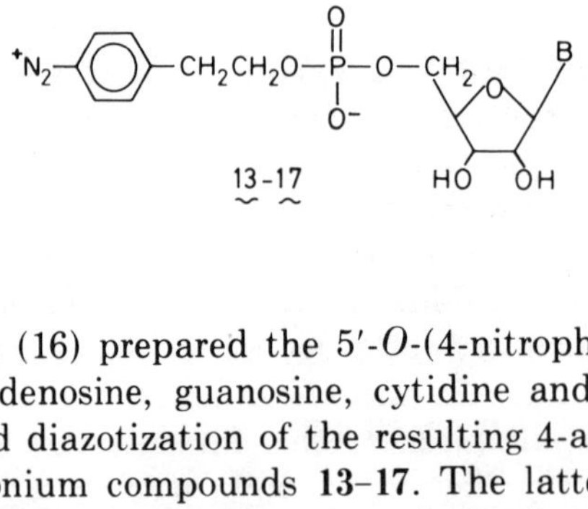

13–17

13, B = ADENINE
14, B = GUANINE
15, B = CYTOSINE
16, B = URACIL
17, B = THYMINE

Lubineau (16) prepared the 5′-*O*-(4-nitrophenylethylphosphoryl) derivatives of adenosine, guanosine, cytidine and uridine. Catalytic hydrogenation and diazotization of the resulting 4-aminophenylethyl esters led to the diazonium compounds **13–17**. The latter were used to covalently substitute bovine serum albumin and the derivatized protein was in turn employed for the induction of antibodies with specificity for the respective nucleotides. 4-Aminophenyl esters of nucleotides which have been reacted with cyanogen-bromide activated Sepharose are listed in Table 7.1.

Eckstein et al. (22) explored a synthetic route, which allows the preparation of guanosine 5′-triphosphate bearing an alkylating function at the terminal phosphate residue. *N*-Carbobenzoxyethanolamine (**18**) was phosphorylated using phosphorus oxychloride in triethylphosphate. Reaction of *N*-carbobenzoxyaminoethyl phosphate (**19**) with diphenyl phosphorochloridate gave P^1-diphenyl P^2-(*N*-carbobenzoxyaminoethyl)pyrophosphate (**20**). Displacement of diphenylphosphate by guanosine 5′-

Table 7.1 Covalent binding of nucleotide 4-aminophenylesters to Sepharose

Compound	Reference
3′-*O*-(4-Aminophenylphosphoryl)2′-deoxythymidine 5′-phosphate	17
5′-*O*-(4-Aminophenylphosphoryl)uridine 2′,(3′)-phosphate	18
Adenosine 5′-triphosphate 4-aminophenylester	19–21
2′-Deoxyadenosine 5′-triphosphate 4-aminophenylester	20, 21
Adenosine 5′-diphosphate 4-aminophenylester	21

diphosphate from **20** afforded P^3-aminoethyl P^1-5′-guanosine triphosphate (**21**). Specific acetylation of **21** was achieved with acetyl-*N*-hydroxysuccinimide ester. The latter reaction could be similarly performed with bromoacetyl-*N*-hydroxysuccinimide ester (**22**). The synthesis of P^3-(6-aminohex-1-*yl*) deoxyguanosine 5′-triphosphate, a compound similar to **21**

was reported by Hoffmann and Blakley (23). The imidazolidate of deoxyguanosine 5′-phosphate (**23**) was reacted with *N*-trifluoroacetyl-6-aminohexanol 1-pyrophosphate (**24**) to give **25** after alkaline hydrolysis of the trifluoroacetyl protecting group. This method is generally applicable to the synthesis of P^3-(ω-aminoalk-1-*yl*)nucleoside triphosphate esters. Compound **25** was coupled to Sepharose by activation of the latter with cyanogen bromide (23). Alkylating derivatives of adenosine 5′-triphosphate P^3-phosphoramidates were described by Zarytova et al. (24). Reaction of adenosine 5′-triphosphate with dicyclohexylcarbodiimide leads to formation of a trimetaphosphate intermediate **27**, which readily forms amidates with amines at the P^3-phosphate moiety. Formation of the trimetaphosphate species **27** could be monitored by ^{31}P-nmr spectroscopy (24). Hampton et al. reported the synthesis of the highly reactive carboxylic-phosphoric acid mixed anhydrides isosteric with AMP and ATP (25, 26). Adenine-9-(β-D-ribofuranosyl)uronic acid (**29**) was activated by diphenyl-phosphorochloridate and the mixed anhydride **30** allowed to react with either tri-*n*-butylammonium phosphate or tri-*n*-butylammonium tripolyphosphate to give the mixed anhydrides **31** and **32** respectively. These mixed anhydrides are extremely reactive. Compound **31** hydrolyzes at least 100 times faster than acetyl phosphate at pH 7.7 (25) and pH 6.5. Hydrolysis seems to be virtually complete in 15 sec. at pH 7.6 and room temperature. The AMP analog **31** irreversibly inactivated rabbit AMP aminohydrolase. Inactivation of the enzyme was prevented in the presence of AMP, thus indicating the specificity of the covalent substitution of the enzyme by **31**. Because of the short half-life of the latter in the reaction medium, the inactivation of AMP aminohydrolase must occur by an extremely rapid reaction (25). Reaction of **31** with adenylosuccinate AMP-lyase from *E. coli* likewise led to specific irreversible inactivation of the enzyme. Attempts to inactivate or inhibit rabbit muscle AMP kinase with **28** were not successful (25), whereas rabbit muscle pyruvate kinase was irreversibly inhibited by the ATP analog **32**. Specificity of substitution was demonstrated by effective protection of the enzyme with ATP, ADP, or phosphoenolpyruvate (25). The aminoacids, which react with the

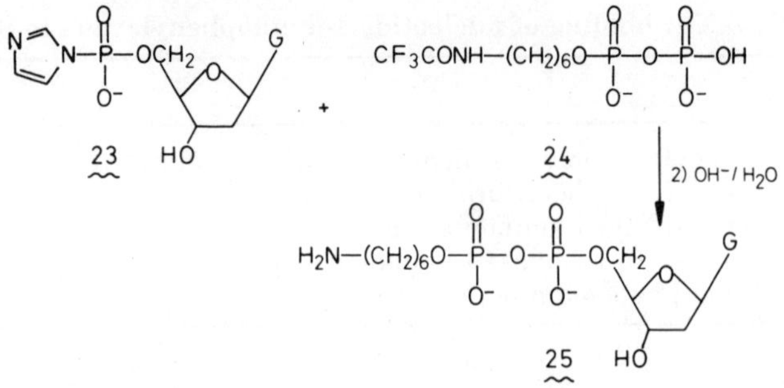

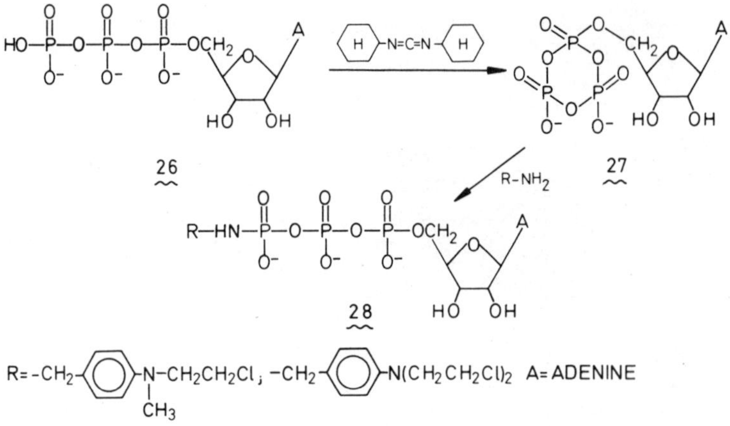

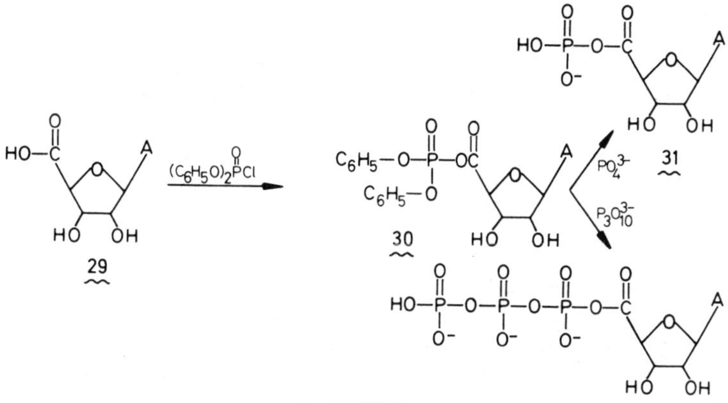

A=ADENINE

mixed anhydrides **31** and **32**, have not yet been identified. It is worthwhile noting that uracil-1-(β-D-ribofuranosyl)uronic acid and thymine-1-(β-D-ribofuranosyl)uronic acid could not be converted to mixed anhydrides analogous to **31** and **32** (25).

7.1.2 Base-Modified Nucleotides

6-Chloropurine-9-(β-D-ribofuranosyl)5′-phosphate (**33**) was the first nucleotide analog bearing a reactive functional group at the heterocyclic

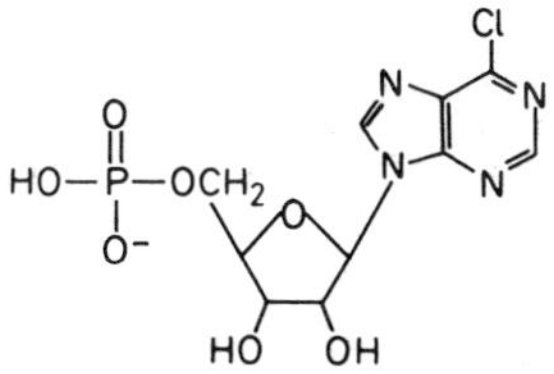

33

substituent to be employed as an "active-site-directed" irreversible inhibitor. The synthesis of **33** by conventional procedures was reported by various authors (27–29). A convenient preparation involves chlorination of 6-mercapto-purine-9-(β-D-ribofuranosyl)5′-phosphate (Chapter 2, Section 2.3; 30). Compound **33** was employed as an "active-site-directed" reagent for the affinity labeling of inosine 5′-phosphate dehydrogenase (31–33) and guanosine 5′-phosphate dehydrogenase (34) of *Aerobacter aerogenes*. It appears that **33** alkylates SH-groups of cysteine residues near the active site of these proteins. Mosbach et al. (29, 35) reacted **33** and other 6-chloropurine nucleotides with 1,ω-bisamino alkanes. The resulting N^6-(ω-aminoalkyl) adenosine nucleotides were coupled to Sepharose activated by cyanogen bromide. Eckstein et al. (28) showed that 6-chloropurine nucleotides such as 6-chloropurine-9-(β-D-ribofuranosyl)5′-triphosphate **34** can

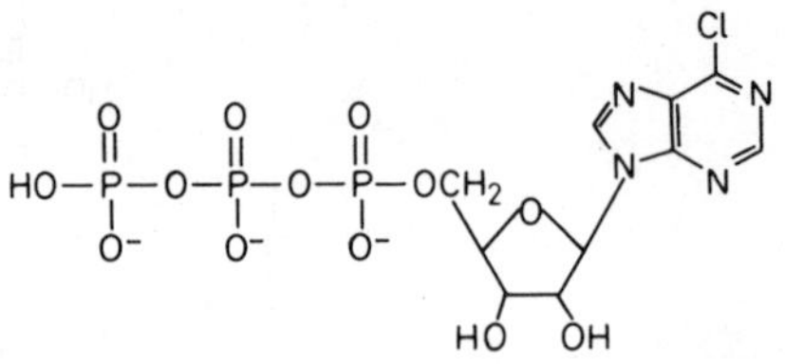

34

be bound to Sepharose directly by reaction of **34** with Sepharose substituted with aminohexyl residues, although efficiency of coupling is low. Far better results were obtained if the chlorine atom at position 6 in compound **34** was replaced by the methylsulfonyl group. 6-Methylsulfonylpurine-9-ribofuranosyl) 5′-triphosphate (**36**) is readily prepared by chlorination of 6-methylthiopurine-9-(β-D-ribofuranosyl) 5′-triphosphate (**35**) (28, 36).

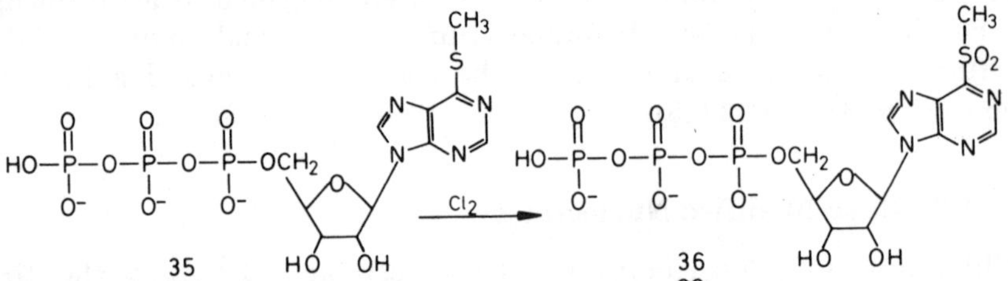

35 $\xrightarrow{Cl_2}$ 36

Compound **36** reacted efficiently at room temperature with ω-aminohexyl-Sepharose (28). This method of coupling proved to be comparable to the reaction of N^6-aminocaproyladenosine 5′-triphosphate with Sepharose activated by cyanogen bromide (26). One wonders whether 6-methylsulfonylpurine nucleotides might not function as "active-site-directed" inhibitors for adenosine nucleotide binding proteins. Lindberg and Mosbach (37) substituted ATP by a N^6-carboxylmethyl group as an alternative means for coupling ATP to an insoluble matrix. The synthesis of N^6-carboxymethyladenosine 5′-triphosphate (45) made use of the Dimroth rearrangement of N^1-substituted adenine derivatives (see Chapter 2, Section 2.1). Alkylation of ATP with iodoacetic acid at pH 6.5 yielded N^1-carboxymethyladenosine 5′-triphosphate (**39**), which in alkaline medium rearranged to **42**. Activation of **42** by carbodiimide and condensation with 1,6-diaminohexane afforded N^6-(6-aminohexyl) carbamoyl methyladenosine 5′-triphosphate (**45**). The procedure outlined above was applied to the preparation of N^6-(6-aminohexyl)carbamoylmethyl adenosine 5′-phosphate (**43**) as well as the 5′-diphosphate (**44**). Compounds **42** and **45** both served as substrates for yeast hexokinase, with the latter being the more effective (37). Compound **45** was coupled via the ω-amino group to cyanogen bromide activated Sepharose. The Sepharose bound ATP functioned as a solid phase substrate for soluble yeast hexokinase (37). The

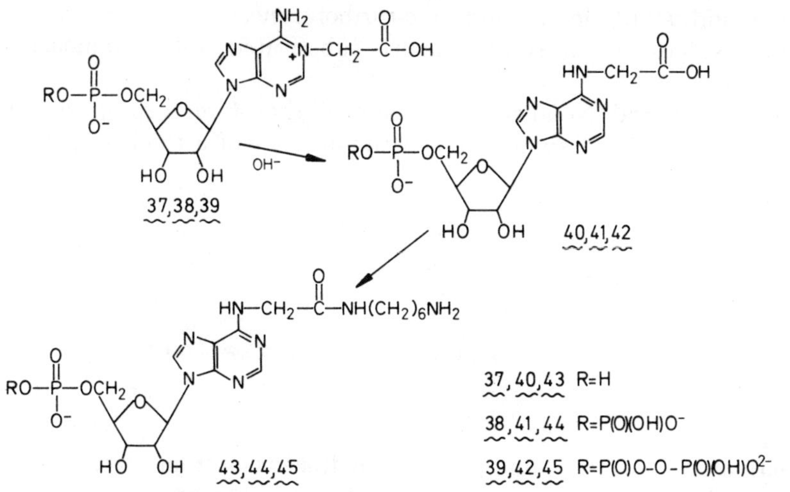

37,38,39

40,41,42

43,44,45

37,40,43 R=H

38,41,44 R=P(O)(OH)O⁻

39,42,45 R=P(O)O-O-P(O)(OH)O²⁻

adenine nucleotides **40–42** are potentially suitable for the preparation of reactive carboxylic acid derivatives which could be useful in the affinity labeling of proteins.

Direct introduction of the N^6-(6-aminohexyl)substituent into adenosine nucleotides in one step was achieved by Yamazaki et al. (38). Reaction of adenosine 5′-diphosphate (**46**) with hexamethylene diisocyanate in hexa-

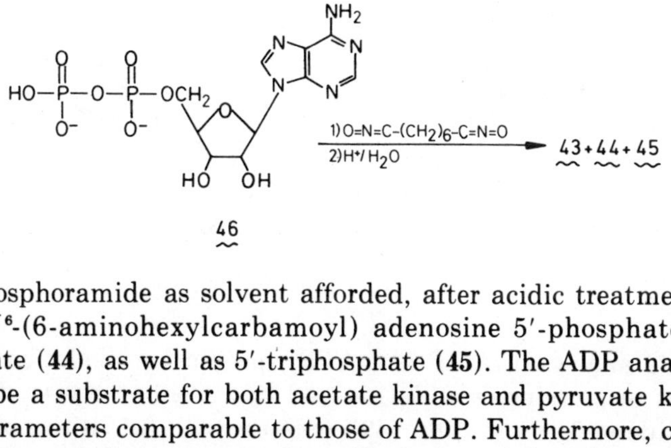

1) O=N=C-$(CH_2)_6$-C=N=O
2) H^+/H_2O

43 + 44 + 45

46

methylphosphoramide as solvent afforded, after acidic treatment, a mixture of N^6-(6-aminohexylcarbamoyl) adenosine 5′-phosphate (**43**), 5′-diphosphate (**44**), as well as 5′-triphosphate (**45**). The ADP analog **44** was found to be a substrate for both acetate kinase and pyruvate kinase with kinetic parameters comparable to those of ADP. Furthermore, compounds **43–45** were coupled to cyanogen-bromide activated Sepharose and the derivatized Sepharose employed in affinity chromatography of various kinases and dehydrogenases (38). The experimental results of Yamazaki et al. confirmed those already published by Lindberg and Mosbach (37).

The susceptibility of the mercapto group in 6-mercaptopurine toward specific alkylation was exploited by Barry and O'Carra (39) for the covalent attachment of analogs of AMP, ATP, and NAD. The procedure involved alkylation of 6-thioinosine 5′-phosphate, 6-thioinosine 5′-triphosphate and *P*[1]-(3-nicotinamide-1-β-D-ribofuranosyl) 5′-*P*[2]-thioinosine) 5′-pyrophosphate by a solid support substituted with bromoacetamido moieties.

Hampton and Nomura (33) observed that 6-mercaptopurine-9-(β-D-ribofuranosyl) 5′-phosphate (**47**) in the absence of a thiol reagent rapidly

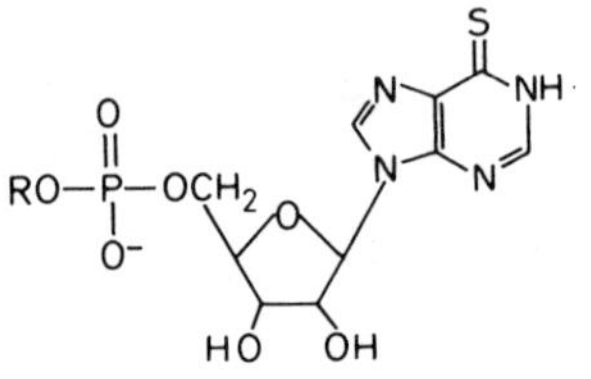

47, R = H

48, R = P(O)(OH)O⁻

47, **48**

inactivated inosine 5′-phosphate dehydrogenase from *Aerobacter aerogenes*. The rate of inactivation was reduced in the presence of inosine 5′-phosphate, thus indicating substitution of the enzyme at the active site. It is suggested that inactivation resulted from disulfide formation between **47** and a cysteine amino acid residue near the active site. A similar behavior was observed with 6-mercapto-purine-9-(β-D-ribofuranosyl) 5′-diphosphate (**48**) and polynucleotide phosphorylase from *M. luteus* (40).

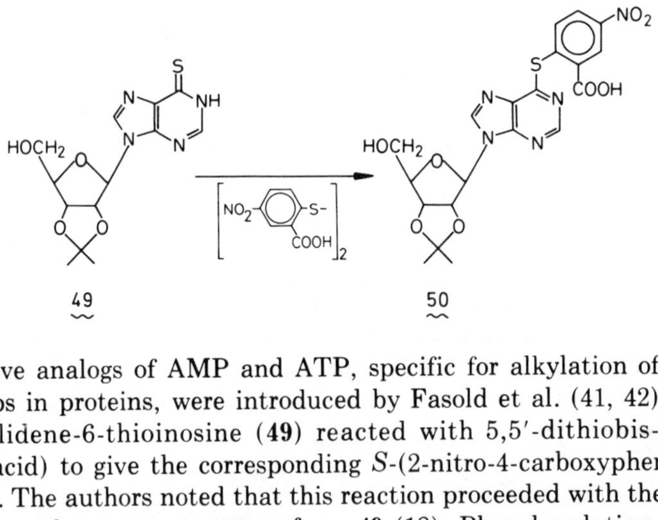

49 **50**

Reactive analogs of AMP and ATP, specific for alkylation of cysteine SH-groups in proteins, were introduced by Fasold et al. (41, 42). 2′,3′-*O*-Isopropylidene-6-thioinosine (**49**) reacted with 5,5′-dithiobis-(2-nitrobenzoic acid) to give the corresponding *S*-(2-nitro-4-carboxyphenyl) derivative **50**. The authors noted that this reaction proceeded with the unusual elimination of a mercapto group from **49** (18). Phosphorylation by phosphorus oxychloride and removal of the isopropylidene group yielded *S*-(2-nitro-4-carboxy-phenyl)-6-mercaptopurine-9-(β-D-ribofuranosyl) 5′-phosphate (**51**). Compound **51** was activated by diphenyl phosphorochloridate for conversion to the 5′-triphosphate (**52**) (43). Starting with dinitrophenylation of **49**, *S*-2,4-dinitrophenyl-6-mercaptopurine-9-(β-D-ribo-

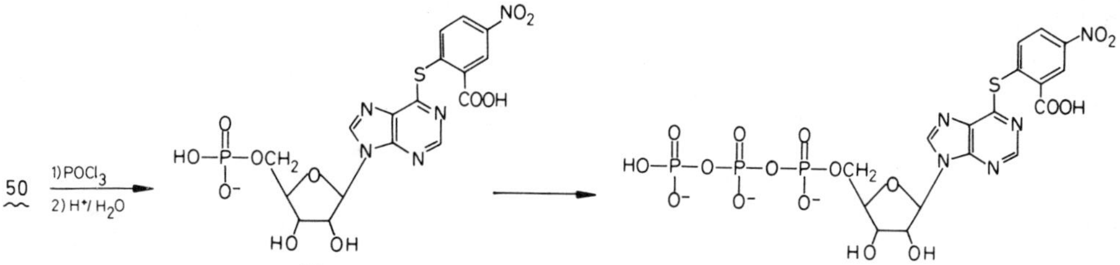

53 R = H

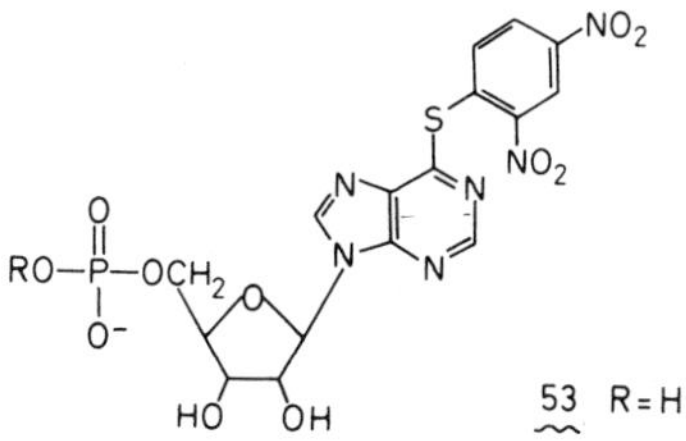

54 R = P(O)O-O-P(O)(OH)O^{2-}

furanosyl) 5′-phosphate (**53**) and 5′-triphosphate (**54**) were prepared by conventional procedures of phosphorylation (42). In nucleophilic reactions with aliphatic SH-groups, the 2-nitro-4-thiobenzoate behaves as a good leaving group. Compound **51** was employed as an "active-site-directed" reagent for phosphorylase b (41). Covalent substitution of the protein by the effector analog **51** led to fractional activation, which could not be increased because of unspecific alkylations of surface SH-groups on prolonged reaction times (41). The ATP analog **52** was employed in affinity-labeling studies with rabbit muscle actin (42). Experiments with **52** as an "active-site-directed" reagent for phosphoribosyl-pyrophosphate-ATPase revealed that the 4-dinitrophenyl group was transfered to nucleophilic functions on the protein (44).

N^6-2- and 4-Fluorobenzoyl derivatives were prepared by Hampton and Slotin (45) as potential active-site labeling reagents. Fluorobenzoyl residues were chosen as the reactive substituents because of the relative ease with which fluorine undergoes nucleophilic displacement from aromatic rings and the smallness of the fluorine atom. Moreover, as indicated by the enhanced alkali lability of N^6-fluorobenzoyladenosine 5′-phosphate compared to N^6-benzoyladenosine 5′-phosphate attack of a nucleophilic function of a protein could occur at the amide carbonyl atom leading to substitution of the protein by a fluorobenzoyl residue. Fluorobenzoylation of adenosine 5′-phosphate (**55**) followed by controlled alkaline hydrolysis gave N^6-fluorobenzoyl derivatives **56** and **57**, which were converted to the respective 5′-diphosphates (**58**, **59**), and 5′-triphosphates (**60**, **61**) by standard procedures. ATP analog **60** was an "active-site-directed" inhibi-

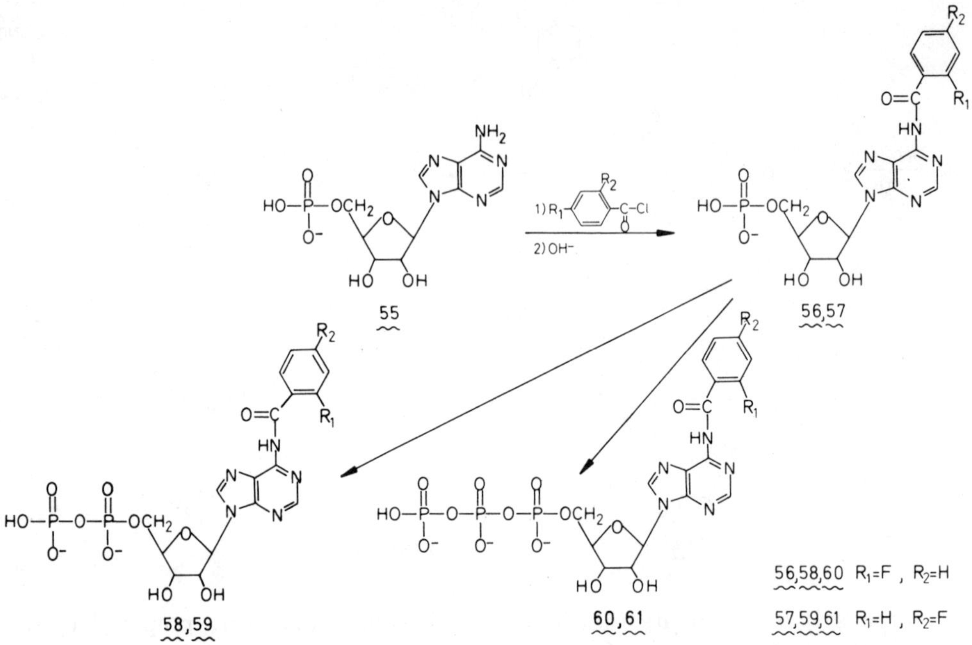

55

56,57

58,59

60,61

56,58,60 R_1=F , R_2=H

57,59,61 R_1=H , R_2=F

tor of adenylate kinases from pig, rabbit, and carp muscle (45). Similarly, analog **61** specifically inactivated rabbit and carp adenylate kinases. Both **60** and **61** were excellent substrates for yeast hexokinase (46).

To synthesize reactive uridine derivatives bearing a functional group at the base moiety, substitution at C(5) of the double bond is the method of choice for the following reasons: (1) introduction of substituents at C(5) does not seemingly effect normal conformation of the parent nucleotide, in contrast to substitution at C(6) which hinders free rotation about the gly-

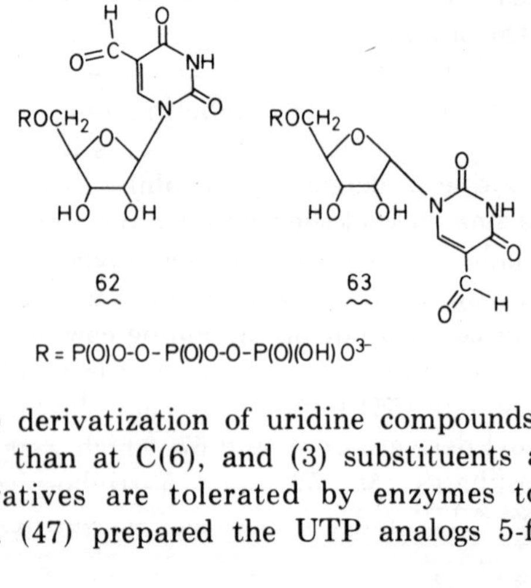

62

63

R = P(O)O-O-P(O)O-O-P(O)(OH) O^{3-}

cosidic bond, (2) derivatization of uridine compounds at C(5) is much easier to achieve than at C(6), and (3) substituents at C(5) in uridine phosphate derivatives are tolerated by enzymes to a large extent. Armstrong et al. (47) prepared the UTP analogs 5-formyl-1-(β-D-ribofuranosyl) uracil 5′-triphosphate (**62**) and 5-formyl-1-(α-D-ribofuranosyl) uracil 5′-triphosphate (**63**) as potential "active-site-directed" inhibitors of *E. coli* DNA-dependent RNA polymerase (see Chapter 2, Section 3.3 for synthesis). Whereas the UTP analog **62** serves as a substrate for RNA polymerase, the α-anomer **63** was found to be a noncompetitive inhibitor of the enzyme. Reduction of a complex between enzyme and analog **63** by sodium borohydride leads to irreversible inactivation of the enzyme, indicating that inhibition of enzymic activity by **63** involves formation of a Schiff base. Competition experiments with normal substrates established the specificity of the inactivation. The substrate **62** also inactivated the enzyme upon treatment of the respective mixture with sodium borohydride, but to a lesser extent.

7.1.3 Sugar-Modified Nucleotides

Ribonucleotides (**64**) with a free 2′,3′-cis diol group are readily oxidized by periodate to the dialdehyde **65** (48). The latter probably exists as the hydrated form (**66**, **67**) in aqueous solution (49, 50). Dialdehydes of the

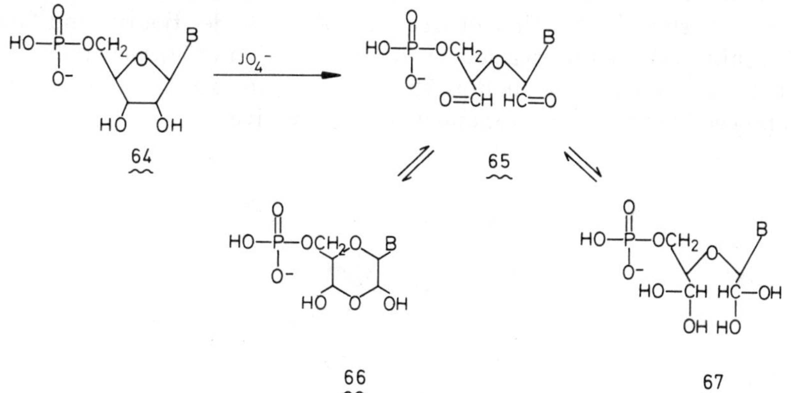

64 65 66 67

type **65** react readily with thiosemicarbazide (51) or nucleophilic primary amines (52) forming the carbinolamines **68** and **69** respectively. Stable morpholino derivatives are formed on reduction of the carbinolamines **68** and **69** with sodium borohydride in aqueous solution (52). Despite the easy access to the dialdehydes of nucleotides and the reactivity of the aldehyde

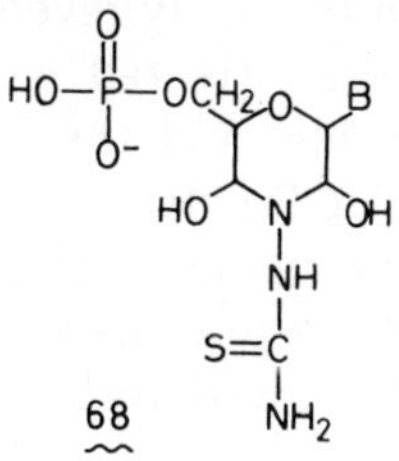

68 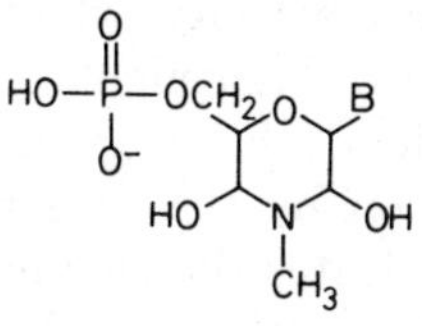 69

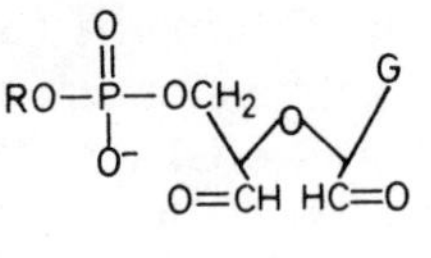

G = GUANINE

70, R = $P(O)(OH)O^-$

71, R = $P(O)O{-}O{-}P(O)(OH)O^{2-}$

70, **71**

groups, only a few attempts have been reported in which nucleotide dialdehydes have been employed as active site directed reagents. Bodley and Gordon studied the effects of periodate oxidized GDP (**70**) and GTP (**71**) on the elongation factor G-stimulated ribosomal GTPase from *E. coli* (53). 2′-*O*-(*R*)-Formyl(guanine-9-*yl*)methyl-3′-diphosphate-3′-deoxy-(*S*)-glyceraldehyde (**70**) (GDP-dialdehyde) and 2′-*O*-(*R*)-formyl(guanine-9-*yl*)-methyl-3′-deoxy-(*S*)-glyceraldehyde (**71**) (GTP-dialdehyde) bound to the ribosomal GTPase. The rate of hydrolysis of GTP-dialdehyde was approximately 0.3% that of GTP. Reduction of complexes between GDP-dialdehyde, elongation factor G and ribosomes by sodium borohydride did not lead to covalent substitution, indicating that binding of **70** did not involve formation of a Schiff base (53). However, the rate of reduction by borohydride of compound **70** within the complex was about one order of magnitude slower than that of free GDP-dialdehyde. Bodley and Gordon (53) confirmed the observation by Khym and Cohn (54) that reduction of a nucleotide dialdehyde such as **70** under certain experimental conditions can proceed via a transient monoalcohol derivative.

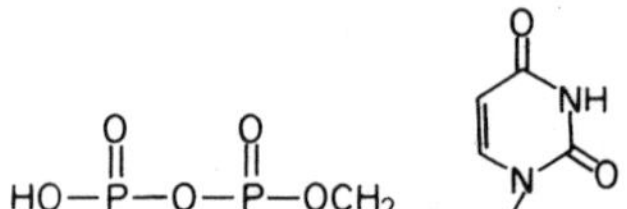

72

The dialdehyde 2′-*O*-(*R*)-formyl(uracil-1-*yl*)methyl-3′-diphosphate-3′-deoxy-(*S*)-glyceraldehyde (**72**) prepared by oxidation of UDP with periodate, has been used as an affinity-labeling reagent for the UDP-galactose/UDP binding site of galactosyltransferase from bovine colostrum (55). Incubation of the enzyme with UDP-dialdehyde (**72**) leads to inactivation of enzymatic activity, which can be slowly reversed by the addition of primary amines. Treatment of the inactivated enzyme with sodium borohydride renders inhibition irreversible, and the enzyme is protected against inactivation by the substrate UDP-galactose. These results indicate that **72** substitutes the enzyme by forming a Schiff base with a lysine amino group at the active site. Indeed the authors succeeded in isolating and characterizing a modified oligopeptide Ser-Gly-Lys-UDP (55).

The successful application of 2′-*O*-(*R*)-formyl(adenine-9-*yl*-)methyl-3′-triphosphate-3′-deoxy-(*S*)-glyceraldehyde (ATP-dialdehyde) as an "active-site-directed" reagent for pyruvate carboxylase was demonstrated by Easterbrook-Smith et al. (56).

Phenol sulfotransferase catalyzes the transfer of a sulfate residue from 3′-*O*-phosphoryladenosine 5′-phosphosulfate to a phenolic acceptor substrate. Adenosine 3′,5′-diphosphate and ADP are competitive inhibitors of this reaction. The periodate oxidation product of the latter, 2′-*O*-(*R*)-formyl(adenine-9-*yl*) methyl-3′-diphosphate-3′-deoxy-(*S*)-glyceraldehyde (ADP-dialdehyde, **73**) has been used by Borchardt et al. (57) as an "active-site-directed" inhibitor for phenol sulfotransferase. This enzyme

$\xrightarrow{NaBH_4}$

73 74

A = ADENINE

was irreversibly inhibited by ADP-dialdehyde (57). The irreversible nature of inactivation became evident as attempts to recover enzymic activity by extensive dialysis failed. The specificity of inactivation has been established by competition experiments with the inhibitor adenosine 3′,5′-diphosphate. It is worthwhile mentioning that the ADP analog **74**, prepared by reduction of ADP-dialdehyde with sodium borohydride was only a weak reversible inhibitor of phenol sulfotransferase.

The reaction of ribonucleotide dialdehydes with the amino groups of proteins followed by subsequent reduction of the Schiff base with borohydride is a general procedure for obtaining protein-nucleotide conjugates. The latter have been successfully used to elicit antibodies specific for the respective nucleotide of the conjugate (58).

A special group of affinity labeling reagents are the so-called k_{cat}-reagents (59). These reagents require an enzyme catalyzed activation leading to the *in situ* generation of the reactive species. Subsequent reaction of the latter with a functional group at the enzyme active site results in either covalent substitution or modification. It is obvious that k_{cat}-reagents represent extremely specific affinity-labeling reagents. 2′-Deoxy-2′-chloronucleoside 5′-diphosphates (**75–77**) and 2′-deoxy-2′-azidonucleoside 5′-diphosphates (**78, 79**) were discovered to be k_{cat}-reagents for ribonucleoside diphosphate reductase from *E. coli* (60). The latter enzyme consists of two protein subunits B1 and B2. Subunit B1 contains dithiols, which are involved in oxidation-reduction processes of electron transport. Subunit B2 possesses a free radical essential for activity, and the

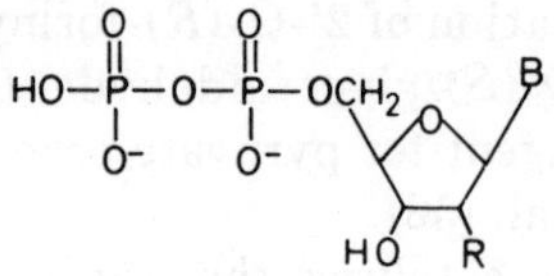

75–79

	B	R
75	Uracil	Cl
76	Cytosine	Cl
77	Adenine	Cl
78	Uracil	N_3
79	Cytosine	N_3

substrates are bound to subunit B1. Interaction of compounds **75–77** with the enzyme causes irreversible inactivation of B1, whereas subunit B2 is unaffected. The specificity of the inactivation is indicated by the following: inactivation requires the presence of subunit B2; it is prevented by the presence of natural substrates; 2′-deoxy-2′-chloronucleoside 5′-monophosphates are ineffective. The reagents **75–77** decompose during the inactivation process into free base, 2-deoxyribose 5′-phosphate, and chloride ion. The inactivation is caused by modification of the dithiols participating in oxidation-reduction, although the chemical nature of this modification is not yet known. The 2′-deoxy-2′-azidonucleoside 5′-diphosphates (**78**, **79**) specifically and irreversibly inactivate subunit B2 without affecting B1. Experimental evidence suggests that the analogs **78**, **79** while being reduced capture and destroy the free radical in B2 by means of the 2′-azido group.

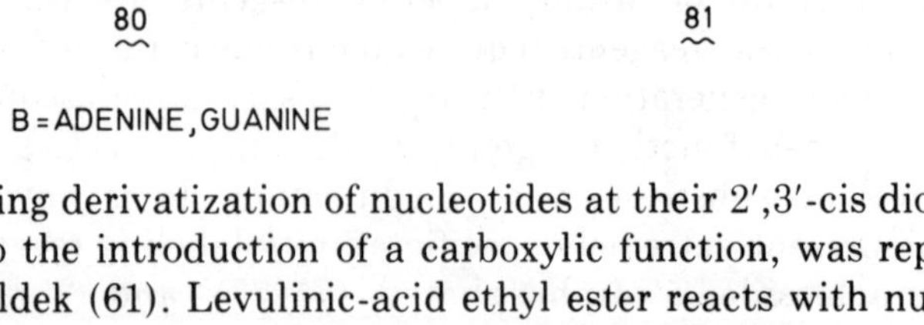

80 → 1) $POCl_3$ 2) OH^- → **81**

B = ADENINE, GUANINE

An interesting derivatization of nucleotides at their 2′,3′-cis diol groups, which leads to the introduction of a carboxylic function, was reported by Seela and Waldek (61). Levulinic-acid ethyl ester reacts with nucleosides under acid catalysis to form the ketal **80**. Phosphorylation, followed by alkaline hydrolysis affords the nucleotide derivatives **81** with a free carboxylic function. Activation of the carboxyl groups in levulinate-$O^{2',3'}$-nucleotide ketals (**81**) with carbodiimides allows specific reactions with nucleophilic groups, for example, amino groups. The above procedure has been employed for the covalent substitution of aminohexyl-Sepharose with AMP or GMP residues respectively (60).

7.2 PHOTOGENERATED REACTIVE DERIVATIVES OF NUCLEOTIDES

7.2.1 Phosphate-Modified Nucleotides

1-(4-Azidophenyl)-2-(5′-guanylyl)pyrophosphate (**83**) has been synthesized as a photoreactive GTP analog suitable for use as an "active-site-directed" reagent with the elongation factor G-dependent GTP binding site on

ribosomes (62). Preparation of the GDP analog **83** was achieved by a standard procedure which involved activation of the monoester phosphate **82** and subsequent reaction with GMP. The GDP analog **83** inhibited the binding of GDP to a complex between elongation factor G and the 50 S ribosomal subunit. Irradiation of the latter complex in the presence of **83** at

wavelengths above 320 nm led to specific covalent substitution of various 50 S-ribosomal proteins supposed to be involved in GDP binding (62).

For the same purpose Ovchinnikov et al. (63) synthesized guanosine 5′-3-(4-azidobenzylamido) triphosphate (**85**). The method employed for the synthesis of **85** is that of Babkina et al. (64). The GTP analog **85** was specifically bound to the binary complex elongation factor G-ribosome as judged from competition with the natural ligand GTP (63, 65). Irradiation of the ternary complex elongation factor G-ribosome GTP analog led to exclusive substitution of the elongation factor G. Modification of ribosomal proteins has been stated to be negligible. Covalent substitution of elongation factor G by the GTP analog **85** can be effectively suppressed by the presence of GTP.

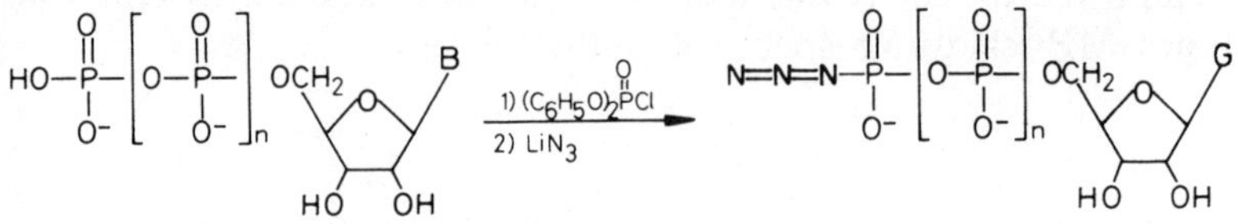

HO—P(=O)(O⁻)—[O—P(=O)(O⁻)]$_n$—OCH$_2$ (ribose, B; HO OH) —1) $(C_6H_5O)_2P(O)Cl$ 2) LiN_3→ N≡N—N—P(=O)(O⁻)—[O—P(=O)(O⁻)]$_n$—OCH$_2$ (ribose, G; HO OH)

86-93 → **94-101**

	n	B		n	B
86, 94	0	Guanine	**90, 98**	1	Adenine
87, 95	1	Guanine	**91, 99**	2	Adenine
88, 96	2	Guanine	**92, 100**	3	Adenine
89, 97	0	Adenine	**93, 101**	4	Adenine

Chladek et al., (66) prepared various nucleoside 5′-phosphorazidates (**94–101**) by the reaction of suitably activated nucleotide precursors (**86–93**) with lithium azide. Although the reaction mixtures have been found to be rather complex because of many side reactions, the phosphoazidates represented the main products of reaction. Photolysis of guanosine 5′-phosphorazidate (**94**) and adenosine 5′-phosphorazidate (**97**) at wavelengths above 280 nm led to 5′-phosphoramidates as the major products. Phosphorazidates are stable close to neutral pH and in 80% aqueous acetic acid, but decompose in strong alkali. Guanosine 5′-*O*-(2-azidodiphosphate) (**95**) competes with GDP in binding to the binary complex elongation factor G-ribosome. Guanosine 5′-*O*-(3-azidotriphosphate) (**96**) inhibits the elongation factor G-stimulated GTPase. Both phosphorazidates should prove useful as photoreactive affinity-labeling reagents in the system described above.

7.2.2 Base-Modified Nucleotides

Muneyama et al. (67) were the first to report the facile displacement of the bromo substituent in 8-bromoadenosine 3′,5′-cyclic phosphate by azide ion. The expectation that the 8-azido group in 8-azidoadenine nucleotides

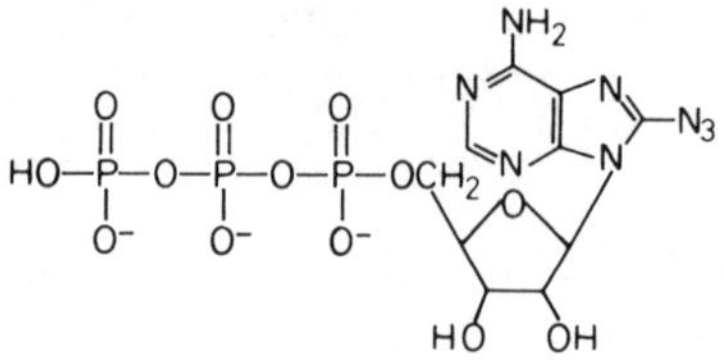

HO—P(=O)(O⁻)—O—P(=O)(O⁻)—O—P(=O)(O⁻)—OCH$_2$ (ribose; HO OH; 8-azidoadenine with NH_2, N_3)

102

might undergo photolysis to a nitrene intermediate stimulated Haley and Hoffman to synthesize 8-azidoadenosine 5′-triphosphate (**102**) as a potential photoreactive affinity-labeling reagent (68). 8-Azidoadenosine 5′-phosphate was prepared from 8-bromoadenosine 5′-phosphate and converted to the 5′-triphosphate by a standard procedure (65) (see also Chapter 2, Section 3.2). The ATP analog **102** serves as substrate for both the sodium potassium ATPase and the Mg^{2+}-ATPase from human red cells (68). Irradiation of red cell membranes in the presence of **102** led to irreversible inactivation of the two types of ATPases. Inactivation is caused by covalent substitution of membrane proteins and is prevented in the presence of sufficiently high concentrations of ATP.

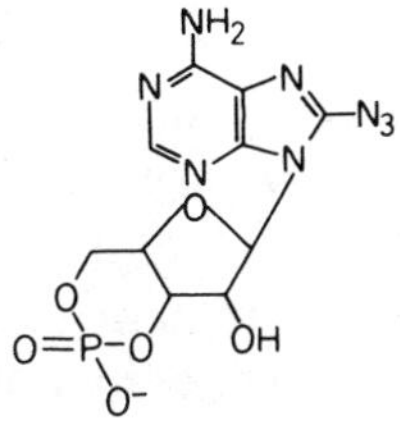

NH2 N N N3 N N O O O=P-O OH O-

103

8-Azidoadenosine 3′,5′-cyclic phosphate (**103**) (64) has been employed by Pomerantz et al. (69) as a photoaffinity label for cAMP-regulated protein kinases from bovine brain. As already shown by Muneyama et al. (67), the analog **103** can substitute for cAMP in stimulation of protein kinases. Irradiation of protein kinase complexed with **103** led to rapid irreversible inactivation, which could be nullified by an excess of cAMP (69). The authors discuss the phenomenon that an excess of a nonreactive ligand with a higher affinity for the protein than the reagent **103** is required to protect the protein. As an explanation a so-called "trapping effect" for the photoreactive agent by the protein active site during photolysis is invoked. The reagent molecules, which are exchanging with the nonreactive ligands during the irradiation period, can be covalently incorporated and this is reflected in a higher apparent affinity for the photoreactive ligand. Thus the concentration of a nonreactive ligand necessary to protect the protein from modification is controlled both thermodynamically and kinetically (69).

The catalytic activity of glutamate dehydrogenase is regulated by purine nucleotides. ADP can be either an activator or inhibitor of this enzyme depending on the experimental conditions. It was therefore tempting to employ the ADP analog 8-azidoadenosine 5′-diphosphate (**104**) as a photoaffinity label. Despite numerous side reactions, compound **104** was obtained in 25% yield by direct reaction of 8-bromoadenosine 5′-diphosphate with azide ions (70). In analogy to ADP, one molecule of **104** has been found to bind per subunit of enzyme. However, stimulation of enzymatic activity was less for the analog than for ADP under saturating conditions. For the first time, photolysis of an 8-azidoadenine nucleotide itself has been studied spectroscopically. Experimental evidence indicates that photolysis of **104** in aqueous solution yields 8-hydroxyaminoadenosine 5′-

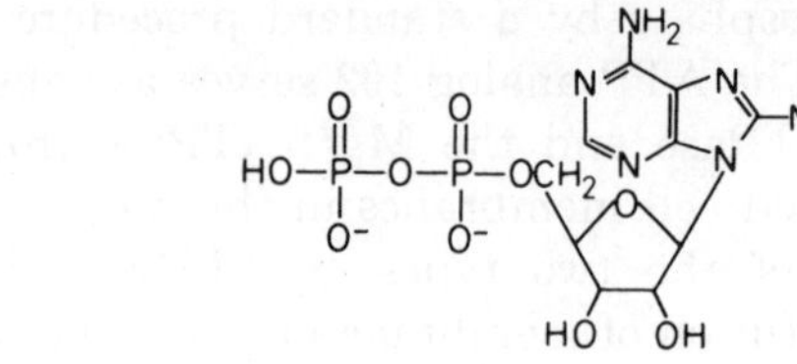

104

diphosphate (**105**).Irradiation of the enzyme in the presence of **104** at the absorption maximum of the latter (282 nm) rapidly inactivated enzymic activity. However, Koberstein et al. (70) noticed that the absorption of compound **104** above 300 nm is sufficiently high to be used in photolysis experiments. Photolysis of glutamate dehydrogenase in the presence of **104**

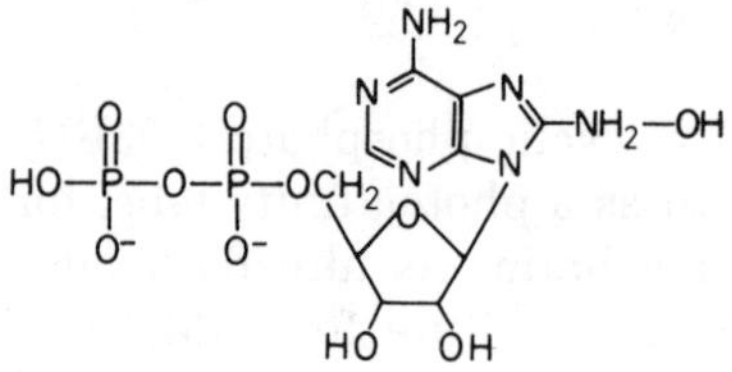

105

performed under these conditions led to covalent substitution of the enzyme. Competition by the natural ligand ADP markedly reduces covalent modification of the protein (70).

Schäfer and Penades (71) utilized 8-azidoadenosine 5′-diphosphate as a photoreactive reagent for the covalent modification of the nucleotide carrier from beef heart mitochondria.

The 8-azidoadenine analogs of NAD^+ and FAD have been synthesized by condensation of nicotinamide or flavine mononucleotide respectively with 8-azidoadenosine 5′-phosphate employing standard phosphorylation procedures (72). Nicotinamide-8-azidoadenine dinucleotide (**106**) has been found to substitute NAD^+ as coenzyme in the case of lactate, glutamate, and alcohol dehydrogenase. Flavin-8-azidoadenine dinucleotide (**107**) replaces FAD as coenzyme for glucose oxidase, although the analog **107** proved to be less effective as an initiator of the reactivation process of the enzyme when compared to FAD. With D-amino acid oxidase, **107** is completely inactive as coenzyme. Absorption difference spectroscopy, however, indicated the formation of a specific complex between the analog **107** and D-amino acid oxidase. Compounds **106** and **107** were studied by absorption spectroscopy, CD and ^1H-nmr spectroscopy. According to the results obtained by the two latter techniques, both analogs appear to prefer a folded conformation in solution. They undergo photolysis upon irradiation as demonstrated by ultraviolet absorption spectroscopy. Thus

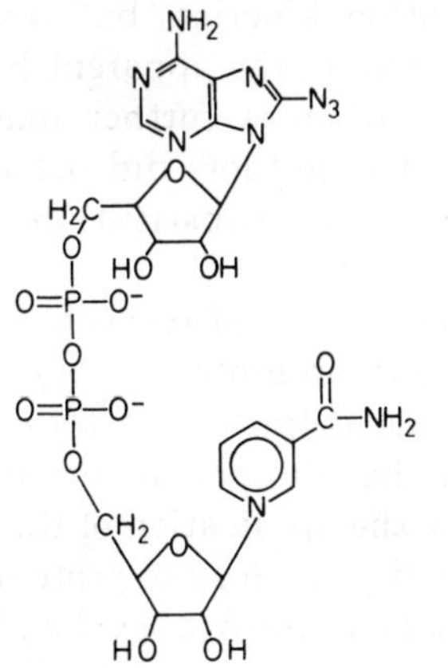

106

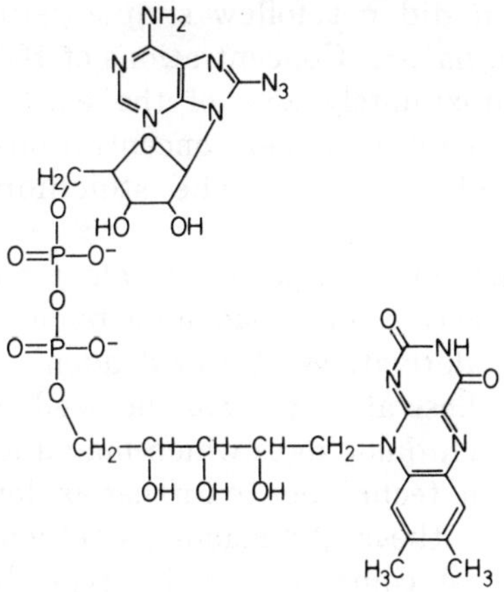

107

the coenzyme analogs **106** and **107** seem to be promising candidates for photoaffinity-labeling studies with NAD^+ or FAD requiring enzymes.

Activation of 8-azidoadenosine 5′-phosphate by diphenyl phosphorochloridate, followed by condensation with *S*-benzoyl-4′-phosphopantetheine yielded *S*-benzoyl (3′-dephospho-8-azido) coenzyme A (**108**) (73). In comparison to benzoyl coenzyme A, analog **108** turned out to be a good substrate for acyl coenzyme A—glycine *N*-acyltransferase from beef liver (73). Photolysis of **108** was demonstrated by measurement of ultraviolet absorption spectra after exposure to light for different times. It is somewhat intriguing that the set of presented absorption spectra does not display an isosbestic point in contrast to similar spectroscopic studies of the photolysis of 8-azidoadenine nucleotides (70, 72). Irradiation of a mixture of compound **108** with the *N*-acyltransferase leads to irreversible inhibition of the enzyme. Experiments with radioactive **108** revealed covalent attachment of the reagent to the enzyme. The specificity of substitution has been demonstrated by effective protection of the enzyme with the substrate benzoyl coenzyme A. The concentration dependence of

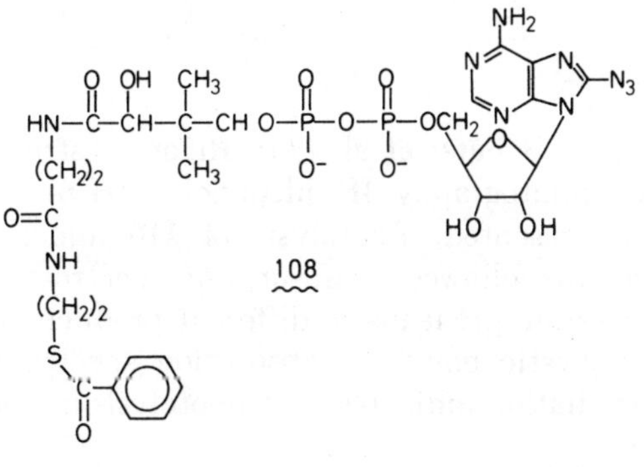

108

inactivation did not follow simple saturation kinetics, but was clearly biphasic in nature. Concentrations of **108** close to the apparent K_m inactivated approximately 25% of the enzyme, whereas further inactivation seemingly requires higher concentrations. The authors did not comment on these observations. The stoichiometry of substitution was not determined.

The efficient and specific covalent modification of respective proteins by photoreactive 8-azidoadenine nucleotides prompted the synthesis of fluorescent derivatives of 8-azidoadenine nucleotides. The rationale that motivated these attempts was the need for the site specific introduction of fluorescent chromophores which may allow the application of fluorescence spectroscopic techniques to further explore the topology of protein binding sites. The synthesis of 8-azido-1,N^6-ethenoadenosine 3′,5′-cyclic phosphate (**109**) the first compound of this type, has been reported by Keeler and

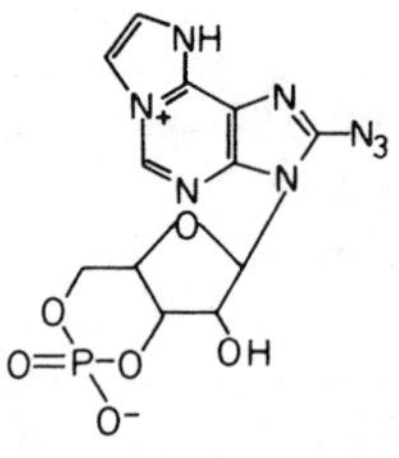

109

Campbell (74). The last step of the preparation involved the quantitative reaction of 8-azidoadenosine 3′,5′-cyclic phosphate with chloroacetaldehyde. Compound **109** has been characterized by absorption and ^{1}H-nmr spectroscopy. Photolysis of **109** leaves the fluorescent chromophore intact.

The preparation of 8-azido-1,N^6-ethenoadenosine 5′-triphosphate (**110**)

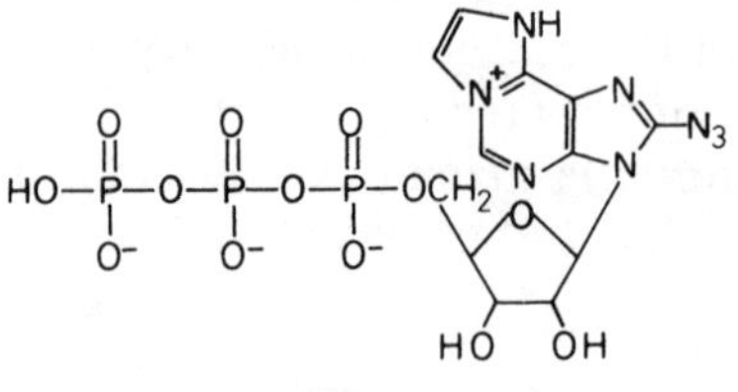

110

has been reported by Schäfer et al. (75). Rigorous structural proof of the ATP analog by chromatography, IR, ultraviolet, ^{1}H-nmr, and fluorescence spectroscopy was presented. Photolysis of **110** under different experimental conditions was followed by absorption spectroscopy. Irradiation at either acidic or alkaline pH leads to different products of photolysis. The observation of isosbestic points in absorption spectra measured at different times of irradiation indicates that photolysis in both cases proceeds via a single set of reactions. The fluorescence emission of **110** upon photolysis increases more than tenfold indicating that the fluorescence quantum yields of **110** and 1,N^6-ethenoadenosine-5′-triphosphate may differ by the same factor.

In attempts to utilize 8-azidoadenine nucleotides as photoreactive reagents for the active-site labeling of polynucleotide phosphorylase and DNA-dependent RNA polymerase, Cartwright et al. (76) found that the above class of compounds reacted in the dark with thiols necessary for the full activity of these enzymes. The reduction of azides by thiols was known to proceed under rather vigorous conditions (77). 8-Azidoadenosine as well as the corresponding 5′-mono- and diphosphate have been found to react rapidly in the dark with a number of dithiols such as dithiothreitol, dithioerythritol, and 1,3-propanedithiol with liberation of nitrogen and formation of the respective 8-aminoadenine derivatives. The second-order rate constants for the reaction of the above 8-azidoadenine nucleotides with dithiothreitol at 25°C and pH 8.9 were determined to be 34.38 (mol^{-1}sec^{-1}), 7.41 (mol^{-1}sec^{-1}) and 6.15 (mol^{-1}sec^{-1}) respectively. The reaction is base catalyzed with an optimum at pH 10. At pH 8.9 and 25°C a 20-fold excess of 2-mercaptoethanol did not react detectably with 8-azidoadenosine 5′-phosphate. However, photolysis of 8-azidoadenosine 5′-triphosphate at pH 8 in the presence of 2.5 mM 2-mercaptoethanol led to formation of the 8-aminoadenine derivative. The authors proposed two alternative mechanisms as outlined in Schemes 7.2 and 7.3 for the reaction of 8-azidoadenine derivatives with dithiols in the dark. In connection

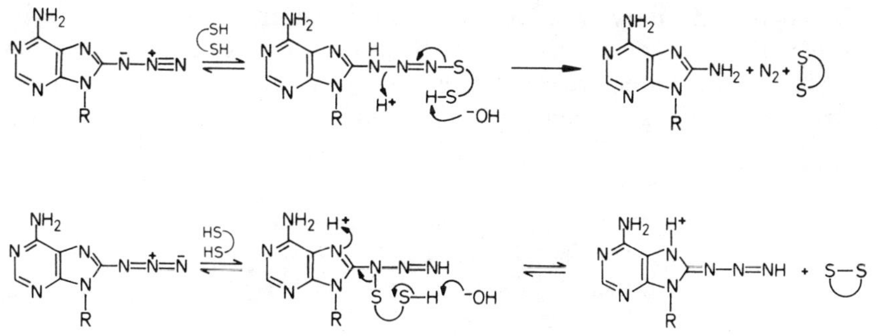

Scheme 7.2

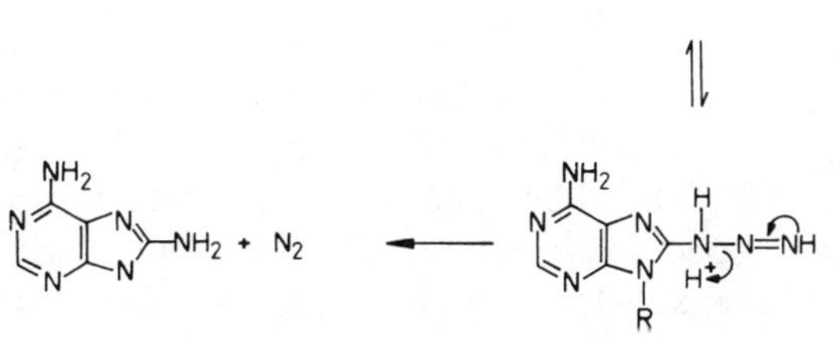

Scheme 7.3

with the findings of Cartwright et al. (76) it appears questionable as to whether 8-azidoadenine nucleotides can be employed as photoaffinity-labeling reagents for those proteins that require thiols as protection against oxidation.

Inactivation of DNA polymerases I and II by irradiation with ultraviolet light follows first-order kinetics (78). Enhanced inactivation was observed in the presence of 5-iodo-2′-deoxyuridine 5′-triphosphate (**111**). Photoinactivation as a function of the concentration of **111** showed

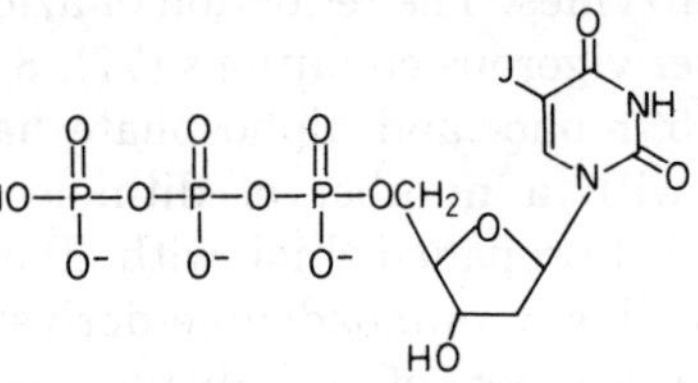

111

saturation kinetics, whereas photoreaction in the presence of 5-iodo-2′-deoxyuridine 5′-phosphate did not. From these observations specific inactivation by **111** was concluded, although the most direct proof for specificity was negative. The natural substrates, 2′-deoxynucleoside 5′-triphosphates, did not protect both enzymes from inactivation. It is assumed that irradiation of **111** produces a radical which inactivates the enzyme.

Photooxidation of 4-thioketopyrimidine nucleotides (**112**) leads to highly electrophilic intermediates (**113**) which readily react with nucleophiles such as primary or secondary amines and ammonia to give derivatives of cytidine (**114**) (see Chapter 2, Section 2.3). The intermediates have not as yet been characterized, but by analogy to the behavior of 6-thioketopurine derivatives, it can be safely concluded that they are identical

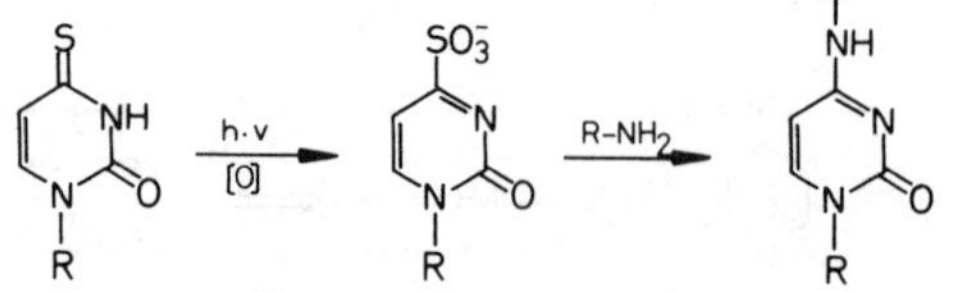

112 113 114

with the respective 4-sulfonates (**113**). The normal oxygen content of solvents or buffers is sufficiently high to allow formation of the intermediates **113**. The above photooxidation reaction was employed (81) in an attempt to specifically modify DNA-dependent RNA polymerase from *E. coli* with the substrate analogy 4-thiouridine 5′-triphosphate (**115**) (79, 80). Irradiation of RNA polymerase at 330–360 nm in the presence of **115** led to irreversible loss of enzymic activity (81). To demonstrate the speci-

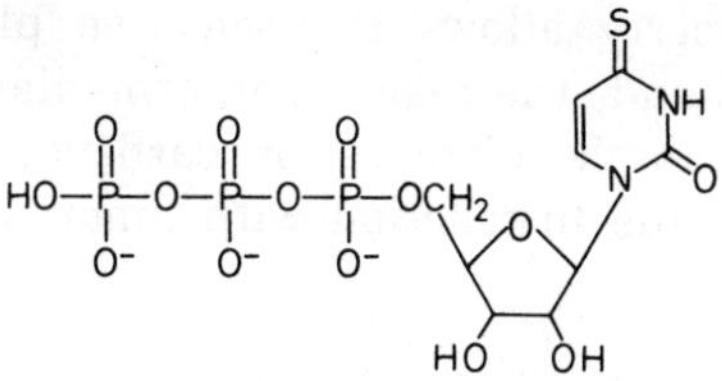

115

ficity of the photochemical reaction, the kinetics of inactivation were investigated. The rate of inactivation by **115** is decreased in the presence of either ATP or UTP, whereas the template poly d(A-T) has very little effect. The enzyme is best protected if both UTP and poly d(A-T) are present. This indicates that RNA polymerase is inactivated by reaction

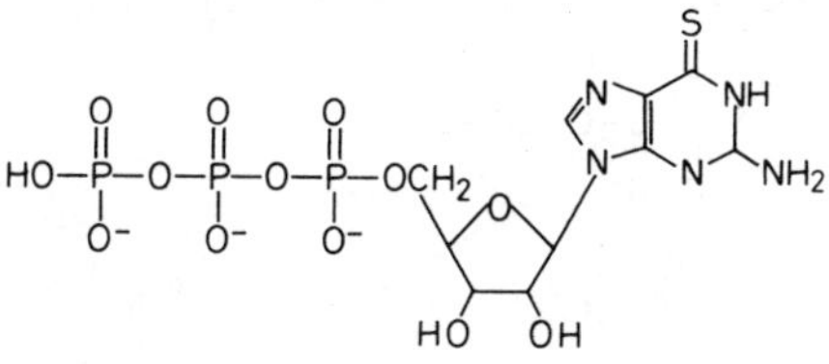

116

with **115** at the substrate binding site. By means of radioactive **115** the covalent substitution of the enzyme by a reactive species derived from **115** was established. The stoichiometry of the chemical modification was determined to be 0.9 (81).

In an analogous experiment DNA-dependent RNA polymerase was irreversibly inactivated by irradiation in the presence of the GTP analog 6-thioketoguanosine 5′-triphosphate (**116**) (82). Compound **116** proved to be less efficient than **115**, which might be explained by the lesser reactivity of the 6-sulfonate intermediate.

Sawada (83) applied the photooxidation of 4-thiouridine 3′-phosphate (**117**) in the presence of ribonuclease A from pancreas for the specific covalent modification of this enzyme. A general advantage of the photooxidation procedure lies in the chemical stability and high wavelength absorption maxima of the thiopyrimidine and thiopurine deriva-

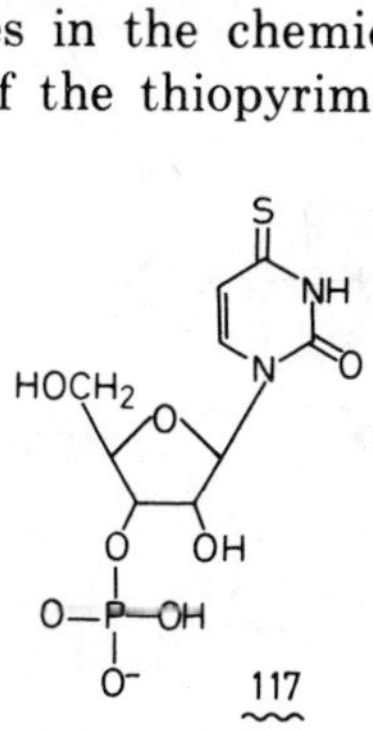

117

tives. The latter property allows the selective photoactivation of the ligand. On the other hand, the reactive intermediates have much longer lifetimes than for example nitrenes or carbenes which may lead to diminished specificity due to reaction with functional groups other than those at the active site.

7.2.3 Sugar-Modified Nucleotides

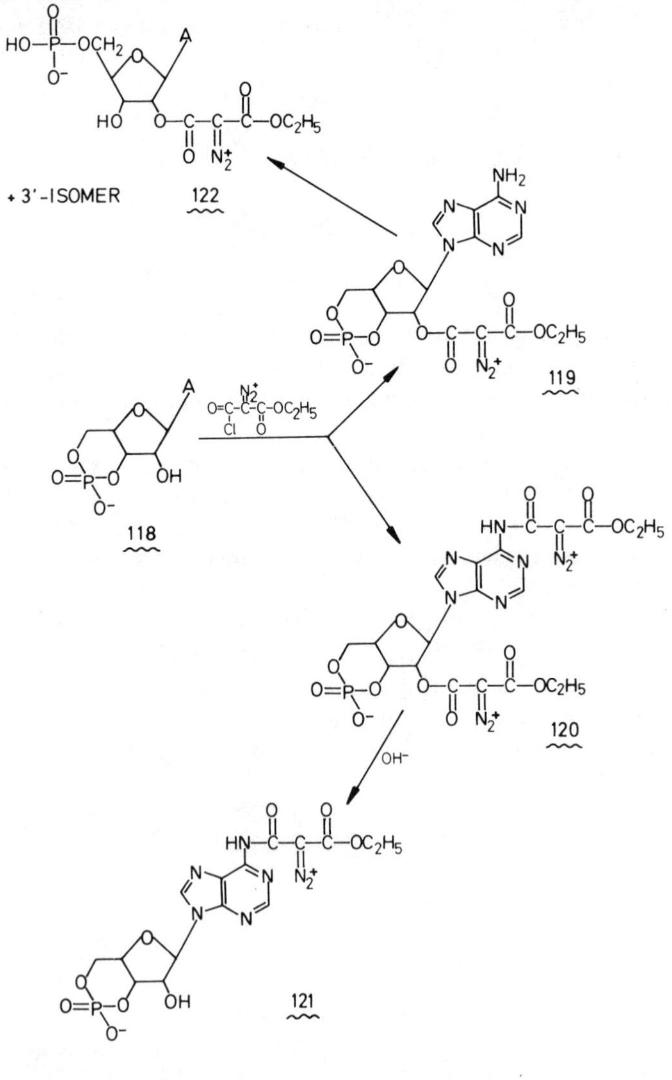

Westheimer and colleagues (84) were the first to introduce a nucleotide analog substituted by a functional group which upon photolysis generated a carbene for the purpose of active-site labeling. The analog was a derivative of NAD^+ carrying a diazoacetate group at the pyridine moiety and was employed as a photoaffinity label for dehydrogenases. Brunswick and Cooperman (85, 86) prepared an ethyl 2-diazomalonyl derivative of adenosine 3′,5′-cyclic phosphate (**118**) and adenosine 5′-phosphate as potential photogenerated "active-site-directed" reagents. Depending on the reaction conditions, acylation of **118** with ethyl 2-diazomalonylchloride in pyridine led to 2′-*O*-(ethyl 2-diazomalonyl)adenosine 3′,5′-cyclic phosphate (**119**) and 2′-*O*-N^6-di-(ethyl 2-diazomalonyl) adenosine 3′,5′-cyclic phosphate (**120**). Controlled alkaline hydrolysis of **120** afforded N^6-(ethyl 2-diazomalonyl)adenosine 3′,5′-cyclic phosphate (**121**). Enzymic hydrolysis of **119** by beef heart phosphodiesterase yielded initially 2′-*O*-(ethyl 2-diazomalonyl) adenosine 5′-phosphate (**122**). Because of the facile

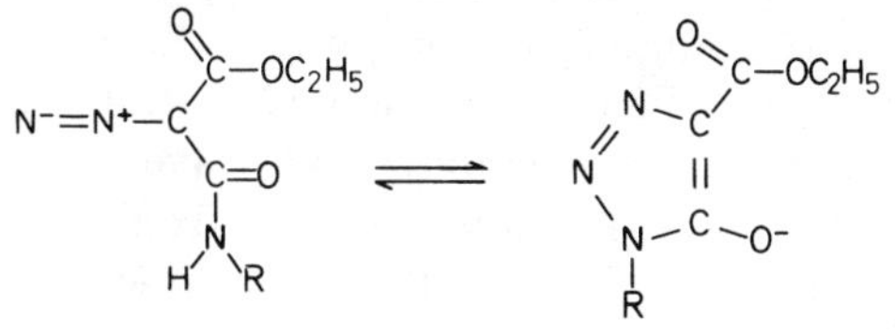

121 ⇌ 123

R=PURINE 3′,5′-CYCLIC RIBOTIDE

acyl migration between cis diol OH groups equilibration occurs giving a mixture of 3′- and 2′-*O*-acylated species (87). The equilibrium constant for the interconversion of 2′,(3′)-*O*-acetyluridine was determined to be 1.7 at pH 7 in favor of 3′-*O*-acetyluridine (87). A similar situation might prevail in the case of compound **122** (86). Irradiation of **119** at 254 nm for 30 second causes 100% degradation, whereas with irradiation at 350 nm for 90 minutes only 80% photolysis had occurred. Compound **119** turned out to be an effector of rabbit muscle phosphofructokinase with a 2.5-fold lesser affinity than cAMP. Irradiation of a mixture of enzyme and **119** for one minute at 254 nm yielded covalently modified enzyme. The rate of modification obeyed saturation kinetics (86). The enzyme can be protected to some extent against substitution by the presence of either ATP or fructose-6-phosphate and almost completely by a 25-fold excess of cAMP. The latter results are in agreement with observations made by others in photoaffinity labelling studies. A large excess of an unreactive ligand of higher affinity is needed to protect protein from specific inactivation by a photoreactive agent of lower affinity.

In contrast to diazoketones the carbenes generated from diazocarboxylic acid esters such as **119** are not converted into the corresponding ketenes by the Wolff rearrangement (88). Brunswick and Cooperman (86) observed that the ethyl 2-diazomalonyl residue in **121** undergoes a pH-dependent Dimroth rearrangement typical for α-diazoamides. The 1,2,3-triazole derivative **123**, which is the predominant form at pH 8 is only poorly photolyzed. This limits the application of derivative **121** as a photoaffinity reagent.

Guillory and coworkers developed a new class of photoreactive nucleotide analogs by acylation of sugar OH groups in nucleotides with 3-*N*-(4-azido-2-nitrophenyl) aminopropionic acid (**124**) (89). The 4-azido-2-nitrophenyl substituent exhibits an absorption maximum at 480 nm. The general procedure employed for acylation of nucleotides with **124** involves the *in situ* conversion of **124** into the imidazolidate by reaction with *N*,*N'*-carbonyldiimidazole. The acylation reaction with the nucleotide is then performed in a mixture of dimethylformamide-water (89, 90). The acyl substituents in 3'-*O*-[*N*-(4-azido-2-nitrophenyl)] amino propionyladenosine 5'-phosphate (**125**) and 5'-triphosphate (**126**) have been found to be rather alkali labile (89). This ester bond is hydrolyzed rapidly at pH's greater than 8 implying that such compounds have to be used at pH's well below 8. The authors claim to have evidence from ^1H-nmr spectroscopy that the acyl substituents in **125** and **126** are located at *O*(3') (89). Irradia-

124 → 1) N,N'-carbonyldiimidazole, 2) NUCLEOTIDE → 125, 126

125 R = H

126 R = $P(O)O\text{-}O\text{-}P(O)(OH)O^{2-}$

A = ADENINE

tion of **125** and **126** at wavelengths above 350 nm caused photolysis, which was monitored by absorption spectroscopy. The observed time-dependent change of the absorption spectra proceeded through a set of isosbestic points (89). Guillory et al., employed the ATP analog **126** as a photogenerated active site reagent for the soluble as well as membrane bound mitochondrial F_1 ATPase (91). Irradiation of F_1 ATPase in the presence of **126** led to covalent modification. The substitution fulfilled all criteria of specific active-site labeling: (1) **126** proved to be a substrate in the dark, (2) the rate of inactivation followed saturation kinetics, and (3) inactivation was suppressed in the presence of ATP. The ATP analog **126** found further application as a photoaffinity label for myosin subfragment 1 ATPase and heavy meromyosin (89, 90).

Guillory and Jeng (90) by means of the above described general procedure prepared the 4-azido-2-nitrophenyl aminobutyric acid ester

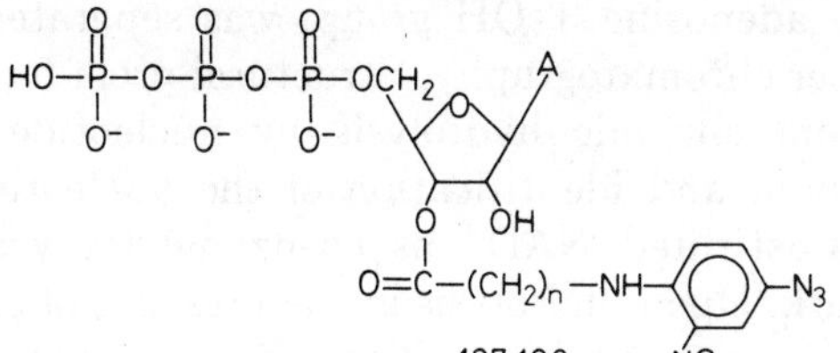

127 n = 3
128 n = 5
A = ADENINE
127,128

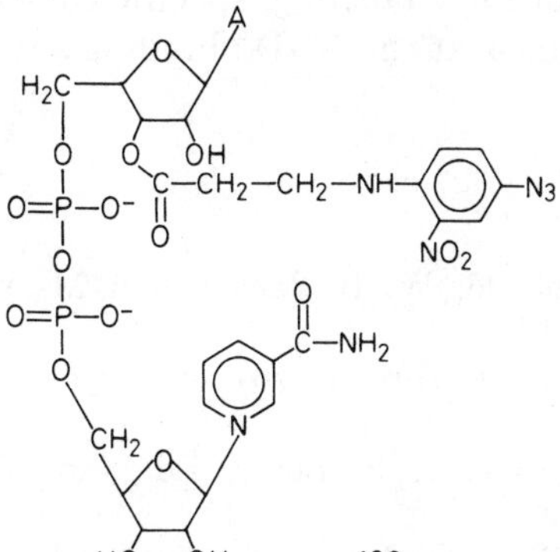

130

(**127**) and the 4-azido-2-nitrophenyl caproic-acid ester of ATP (**128**), in which the photoreactive group is separated from the nucleotide by longer spacer arms.

Schäfer et al. (92) reported the acylation of ADP with 4-*N*-(4′-azido-2′-nitrophenyl)amino butyryl imidazolidate. The resulting 3′-*O*-[-*N*-(4-azido-

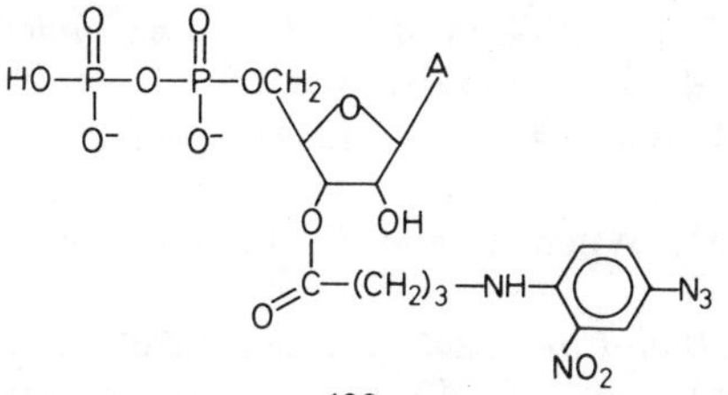

129
A = ADENINE

2-nitrophenyl)amino]butyryladenosine 5′-diphosphate (**129**) was employed as an affinity-labeling reagent for chloroplast ATP-synthetase (92). In the dark, the ADP analog is not phosphorylated but functions as a competitive inhibitor. It is worth to mention that 8-azidoadenosine 5′-diphosphate did not serve as a photoaffinity labelling reagent for the chloroplast ATP-synthetase, probably because of sterical restrictions (92).

The 3′-*O*-[3-*N*-(4-azido-2-nitrophenyl)amino]propionyl derivative of NAD^+ (**130**) has been prepared by acylation of NAD^+ with 4-azido-2-nitrophenylaminopropionyl imidazolidate as described above (93). The NAD^+

analog, acylated at the adenosine 3′-OH group, was separated from the reaction mixture by paper chromatography. Structural proof for the NAD$^+$ analog **130** resulted from enzymic hydrolysis by nucleotide pyrophosphatase from snake venom and identification of the nucleotides formed (93). Compound **130** substituted NAD$^+$ as coenzyme for yeast alcohol dehydrogenase in the dark. Upon photolysis in the presence of enzyme, the analog **130** becomes a potent inhibitor. The photoreaction caused, as judged by common criteria of affinity labeling, specific covalent modification of yeast alcohol dehydrogenase at the NAD$^+$ binding site (93).

REFERENCES

1. *Methods in Enzymology*, Vol. 46, W. B. Jacoby and M. Wilchek, Eds., Academic Press, New York, 1977.
2. B. R. Baker, *Design of Active-Site Directed Irreversible Enzyme Inhibitors*, Wiley, New York, 1967.
3. E. Shaw, in *The Enzymes*, Vol. 1, P. Boyer, Ed., 3rd ed., pp 91–146, Academic Press, New York, 1970.
4. T. L. Gilchrist and C. W. Rees, *Carbenes, Nitrenes and Arynes*, Nelson, London, 1969.
5. W. Kirmse, *Carbene Chemistry*, Academic Press, New York, 1964.
6. A. Singh, E. R. Thornton, and F. Westheimer, *J. Biol. Chem.* **237,** PC 3006 (1962).
7. J. Shafer, P. Baranowsky, R. Laursen, F. Finn, and F. Westheimer, *J. Biol. Chem.* **241,** 421–427 (1966).
8. J. R. Knowles, *Acc. Chem. Res.* **5,** 155–160 (1972).
9. H. Bayley and J. R. Knowles, in *Methods in Enzymology* Vol. 46, pp 69–114, W. B. Jacoby and M. Wilchek, Eds., Academic Press, New York, 1977.
10. P. Cuatrecasas, M. Wilchek, and C. B. Anfinsen, *Biochemistry* **8,** 2277–2284 (1969).
11. P. Cuatrecasas, M. Wilchek, and C. B. Anfinsen, *J. Biol. Chem.* **244,** 4316–4329 (1969).
12. P. Cuatrecasas, *J. Biol. Chem.* **245,** 574–584 (1970).
13. R. P. Glinski and M. B. Sporn, *Biochemistry* **11,** 405–413 (1972).
14. M. B. Sporn, D. M. Berkowitz, R. P. Glinski, A. B. Ash, and C. L. Stevens, *Science* **164,** 1408–1410 (1969).
15. A. W. Schrecker, M. B. Sporn, and R. C. Gallo, *Cancer Res.* **32,** 1547–1553 (1972).
16. A. Lubineau, *Bull. Soc. Chim. France* **4,** 1569–1573 (1972).
17. P. Cuatrecasa, M. Wilchek, and C. B. Anfinsen, *Proc. Nat. Acad. Sci.* (*US*) **61,** 636–643 (1969).
18. M. Wilchek and M. Gorecki, *Eur. J. Biochem.* **11,** 491–494 (1969).
19. H. Sternbach, F. v. d. Haar, E. Schlimme, E. Gaertner, and F. Cramer, *Eur. J. Biochem.* **22,** 166–172 (1971).
20. O. Berglund and F. Eckstein, *Eur. J. Biochem.* **28,** 492–496 (1972).

21. O. Berglund and F. Eckstein, in *Methods in Enzymology* Vol. 34, pp 253–261, W. B. Jacoby and M. Wilchek, Eds., Academic Press, New York, 1974.
22. F. Eckstein, W. Bruns, and A. Parmeggiani, *Biochemistry* **14,** 5225–5232 (1975).
23. P. J. Hoffmann and R. L. Blakley, *Biochemistry* **14,** 4804–4812 (1975).
24. V. F. Zarytova, D. G. Knorre, V. A. Kurbatov, A. V. Lebedev, V. V. Samukov, and G. V. Shishkin, *Bioorg. Chem.* (*USSR*) **1,** 793–798 (1975).
25. A. Hampton, P. J. Harper, T. Sasaki, P. Howgate, and R. K. Preston, *Biochem. Biophys. Res. Commun.* **65,** 945–950 (1975).
26. A. Hampton, P. J. Harper, and P. Howgate, in *Methods in Enzymology* Vol. 46, pp. 302–307, W. B. Jacoby and M. Wilchek, Eds., Academic Press, New York, 1977.
27. A. Hampton and M. H. Maguire, *J. Am. Chem. Soc.* **83,** 150–157 (1961).
28. F. Eckstein, M. Goumet, and R. Wetzel, *Nucleic Acids Res.* **2,** 1771–1775 (1975).
29. H. Guilford, P. O. Larsson, and K. Mosbach, *Chem. Scr.* **2,** 165–170 (1972).
30. A. Hampton, in *Methods in Enzymology*, Vol. 46, pp. 299–302, W. B. Jacoby and M. Wilchek, Eds., Academic Press, New York, 1977.
31. A. Hampton, *J. Biol. Chem.* **238,** 3068–3074 (1963).
32. L. W. Brox and A. Hampton, *Biochemistry* **7,** 2589–2596 (1968).
33. A. Hampton and A. Nomura, *Biochemistry* **6,** 679–689 (1967).
34. L. W. Brox and A. Hampton, *Biochemistry* **7,** 398–405 (1968).
35. P. Brodelius, P. O. Larsson, and K. Mosbach, *Eur. J. Biochem.* **47,** 81–89 (1974).
36. R. Wetzel and F. Eckstein, *J. Org. Chem.* **40,** 658–660 (1975).
37. M. Lindberg and K. Mosbach, *Eur. J. Biochem.* **53,** 481–486 (1975).
38. Y. Yamazaki, H. Maeda, and H. Suzuki, *Eur. J. Biochem.* **77,** 511–520 (1977).
39. S. Barry and P. O'Carra, *FEBS-Lett.* **37,**134–139 (1973).
40. J. A. Carbon, *Biochem. Biophys. Res. Commun.* **7,** 366–369 (1962).
41. F. W. Hulla and H. Fasold, *Biochemistry* **11,** 1056–1061 (1972).
42. U. Faust, H. Fasold, and F. Ortanderl, *Eur. J. Biochem.* **43,** 273–279 (1974).
43. H. Fasold, F. W. Hulla, F. Ortanderl, and M. Rack, in *Methods in Enzymology* Vol. 46, pp. 289–295, W. B. Jacoby and M. Wilchek, Eds. Academic Press, New York, 1977.
44. T. Dall-Larsen, H. Fasold, C. Klungsoyr, H. Kryvi, C. Meyer, and F. Ortanderl, *Eur. J. Biochem.* **60,** 103–107 (1975).
45. A. Hampton and L. A. Slotin, *Biochemistry* **14,** 5438–5444 (1975).
46. A. Hampton and L. Slotin, in *Methods in Enzymology*, Vol. 46, pp. 295–299, W. B. Jacoby and M. Wilchek, Eds. Academic Press, New York, 1977.
47. V. W. Armstrong, H. Sternbach, and F. Eckstein, *Biochemistry* **15,** 2086–2091 (1976).
48. B. Lythgoe and A. R. Todd, *J. Chem. Soc.* **1944**, 592.
49. N. K. Kochetkov and E. I. Budovskii Eds., *Organic Chemistry of Nucleic Acids*, pp. 467–468, Plenum Press, London, 1972.
50. D. H. Rammler, *Biochemistry* **10,** 4699–4705 (1971).
51. R. D. Dulbecco and J. D. Smith, *Biochim. Biophys. Acta* **39,** 358–361 (1960).

52. J. X. Khym, *Biochemistry* **2,** 344–350 (1963).
53. J. W. Bodley and J. Gordon, *Biochemistry* **13,** 3401–3405 (1974).
54. J. X. Khym and W. E. Cohn, *J. Am. Chem. Soc.* **82,** 6380–6386 (1960).
55. J. T. Powell and K. Brew, *Biochemistry* **15,** 3499–3505 (1976).
56. S. B. Easterbrook-Smith, J. L. Wallace, and D. B. Keech, *Eur. J. Biochem.* **62,** 125–130 (1976).
57. R. T. Borchardt, S. E. Wu, and C. S. Schasteen, *Biochem. Biophys. Res. Commun.* **81,** 841–849 (1978).
58. B. F. Erlanger and S. M. Beiser, *Proc. Nat. Acad. Sci. (US)* **52,** 68–74 (1964).
59. F. Wold, in *Methods in Enzymology*, Vol. 46, pp. 3–14, W. B. Jacoby and M. Wilchek, Eds. Academic Press, New York, 1977.
60. L. Thelander, B. Larsson, J. Hobbs, and F. Eckstein, *J. Biol. Chem.* **251,** 1398–1405 (1976).
61. F. Seela and S. Waldek, *Nucleic Acids Res.* **2,** 2343–2356 (1975).
62. J. A. Massen and W. Moller, *Proc. Nat. Acad. Sci. (US)* **71,** 1277–1280 (1974).
63. A. S. Ghirshovich, V. A. Podnyakov, and Y. A. Ovchinnikov, in *Methods in Enzymology*, Vol. 46, pp. 656–660, W. B. Jacoby and M. Wilchek, Eds., Academic Press, New York, 1977.
64. G. T. Babkina, V. F. Zarytova, and D. G. Knorre, *Bioorg. Chem. (USSR)* **1,** 611 (1975).
65. A. S. Ghirshovich, V. A. Podnyakov, and Y. A. Ovchinnikov, *Eur. J. Biochem.* **69,** 321–328 (1976).
66. S. Chladek, K. Quiggle, G. Chinali, J. Kohut III, and J. Ofengand, *Biochemistry* **16,** 4312–4319 (1977).
67. K. Muneyama, R. J. Bauer, D. A. Shuman, R. K. Robins, and L. N. Simon, *Biochemistry* **10,** 2390–2395 (1971).
68. B. E. Haley and J. F. Hoffman, *Proc. Nat. Acad. Sci. (US)* **71,** 3367–3371 (1974).
69. A. H. Pomerantz, S. A. Rudolph, B. E. Haley, and P. Greengard, *Biochemistry* **14,** 3858–3862 (1975).
70. R. Koberstein, L. Cobianchi, and H. Sund, *FEBS-Lett.* **64,** 176–180 (1976).
71. G. Schäfer and S. Penades, *Biochem. Biophys. Res. Commun.* **78,** 811–816 (1977).
72. R. Koberstein, *Eur. J. Biochem.* **67,** 223–229 (1976).
73. E. P. Lau, B. E. Haley, and R. E. Barden, *Biochem. Biophys. Res. Commun.* **76,** 843–849 (1977).
74. E. K. Keeler and P. Campbell, *Biochem. Biophys. Res. Commun.* **72,** 575–580 (1976).
74. H. J. Schäfer, P. Scheurich, G. Ratgeber, and K. Dose, *Nucleic Acids Res.* **5,** 1345–1351 (1978).
76. I. L. Cartwright, D. W. Hutchinson, and V. W. Armstrong, *Nucleic Acids Res.* **3,** 2331–2339 (1976).
77. T. Saegusa, Y. Ito, and T. Shimizu, *J. Org. Chem.* **35,** 2979–2982 (1970).
78. K. Y. Ku and W. H. Prusoff, *J. Biol. Chem.* **249,** 1239–1246 (1974).
79. H. R. Rackwitz and K. H. Scheit, *Eur. J. Biochem.* **72,** 191–200 (1977).
80. F. Cramer, E. M. Gottschalk, H. Matzura, K. H. Scheit, and H. Sternbach, *Eur. J. Biochem.* **19,** 379–385 (1971).

81. A. M. Frischauf and K. H. Scheit, *Biochem. Biophys. Res. Commun.* **53,** 1227–1233 (1973).
82. K. H. Scheit, unpublished results.
83. F. Sawada, *Biochem. Biophys. Res. Commun.* **64,** 311–316 (1975).
84. D. T. Browne, S. S. Hixson, and F. H. Westheimer, *J. Biol. Chem.* **246,** 4477–4484 (1971).
85. D. J. Brunswick and B. S. Cooperman, *Proc. Nat. Acad. Sci.* (*US*) **68,** 1801–1804 (1971).
86. D. J. Brunswick and B. S. Cooperman, *Biochemistry* **12,** 4074–4078 (1973).
87. B. E. Griffin, M. Jarman, C. B. Reese, J. E. Sulston, and D. R. Trentham, *Biochemistry* **5,** 3638–3649 (1966).
88. E. S. Gould, *Mechanismus und Struktur in der Organischen Chemie*, pp. 760–761, Verlag Chemie, Weinheim, 1962.
89. S. J. Jeng and R. J. Guillory, *J. Supramol. Struct.* **3,** 448–468 (1975).
90. R. J. Guillory and S. J. Jeng, in *Methods in Enzymology*, Vol. 46, pp. 259–288, W. B. Jacoby and M. Wilchek, Eds. Academic Press, New York, 1977.
91. J. Russel, S. J. Jeng, and R. J. Guillory, *Biochem. Biophys. Res. Commun.* **70,** 1225–1234 (1976).
92. G. Schäfer, G. Onur, K. Edelmann, S. Bickel-Sandkötter, and H. Strotmann, *FEBS-Lett.* **87,** 318–322 (1978).
93. S. Chen and R. J. Guillory, *J. Biol. Chem.* **252,** 8990–9001 (1977).

Author Index

Numbers in parentheses indicate where the author's name is cited in references listed at the end of an individual chapter.

Acs, G., 67 (88)
Adamczyk, D. L., 62, (86)
Adelman, R. C., 161, (191)
Adler, J., 49, (84)
Agarwal, K. C., 62, 63, (86, 87)
Agarwal, S. C., 36, (82)
Aivasashvili, V. A., 172, (193)
Akerfeldt, S., 112, (139)
Albrecht, H. P., 129, (140), 186, (194)
Alderfer, J. L., 171, (193)
Alegria, A. H., 50, (84)
Allen, L. B., 68, 70, 71, (88), 144, (190)
Altwerger, L., 67, (88)
Anai, K., 206, (217)
Anfinsen, C. B., 221, 225, (250)
Angell, G. L., 1, (11)
Anisuzzaman, A. K. M., 175, (193)
Aradi, J., 53, (85)
Arbor, A., 143, (189)
Armanath, V., 20, (78)
Armstrong, V. W., 38, 56, (82, 85), 106, 135, 136, (139, 141), 163, 169, (192), 232, 243, 244, (251, 252)
Arnott, S., 6, (12)
Arora, S. K., 32, (81)
Asai, M., 153, 154, (191)
Ash, A. B., 223, (250)
Atkinson, M. R., 101, (139), 164, 165, (192)
Auerbach, C., 14, (77)
Aurbach, G. D., 97, 99, (138)
Avison, A. W. D., 203, (216)

Babcock, D., 97, 98, 99, (138)
Babkina, G. T., 237, 239, (252)
Baddiley, J., 211, (217)
Baer, H. H., 186, (194)
Bagdasarian, A., 126, (140)
Bähr, W., 30, 33, (80, 81)
Baker, B. R., 143, 162, (189, 192), 220, (250)
Bakina, G. T., 136, (141)
Bald, R., 57, 58, 59, 73, (85, 86), 92, (95)
Ballantyne, W., 97, 98, 99, (138)
Ballard, F. J., 161, (191)
Balny, C., 32, (81)
Bär, H. P., 103, 118, (139, 140)
Baranowsky, P., 220, (250)
Barden, R. E., 241, (252)
Bardos, T. J., 53, (85)
Barfknecht, R. L., 57, (85)
Barrio, J. R., 21, 22, 23, 33, 49, 69, 70, (79, 81, 88)
Barry, S., 229, (251)

Bartholomew, D. G., 68, (88)
Bârzu, O., 44, (83)
Bass, L. W., 1, (11)
Batt, R. D., 60, (86)
Bauer, R. J., 238, 239, (252)
Bax, P. C., 125, 126, 127, (140)
Bayer, M., 113, (139), 197, (215)
Bayley, H., 221, (250)
Beezley, D. N., 14, (77)
Beiser, S. M., 235, (252)
Bendich, A., 28, (80)
Benitez, A., 143, (189)
Bennett, G. N., 100, (138), 172, (193)
Bentley, H. R., 161, (191)
Beranek, J., 75, (89)
Berg, P., 49, (84)
Berglund, O., 225, (250, 251)
Bergmann, W., 145, (190)
Berkowitz, D. M., 223, (250)
Bessman, M. J., 49, (84)
Bibilashvili, R. S., 172, (193)
Bickel-Sandkötter, S., 249, (253)
Blackburn, B. J., 178, 179, (194)
Blackburn, G. M., 203, (216)
Blakley, R. L., 225, (251)
Block, A., 174, (193)
Bobst, A. M., 168, (192)
Boder, G. B., 91, (95)
Bodley, J. W., 234, (252)
Boezi, J. A., 161, 172, (191, 193)
Bojanovski, D., 184, (194)
Boos, K. S., 179, 184, (194)
Borchardt, R. T., 235, (252)
Borden, K. F., 198, 205, (216)
Boswell, K. H., 65, 71, (87, 88)
Bouquet, M., 162, (191)
Boxer, G. E., 161, 162, 163, (191, 192)
Bradbury, E., 112, (139)
Bradshaw, T. K., 46, 52, (84)
Brdar, B., 67, (88)
Brennan, T., 67, (88)
Brent, T. P., 150, 151, (191)
Brentnall, H. J., 39, (82)
Brew, K., 234, (252)
Brigl, P., 198, (215)
Brimacombe, R. L. C., 16, (78)
Brodelius, P., 227, (251)
Brookes, P., 14, 15, 19, (77, 78)
Broom, A. D., 20, (78), 169, 170, (193)
Broomhead, J. M., 2, (12)
Broude, N. E., 21, (79)
Brown, D. M., 1, (11)
Brown, G. B., 18, 19, 28, (78, 80)
Brown, M. J., 203, (216)
Browne, D. T., 246, (253)
Brox, L. W., 113, (139), 197, (215), 227, (251)
Bruns, W., 135, 136, (141), 224, (251)
Brunswick, D. J., 247, (253)
Budowski, E. I., 15, 17, 19, 21, 30, 31, 36, 59, 60, 61, (78, 79, 80, 82, 86), 223, (251)
Bugg, C. E., 2, (12), 26, (80)
Bunick, G., 150, (190)
Burchenal, J. H., 24, 49, (79, 84)
Burger, A., 125, (140)
Burgers, P. M. J., 108, (139)
Burgi, E., 24, (79)
Burke, D. C., 145, (190)

Campbell, P., 242, (252)
Carbon, J. A., 25, (80), 230, (251)
Cardeilhac, P. T., 146, (190)
Cargill, R. L., 202, (216)
Carpenter, J. M., 56, (85)
Cartwright, I. L., 38, (82), 243, 244, (252)
Cashel, M., 162, (191)
Catlin, J. C., 168, (192), 200, 201, (216)
Cerami, A., 67, (88)
Cerutti, P. A., 33, (81)
Chachaty, C., 8, (12)
Chambers, R. W., 30, 93, 95, (80, 95), 212, (218)
Chang, P. K., 50, (84)
Chassy, B. M., 161, (191)
Chedda, G. B., 19, 20, (78)

Chen, S., 249, 250, (253)
Cheong, L., 28, (80)
Cheung, C. P., 14, (77)
Chinali, G., 136, (141), 238, (252)
Chinault, A. C., 138, (141)
Chladek, S., 112, 136, (139, 141), 174, (193), 238, (252)
Chong, K. J., 205, (217)
Christensen, L. F., 41, 71, (82, 88)
Chu, M.-Y., 38, 39, (82)
Chu, S. H., 34, 38, 39, (81, 82)
Cihak, A., 75, 76, (89)
Clark, V. M., 195, (215)
Clarke, D. A., 27, (80)
Clauwaert, J., 4, (12)
Cleland, W. W., 111, (139)
Cline, J. C., 91, (95)
Cobianchi, L., 37, (82), 239, 240, 241, (252)
Cohen, B. I., 21, (79)
Cohen, S. S., 143, 144, 145, 146, 150, 151, 164, 165, (189, 190, 192)
Cohn, M., 106, 110, 111, (139)
Cohn, P., 187, (194)
Cohn, W. E., 59, 94, (86, 95), 234, (252)
Cole, Q. P., 62, (86)
Comstock, P., 171, (193)
Cook, A. F., 128, (140)
Cook, P. D., 66, (87)
Cooperman, B. S., 247, (253)
Cornelius, R. D., 111, (139)
Cosgrove, D. J., 202, (216)
Coulter, C. L., 32, (81), 112, (139)
Cramer, F., 17, 18, 30, 37, 42, 44, 67, (78, 80, 82, 83, 88), 112, 116, 117, 118, (139, 140), 161, 184, (191, 194), 198, 208, 214, (215, 217, 218), 225, 244, (250, 252)
Crea, R., 213, (218)
Creasy, W. A., 150, (190)
Cuatrecasas, P., 99, (138), 221, 223, 225, (250)
Cunningham, K. G., 161, (191)
Cusack, N. J., 70, (88)
Dahl, J. L., 60, 61, 62, (86)
Dale, R. M. K., 53, 54, (85)
Dall-Larsen, T., 231, (251)
Dammann, L. G., 22, 23, 49, (79)
Danenberg, P. V., 57, (86)
Danielzadeh, A. B., 97, (138)
Dansoreanu, M., 44, (83)
Darlix, J. L., 25, 62, 63, 67, (80, 86)
Darnell, J. E., 14, (77)
Dash, B., 19, (78)
Dattagupta, J. K., 56, (85)
Davis, J. C., 1, (11)
Davoll, J., 42, 44, (83)
Dawson, W. H., 202, (216)
Dea, P., 66, 68, 71, (87, 88)
DeClercq, E., 65, (87), 183, (194)
Dekker, C. A., 2, (11), 170, (193)
Demushkin, V. P., 31, 59, 60, (80)
Descamps, J., 183, (194)
Desrosiers, R. C., 14, (77)
DeSomer, P., 183, (194)
de Wachter, R., 54, (85)
Deutscher, M. P., 164, 165, (192)
Dewey, V. C., 62, (86)
Deyrup, J. A., 90, (95)
Dietz, A., 76, (89)
Dimroth, K., 198, (215)
Dimroth, O., 15, (77)
Dirheimer, G., 136, 137, (141)
Distler, J. J., 213, (218)
Dixon, H. B. F., 137, 138, (141)
Doerr, I. L., 24, 26, 27, 28, 43, (79, 80)
Doherty, D. G., 59, (86)
Dombrowski, J., 19, (78)
Dondon, J., 49, 50, (84)
Dondon, L., 94, (95)
Donohue, J., 6, (12)
Dose, K., 37, (82), 242, (252)
Doskocil, J., 76, (89)
Doty, P., 21, (78)
Downs, R. W., Jr., 97, 99, (138)

Drushinina, T. N., 60, 61, (86)
Duke, J. A., 25, (80)
Dulaney, E. L., 161, (191)
Dulbecco, R. D., 233, (251)
Dumas, L. B., 115, (140), 203, (216)
Dunaway-Mariano, D., 101, (139)
Dunlap, B. E., 171, 172, (193)
Dunlap, R., 202, (216)
Dunn, D. B., 42, (83)
Duschinsky, R., 49, (84)

Easterbrook-Smith, S. B., 235, (252)
Eaton, M. A. W., 50, (84)
Ebel, J. P., 136, 137, (141)
Eckstein, F., 24, 29, 56, (80, 85), 98, 99, 101, 102, 103, 104, 105, 106, 107, 109, 110, 112, 135, 136, (138, 139, 141), 163, 167, 168, 169, (192, 193), 198, 199, 203, (216, 217), 224, 225, 227, 228, 232, 235, 237, (250, 251, 252)
Edelheit, E. B., 161, (191)
Edelman, B., 95, (95)
Edelmann, K., 249, (253)
Eidinoff, M. L., 28, (80)
Eisenberg, P., 5, (12)
Elion, G. B., 24, 73, (79, 89)
Eliseeva, G. I., 55, 60, 61, (85, 86)
Ellis, D. B., 153, (191)
Ellison, R. R., 24, (79)
Engel, M. L., 136, (141)
English, J. P., 26, (86)
Erdenreich, E. S., 102, 104, (139)
Eriksson, B., 71, (88)
Erlanger, B. F., 235, (252)
Ernst, L., 119, 120, (140)
Eto, M., 204, (217)
Etzold, G., 165, 166, (192)
Evans, F. E., 38, 40, 63, (82, 87)
Evans, J. S., 76, (89)

Faerber, P., 29, 30, 31, 32, 33, 37, (80, 81)
Farier, J., 2, (12)
Fasold, H., 230, 231, (251)
Faust, U., 230, 231, (251)
Fecher, R., 49, (84)
Feldmann, W., 202, (216)
Field, A. K., 167, (192)
Fikus, M., 44, (83)
Finn, F., 220, (250)
Fischer, E., 201, (216)
Fischer, H. O. L., 186, (194)
Fisher, L. V., 144, (190)
Flesher, J. W., 97, 103, (138)
Fleysher, M. H., 17, 56, (78, 85)
Florentjev, V. L., 172, 184, (193, 194)
Folayan, J. O., 38, 49, 52, (82, 84, 85)
Folkers, K., 161, (191)
Follmann, H., 144, 150, (190)
Forrest, H. S., 90, (95)
Fox, J. J., 1, 3, (11, 12), 24, 26, 27, 28, 43, 49, (79, 80, 84), 186, (194)
Fraenkel-Conrat, H., 21, (78)
Franke, A., 199, (216)
Fraser, T. H., 163, (192)
Frazier, J., 37, 44, 50, (82, 83, 84)
Freese, E., 43, (83)
Freist, W., 37, 42, (82, 83), 116, (140), 208, (217)
Fresco, J. R., 33, (81)
Friderici, K. H., 14, (77), 171, 172, (193)
Friedman, O. M., 19, (78)
Frihart, C. R., 22, (79)
Frischauf, A. M., 244, 245, (253)
Fromageot, P., 25, 62, 63, 67, (80, 86)
Frye, R., 66, (87)
Fuertes, M., 130, (141)
Fujii, T., 90, (95)
Fukui, T., 35, 36, 38, 42, 66, (81, 82, 83, 87), 167, 169, 172, (192, 193)
Fuller, W., 63, 67, (87)
Fulmor, W., 142, (189)

Furlong, N. B., 150, (190)
Furth, J. J., 143, 144, 150, 151, (189, 190)
Furuichi, Y., 14, (77)
Furukawa, Y., 72, (88), 170, (193), 202, (216)
Furusawa, K., 212, (218)
Futai, M., 160, (191)

Gabrielyan, N. D., 60, 61, (86)
Gaertner, E., 225, (250)
Gallo, R. C., 223, 224, (250)
Gassen, H. G., 52, 66, (85, 87)
Gatlin, L., 1, (11)
George, P., 5, (12)
Gerard, G. F., 172, (193)
Gerber, N. N., 162, (191, 192)
Gerchman, L. L., 19, (78)
Gershowitz, A., 14, 15, (77)
Gerzon, K., 91, (95)
Ghirshovich, A. S., 237, 239, (252)
Gilchrist, T. L., 220, (250)
Gilham, P. T., 100, (138), 172, (193)
Gin, J. B., 170, (193)
Gindl, H., 104, 105, (139)
Gitterman, C. O., 161, (191)
Giziewicz, J., 56, (85)
Glinski, R. P., 168, (192), 223, (250)
Goia, I., 44, (83)
Golas, T., 44, (83)
Goldberg, I. H., 95, (95)
Goldberg, N. D., 60, 61, (86)
Goldstein, J. H., 1, (11)
Gomez-Guillen, M., 184, (194)
Goodman, L., 1, (11), 143, 144, 152, (189, 190)
Goody, R. S., 105, 106, 107, 109, 112, (139)
Gordon, C. N., 161, (191)
Gordon, J., 234, (252)
Gorecki, M., 225, 230, (250)
Gottschalk, E. M., 244, (252)
Gough, G. R., 42, (83), 100, (138)
Gould, E. S., 247, (253)
Goumet, M., 227, 228, (251)
Grachev, M. A., 28, 31, 59, 60, (80), 136, (141)
Greene, G. L., 115, (139)
Greenfield, J. C., 23, (79)
Greengard, P., 239, (252)
Gresham, C., 150, (190)
Griffin, B. E., 15, 19, 56, (77, 85), 247, (253)
Grimm, W. A. H., 19, 20, (78)
Groner, Y., 13, 14, 19, (77)
Gruber, B. A., 22, 23, (79)
Grunberger, D., 63, 67, (87, 88)
Grunberg-Manago, M., 21, 49, 50, 94, (78, 84, 94)
Gueron, M., 8, (12)
Guilford, H., 227, (251)
Guillory, R. J., 248, 249, 250, (253)
Gumport, R. I., 23, (79), 161, (191)
Guschlbauer, W., 168, (192)

Haar, W., 101, 103, (139)
Hachmann, J., 21, (79)
Haeffner, E. W., 202, (216)
Hagenberg, L., 66, (87)
Haines, J. A., 14, 15, 16, 19, (77)
Haley, B., 134, 135, (141)
Haley, B. E., 37, 38, (82), 238, 239, 241, (252)
Hall, R. H., 13, 14, 17, 56, 92, (77, 78, 85, 95), 170, (193)
Hamel, E., 162, (191)
Hampton, A., 24, 26, 27, 43, (79, 80), 113, 130, 131, 132, 133, (139, 141), 177, 178, 187, 188, 189, (194), 197, 204, (215, 217), 226, 227, 228, 230, 231, 232, (251)
Handschuhmacher, R. E., 72, 73, 74, (88, 89)
Hanka, L. J., 76, (89)
Hapke, B., 116, (140)
Harada, F., 65, 67, (87, 88)

Harper, P. J., 130, 131, 132, (141), 178, (194), 226, 227, 228, (251)
Harris, M. R., 203, (216)
Harris, R., 144, (190)
Haschemeyer, A. E. V., 6, (12)
Haselkorn, R., 21, (78)
Hashimoto, M., 206, 209, (217)
Haskell, T. H., 143, (189)
Hata, T., 200, 205, 208, 210, 212, (216, 217, 218)
Hatfield, D., 90, (95)
Hattori, M., 37, 42, (82, 83)
Hawkinson, S. W., 32, (81)
Hawrelak, S. D., 66, (87)
Hayatsu, H., 30, 33, (80, 81)
Heavner, G. A., 115, (139)
Hebborn, P., 19, 20, (78)
Hecht, S. M., 66, (87), 129, 138, (140, 141)
Heidelberger, C., 48, 49, 57, (84, 86), 166, (192)
Heinlein, K., 170, 171, (193)
Helgstrand, E., 71, (88)
Henry, P., 62, (87)
Herriott, R. M., 14, (77)
Hershey, J. W. B., 97, 99, (138)
Hes, J., 198, 199, (216)
Hess, V. F., 73, (89)
Hettler, H., 113, 114, 115, 117, 118, (139, 140), 198, (215)
Hewson, K., 42, 44, 62, (83, 86), 128, (140)
Hieda, H., 153, 154, (191)
Hilleman, M. R., 167, (192)
Hillen, W., 52, (85)
Hintsche, R., 165, 166, (192)
Hirai, K., 173, (193)
Hiratsuka, T., 176, (194)
Hitchings, G. H., 24, 73, (79, 89)
Hixson, S. S., 246, (253)
Ho, Y. K., 53, (85)
Hoard, D. E., 21, (78), 214, (218)
Hobbs, J., 167, 168, 169, (192, 193), 235, 237, (252)
Hoeksema, H., 143, (189)
Hoffman, D. J., 174, (193)
Hoffman, J. F., 37, 38, (82), 238, 239, (252)
Hoffmann, P. J., 225, (251)
Hogenkamp, H. P. C., 144, 150, (190)
Holmes, R. E., 36, (81)
Holmes, W. L., 49, (84)
Holy, A., 15, 56, 57, 58, 59, 62, 63, 73, 75, 77, (77, 85, 86, 87, 89), 92, (95), 122, 123, 124, 125, 126, 133, (140, 141), 147, 151, 152, 156, 157, 158, 159, 160, 180, 181, 182, 183, 187, (190, 191, 194), 207, (217)
Hong, C. I., 19, 20, (78)
Honjo, M., 38, 42, 72, (82, 83, 88), 170, 186, (193, 194), 202, 208, (216, 217)
Hopkins, G. R., 3, (12)
Hosai, T., 204, (217)
Howard, F. B., 2, (12), 44, 50, (83, 84)
Howgate, P., 226, 227, 228, (251)
Hruby, D. E., 14, (77)
Hruska, F. E., 74, (89)
Huang, G. F., 55, (85)
Hubert-Habart, M., 144, 152, (190)
Huennekens, F. M., 22, (79)
Huffman, J. H., 70, (88), 144, (190)
Hughes, T. R., 106, (139)
Hulla, F. W., 230, 231, (251)
Humphreys, D. A., 21, (78)
Hurwitz, J., 62, (86), 95, (95)
Hutchinson, D. W., 38, 39, 46, 49, 50, 52, (82, 84, 85), 195, (215), 243, 244, (252)

Ichino, M., 50, (84)
Iida, Y., 29, (80)
Ikeda, K., 29, (80)
Ikehara, M., 35, 36, 37, 38, 39, 40, 42, 65, 67, (81, 82, 83, 87, 88), 144, 149, 167, 169, 172, 176, 178, 179, (190, 192, 193, 194)
Illiano, G., 99, (138)

Imai, K. I., 42, (83), 186, (194), 202, (216)
Imazawa, M., 29, 59, (80, 86)
Imoto, M., 179, 180, (194)
Inoue, H., 59, (86)
Ishii, F., 32, (80)
Ishikawa, F., 44, (83)
Ito, Y., 243, (252)
Ivanova, G. S., 180, 181, (194)
Ivanovics, G. A., 63, 66, (87)

Jacob, S. T., 150, (190)
Jaffe, E. K., 110, 111, (139)
Jaffe, J. J., 73, (88)
Janik, B., 167, (192)
Janion, C., 44, (83), 171, (193)
Jarman, M., 247, (253)
Jastorff, B., 42, (83), 113, 114, 115, 116, 117, 118, 119, (139, 140), 198, 208, (215, 217)
Jebeleanu, G., 44, (83)
Jelinek, W., 14, (77)
Jeng, S. J., 248, (253)
Jensen, L. H., 2, (12)
Johansson, N. G., 71, (88)
Johnson, D., 144, (190)
Johnson, I. S., 91, (95)
Johnson, K. L., 170, 171, (193)
Johnson, P. W., 134, (141)
Jolly, J. F., 161, (191)
Jones, G. H., 127, 128, 129, 130, (140), 198, (215)
Jones, J. W., 15, (77), 90, (95)
Jordan, D. O., 1, (11)
Jurwitz, J., 13, 14, 19, (77)

Kaczka, E. A., 161, (191)
Kahan, F. M., 62, (86), 95, (95)
Kakiuchi, N., 167, 169, (192, 193)
Kalamas, R. L., 168, (192)
Kaleja, R., 44, 45, (83)
Kalman, T. I., 53, (85)
Kamimura, A., 152, 153, (191)
Kampe, W., 52, (85), 206, 215, (217, 218)
Kampf, A., 57, (85)
Kanai, Y., 170, (193)
Kanaoka, Y., 33, (81)
Kaneko, M., 144, (190)
Kaplan, N. O., 38, 40, (82)
Kappenberger, H., 198, (215)
Kappler, F., 132, 133, (141), 187, 188, 189, (194), 204, (217)
Kapuler, A. M., 36, 37, 40, 58, (82)
Karnofsky, D. A., 24, (79)
Karon, M., 62, (87)
Kasai, T., 25, (80)
Kato, M., 45, 46, 64, (83, 87)
Kato, T., 201, 202, (216)
Kawada, I., 50, (84)
Kawai, I., 202, (216)
Kawana, M., 63, (87)
Kawazoa, Y., 41, (82)
Kazimierczuk, Z., 44, (83)
Keech, D. B., 235, (252)
Keeler, E. K., 242, (252)
Keilova, H., 73, (88)
Kemp, A., 37, (82)
Kenyon, G. L., 97, 115, (140), 203, (216)
Kettler, M., 98, 99, (138)
Kezdi, M., 44, (83)
Khan, A., 171, (193)
Khan, M. K., 171, (193)
Khare, G. P., 70, (88)
Khorana, H. G., 195, 200, 203, 204, 213, (215, 216, 217, 218)
Khurshid, M., 171, (193)
Khwaja, T. A., 67, 71, (88), 144, 147, 166, 170, (190, 192, 193), 198, 199, (216)
Khym, J. X., 233, 234, (252)
Kidder, G. W., 62, (86)
Kielanoeska, M., 171, (193)
Kigwana, L. J., 71, (88)
Kikuchi, Y., 173, (193)
Kikugawa, K., 50, (84)
Kim, S. H., 68, (88)
Kimura, M., 144, (190)
Kinoshita, M., 179, 180, (194)
Kirby, A. J., 195, (215)

Kirkegaard, L. H., 19, 20, 23, 49, (78, 79)
Kirmse, W., 220, (250)
Kissman, H. M., 162, (192)
Kita, M., 32, (81)
Kitano, S., 66, (87)
Klenow, H., 161, 162, 163, (191, 192)
Klungsøyr, C., 231, (251)
Knoll, J. E., 28, (80)
Knorre, D. G., 136, (141), 225, 237, 239, (251, 252)
Knowles, J. R., 221, (250)
Kobayashi, K., 42, (83), 170, (193), 202, 204, (216, 217)
Koberstein, R., 37, 39, (82), 239, 240, 241, (252)
Koch, R. E., 43, (83)
Kochetkov, N. K., 15, 17, 19, 21, 30, 31, 36, 55, 59, 60, 61, (78, 79, 80, 82, 85, 86), 233 (251)
Kohn, B. D., 187, (194)
Kohut, J., III, 136, (141), 238, (252)
Kokko, J. P., 1, (11)
Kornberg, A., 49, (84), 164, 165, (192)
Kost, A. A., 21, (79)
Kotick, M. P., 167, (192)
Kowollik, G., 165, 166, (192)
Kozarich, J. W., 138, (141)
Krakoff, I., 24, (79)
Kram, D. C., 62, (86)
Krebs, T., 118, (140)
Kreis, W., 147, (190)
Kreiser, T. H., 167, (192)
Kreishman, G. P., 71, (88)
Kritsyn, A. M., 184, (194)
Krokan, H., 150, (190)
Krug, F., 99, (138)
Krygier, V., 150, 151, (191)
Kryvi, H., 231, (251)
Ku, K. Y., 244, (252)
Kućan, I., 95, (95)
Kućerova, Z., 59, (86), 134, (141)
Kula, M. R., 119, (140)
Kulikowski, T., 56, (85)
Kumashiro, I., 18, 42, (78, 83), 152, 153, (191)
Kurbatov, V. A., 225, (251)
Kurkov, V., 93, (95), 212, (218)
Kusashio, K., 201, (216)
Kusmierek, J. T., 16, 21, (78), 171, (193)
Kuwada, Y., 186, (194)
Kwan, S. W., 63, (87)

Labitan, A., 150, 151, (191)
Lagowski, J. M., 90, (95)
Lake, W. C., 175, (193)
Lampen, J. O., 62, (86)
Lamprecht, W., 184, (194)
Lampson, G. P., 167, (192)
Lancaster, J. E., 142, (189)
Lang, R. A., 35, 36, (81)
Langen, P., 165, 166, (192)
Larsen, M., 98, (138)
Larsson, A., 71, (88)
Larsson, B., 235, 237, (252)
Larsson, P. O., 227, (251)
Lau, E. P., 241, (252)
Laursen, R. A., 90, 91, 92, (95), 156, (191), 220, (250)
Lavallee, D. K., 112, (139)
Lavayre, J., 38, 39, (82)
Lawley, P. D., 14, 15, 19, (77, 78)
Lebedev, A. V., 225, (251)
Lechevalier, H. A., 162, (191, 192)
Ledis, S. L., 200, 201, (216)
Lee, C. H., 63, (87)
Lee, T. T., 76, (89)
Lee, W. W., 143, 144, (189, 190)
Lehman, I. R., 49, (84)
Leng, M., 38, 39, (82)
Lengyel, P., 30, (80)
Leonard, N. J., 19, 20, 21, 22, 23, 33, 49, 68, 69, 70, (78, 79, 81, 88), 90, 91, 92, (95), 156, (191)
LePage, G. A., 143, 150, 153, (189, 190, 191)
Letsinger, R. L., 115, 116, (139, 140), 203, (216)

Letters, R., 49, (84)
Levene, P. A., 1, (11)
Levin, D. H., 62, 63, (86, 87)
Lewin, S., 21, (78)
Lichtenthaler, F. W., 186, (194)
Lichtfield, G. J., 70, (88)
Lin, G. H. Y., 32, (81)
Lindberg, M., 228, 229, (251)
Littlefield, J. W., 42, (83)
Livingston, D. C., 53, 54, (85)
Lo, K. W., 200, 201, (216)
Lohrmann, R., 204, (217)
Long, R. A., 147, (190)
Lowe, C. R., 38, (82)
Lowy, B. A., 42, 44, (83)
Lozzio, C. B., 133, (141)
Lubineau, A., 224, (250)
Ludlum, D. B., 19, (78)
Lührmann, R., 52, (85)
Luyten, W. C., 213, (218)
Lythgoe, B., 42, (83), 233, (251)

McCarthy, J. R., Jr., 164, (192)
McClure, W. R., 44, (83)
McCormick, D. B., 36, (82)
McCoss, M., 49, (84), 144, (190)
Mackey, J. K., 172, (193)
Maeda, H., 229, (251)
Maeda, M., 41, (82)
Maelicke, A., 67, (88)
Maguire, M. H., 24, 42, (80, 83), 227, (251)
Mahapatra, G. N., 19, (78)
Maley, F., 50, 60, (84, 86)
Maley, G. F., 60, (86)
Mandel, H. G., 62, (86)
Mandel, J. A., 1, (11)
Mandell, L., 1, (11)
Mantsch, H. H., 44, (83)
Marcus, A., 14, 19, (77)
Marecek, J. F., 205, (217)
Markham, R., 62, (86)
Marsh, R. E., 2, (12)
Martin, D. M. G., 170, (193)
Martin, E., 53, 54, (85)
Martin, K. J., 60, (86)
Marumoto, R., 42, (83), 208, (217)
Mason, D. J., 76, (89)
Massen, J. A., 237, (252)
Matthaei, H., 66, (87)
Matzura, H., 244, (252)
Maurer, H., 21, (79)
Maurer, W., 101, 103, (139)
Mautner, H. G., 32, 35, (81)
Meissner, L., 63, (87)
Meloni, M. L., 162, 163, (192)
Mermann, A. L., 24, (79)
Mertes, M. P., 57, 75, (85, 89), 198, 199, (216)
Meyer, C., 62, (87), 231, (251)
Meyer, H., 198, (215)
Meyer, R. B., Jr., 41, 46, 66, 71, (82, 83, 87, 88)
Meyer, W. E., 142, (189)
Mian, A. M., 66, (87), 144, (190)
Michal, F., 42, (83)
Michelson, A. M., 1, (11), 16, 21, 28, 32, 40, 49, 50, 58, (78, 81, 82, 84), 91, 92, 94, (95), 113, (139), 195, 211, (215, 217, 218)
Mikhailov, S. N., 184, (194)
Miles, H. T., 1, 2, (11, 12), 37, 44, 50, (82, 83, 84)
Miller, E. E., 200, 201, (216)
Miller, J. P., 41, 65, 70, 71, (82, 87, 88)
Miller, R. L., 62, (86)
Milne, G. H., 34, 35, (81)
Mirbach, H., 198, (215)
Misiorny, A., 71, (88)
Mitsugi, K., 152, 153, (191)
Mittelman, A., 19, 20, (78)
Miura, K., 25, 29, (80)
Miyaki, M., 72, (88), 155, 156, (191)
Miyashita, O., 72, (88)
Mizuno, Y., 29, 66, (80, 87)
Moffatt, J. G., 59, (86), 127, 128, 129, 130, (140), 164, 165, 167, 168, (192), 198, 203, 212, 213, (215, 216, 218)
Moller, W., 237, (252)

Momparler, R. L., 76, (89), 150, 151, (190, 191)
Monastirskaja, G. S., 15, (78)
Monny, C., 40, 50, 58, (82, 84), 91, 92, (95)
Monro, R. E., 97, 99, (138)
Montgomery, J. A., 24, 42, 44, 62, (79, 80, 83, 86), 128, (140)
Morgan, M., 14, (77)
Morr, M., 119, 120, (140)
Morrice, A. G., 68, 70, (88)
Morris, F., 126, (140)
Morton, G. D., 142, (189)
Mosbach, K., 227, 228, 229, (251)
Moss, B., 14, 15, (77)
Mukaiyama, T., 200, 206, 209, (216, 217)
Müller, H., 198, (215)
Muneyama, K., 65, (87), 238, 239, (252)
Mungall, W. S., 115, (139)
Murao, K., 37, 67, (82, 87, 88)
Murayama, A., 117, 118, (140), 198, (215)
Murphy, A. J., 25, (80)
Murphy, M. L., 24, (79)
Murray, A. W., 101, (139)
Murray, J., 60, (86)
Mushika, Y., 200, 206, (216, 217)
Myers, T. C., 97, 103, 133, (138, 141)
Myers, W. P. L., 24, (79)
Mynott, R. J., 74, (89)

Nagura, T., 176, 178, (194)
Nagyvary, J., 112, (139), 149, 150, (190)
Naik, S. R., 49, (84)
Naka, T., 38, (82)
Nakajima, Y., 210, (217)
Nakamura, K., 97, 103, (138)
Narindrasorasak, S., 144, (190)
Nathan, N., 73, (89)
Neal, S. N., 14, 19, (77)
Neef, V. G., 22, (79)
Nemes, M. M., 167, (192)
Neuberg, C., 201, (216)
Neunhoeffer, H., 214, (218)
Niedballa, U., 171, (193)
Nirenberg, M., 63, (87)
Nishimura, S., 65, 67, (87, 88)
Nishimura, T., 155, 156, 160, (191), 208, (217)
Nohara, A., 186, (194), 202, (216)
Nomura, A., 66, (87), 227, 230, (251)
Noren, J. O., 71, (88)
Novak, J. J. K., 162, (191)
Nushi, K., 41, (82)
Nutter, R. L., 3, (12)

O'Carra, P., 229, (251)
Ochiai, H., 33, (81)
Odajima, K., 29, (80)
Öberg, B., 71, (88)
Ofengand, J., 136, (141), 238, (252)
Ogilvie, K. K., 176, (194)
Ohkawa, H., 204, (217)
Ohtsuka, E., 38, 39, 65, (82, 87), 172, 176, 178, (193, 194)
Ojala, D., 97, 98, 99, (138)
Okazaki, H., 205, (217)
Olfermann, G., 186, (194)
O'Neal, C., 63, (87)
Onoue, Y., 46, (83)
Onur, G., 249, (253)
Opar, G. E., 36, (82)
Ortanderl, F., 230, 231, (251)
Ortiz, P. J., 150, (190)
Osaki, S., 57, (85)
Ott, D. G., 214, (218)
Ouchi, S., 201, (216)
Ovchinnikov, Y. A., 237, 239, (252)
Owen, G. R., 167, (192)
Oyen, T. B., 77, (89)
Øzer, I., 161, 163, (191)

Paces, V., 76, (89)
Pal, B. C., 171, (193)
Pal, C. P., 35, (81)

Parikh, I., 99, (138)
Parikh, J. R., 125, (140)
Parks, R. E., Jr., 60, 61, 62, 63, (86, 87)
Parmeggiani, A., 98, 99, 135, 136, (138, 141), 224, (251)
Paul, A. V., 125, 126, 127, (140)
Paul, B., 132, 133, (141)
Paul, I. C., 22, 23, (79)
Penades, S., 37, (82), 240, (252)
Penzer, G. R., 22, (79)
Perini, F., 24, (80), 130, 131, 132, 133, (141)
Pettit, G. R., 1, (11)
Pfitzner, K. E., 127, (140), 164, (192)
Pfleidere, G., 66, (87)
Phillips, R., 5, (12)
Philipson, L., 71, (88)
Pischl, H., 77, (89)
Piskala, A., 75, (89)
Pizer, L. I., 145, (190)
Pleiss, M. G., 33, (81)
Ploeser, J. McT., 60, (86)
Plunkett, W., 150, (190)
Pochon, F., 16, 28, 32, (78, 81), 94, (95)
Podnyakov, V. A., 237, 239, (252)
Pollak, H., 201, (216)
Pomerantz, A. H., 239, (252)
Powell, J. T., 234, (252)
Prasolov, V. S., 184, (194)
Preston, R. K., 226, 227, (251)
Provenzale, R. G., 149, 150, (190)
Prusiner, P., 67, (88)
Prusoff, W. H., 48, 49, (84), 244, (252)
Prydz, H., 150, (190)
Ptak, M., 38, 39, (82)
Pugh, L. H., 162, (192)

Quiggle, K., 136, (141), 238, (252)

Rabinowitz, M., 95, (95)
Rack, M., 230, (251)
Rackwitz, H. R., 25, 27, 28, 30, 33, 44, 51, (80, 83, 84), 244, (252)
Rahman, A., 6, (12)
Raleigh, J. A., 178, 179, (194)
Ramani, G., 49, (84)
Ramirez, F., 205, (217)
Rammler, D. H., 125, 126, 127, (140), 233, (251)
Randerath, K., 18, (78)
Ratgeber, G., 242, (252)
Rees, C. W., 220, (250)
Reese, C. B., 14, 15, 16, 19, (77, 78), 170, (193), 198, 199, (216), 247, (253)
Reich, E., 25, 36, 37, 43, 44, 58, 62, 63, 67, (80, 82, 83, 86, 87, 88)
Reichard, P., 150, (191)
Reist, E. J., 143, (189)
Remy, D. C., 49, (84)
Remy, P., 136, 137, (141)
Revankar, G. R., 68, (88), 144, (190)
Revel, M., 136, (141)
Reverman, L. F., 167, (192)
Reyes, P., 57, (86)
Rhaese, H. J., 19, (78)
Rich, A., 6, (12), 56, (85), 163, (192)
Richardson, D. J., 102, (139)
Rimpler, M., 37, (82)
Rinehart, R. R., 90, (95)
Roberts, M., 52, (85)
Robins, M. J., 49, (84), 164, (192)
Robins, R. K., 15, 32, 35, 36, 41, 46, 63, 65, 66, 68, 70, 71, (77, 81, 82, 83, 87, 88), 90, (95), 130, (141), 144, 147, 164, 169, 170, (190, 192, 193), 238, 239, (252)
Roblin, R. O., Jr., 62, (86)
Roboz, J., 71, (88)
Robson, J. M., 14, (77)
Roe, J., 112, (139)
Roesler, G., 119, (140)
Rose, K. M., 150, (190)
Rose, S. D., 23, 24, (79)
Roseman, S., 213, (218)

Rottman, F. M., 14, (77), 170, 171, 172, (193)
Rousseau, R. J., 63, 66, (87)
Rowohl-Quisthoudt, G., 37, (82)
Roy, J. K., 62, (86)
Roy-Burman, P., 24, 50, 52, 59, 60, 61, 64, 65, 67, (79, 84, 85, 86, 87), 150, (190)
Roy-Burman, S., 52, 60, 61, (85, 86)
Rudolph, S. A., 239, (252)
Russel, A. F., 164, 165, (192)
Russel, J., 248, (253)
Rüterjans, H., 101, 103, (139)
Rutman, R. J., 5, (12)

Saegusa, T., 243, (252)
Saenger, W., 5, 6, 7, 8, 9, (12), 32, 56, 58, 74, (81, 85, 86, 89), 102, 106, (139)
Saffhill, R., 203, (216)
Saito, A., 72, (88), 155, 156, (191)
Saito, T., 42, (83)
Sakaguchi, K., 173, (193)
Sakai, T. T., 57, (86)
Salditt-Georgieff, M., 14, (77)
Samejima, T., 32, (89)
Sami Khan, M., 168, (192)
Sampson, S. D., 162, 163, (192)
Samukov, V. V., 225, (251)
Saneyoshi, M., 24, 28, 29, 32, (80, 81)
Sanno, Y., 202, (216)
Santi, D. V., 57, (86)
Sarma, R. H., 63, 74, (87, 89)
Sasaki, T., 132, 133, (141), 177, 178, (194), 226, 227, (251)
Sato, E., 33, (81)
Sato, T., 42, (83)
Sattsangi, P. D., 22, 23, (79)
Sawada, F., 28, 32, (80, 81), 245, (253)
Schäfer, E. A., 18, (78)
Schäfer, G., 37, (82), 240, 249, (252, 253)
Schäfer, H. J., 37, (82), 242, (252)
Schaller, H. 214, (218)
Schasteen, C. S., 235, (252)
Schattka, K., 116, 117, (140), 208, (217)
Schaub, R. E., 162, (192)
Scheit, K. H., 6, (12), 15, 17, 21, 25, 27, 28, 29, 30, 31, 32, 33, 37, 43, 44, 51, 52, 56, 58, (77, 78, 80, 81, 83, 84, 85), 113, (139), 161, 163, 171, (191, 193), 197, 198, 199, 214, 215, (215, 216, 218), 244, 245, (252, 253)
Scheurich, P., 37, (82), 242, (252)
Schimmelschmidt, K., 198, (215)
Schindler, R., 73, (89)
Schirmer, R. H., 107, 109, (139)
Schlimme, E., 179, 184, (194), 225, (250)
Schlingloff, G., 44, (83)
Schmidt, A., 14, 19, (77)
Schmidt, D. G., 171, (193)
Schmitz, R. Y., 19, 20, (78)
Schofield, J. A., 120, 122, (140), 207, (217)
Scholten, M. B., 65, (87)
Schrader, E., 37, (82)
Schrecker, A. W., 223, 224, (250)
Schroeder, W., 143, (189)
Schulz, G., 30, (80)
Schulz, H. H., 101, 103, (139)
Schwalbe, C. H., 106, (139)
Schwarz, U., 52, (85)
Schweizer, M. P., 32, 71, (81, 88)
Scopes, D. I. C., 69, 70, (88)
Secrist, J. A., III, 21, 22, 23, 33, (79, 81)
Seela, F., 17, (78), 236, (252)
Seibert, W., 72, (88)
Seidel, H., 18, (78)
Seita, T., 179, 180, (194)
Sekine, M., 208, 212, (217, 218)
Setondji, J., 137, (141)
Shafer, J., 220, (250)
Shaffer, P. J., 57, (85)
Shapiro, P., 212, (218)
Shapiro, R., 21, 36, (79, 82), 93, (95)
Shapshak, P., 162, (191)

Sharma, O. K., 14, (77)
Shatkin, A. J., 14, (77)
Shaw, E., 220, (250)
Shaw, G., 56, 70, (85, 88)
Shdanov, G. L., 60, 61, (86)
Shefter, E., 6, (12), 32, (81)
Shibaev, V. N., 15, 17, 19, 21, 28, 30, 31, 36, 55, 59, 60, 61, (77, 78, 79, 80, 82, 85, 86)
Shibaeva, R. P., 15, (78)
Shigeura, H. T., 161, 162, 163, (191, 192)
Shimada, Y., 206, 210, (217)
Shimizu, B., 72, (88), 144, 153, 154, 155, 156, 160, (190, 191)
Shimizu, T., 243, (252)
Shire, D., 203, (216)
Shishkin, G. V., 225, (251)
Shiue, C. Y., 34, 38, 39, (81, 82)
Shiuey, S. J., 21, (79)
Shugar, D., 1, 3, (11, 12), 16, 17, 21, 44, 56, (78, 83, 85), 171, (193)
Shulman, L. D. H., 95, (95)
Shuman, D. A., 46, 65, (83, 87), 238, 239, (252)
Sidwell, R. W., 66, 70, 71, (87, 88), 144, 147, (190)
Simon, L. N., 41, 65, 66, 70, 71, (82, 87, 88), 133, (141), 147, (190), 238, 239, (252)
Simonson, L. P., 103, (139)
Simpson, P. J., 91, (95)
Sims, E. S., 49, (84)
Simuth, J., 159, 171, (191, 193)
Singer, B., 13, 14, 21, (77)
Singh, A., 220, (250)
Sinsheimer, R. L., 3, (12)
Skoda, J., 72, 73, 74, 75, 76, (88, 89), 134, (141)
Skoog, F., 19, 20, (78)
Slapikoff, S., 49, (84)
Slomp, G., 143, (189)
Slotin, L. A., 132, 133, (141), 176, (194), 195, (215), 228, 231, 232, (251)
Smith, H. W., 19, (78)
Smith, J. D., 233, (251)
Smith, M., 134, (141), 198, 205, (216)
Smrt, J., 50, 62, 75, (84, 86, 89), 122, 124, 125, (140), 146, (190), 200, 201, 207, (216, 217)
Sobell, H. M., 36, 39, (82)
Solotorovsky, M., 162, (192)
Sommer, H., 33, (81)
Sommer, R. G., 167, (192)
Son, T. D., 8, (12)
Sono, M., 33, (81)
Sorm, F., 50, 58, 63, 72, 73, 74, 75, 76, (84, 86, 87, 88, 89), 122, 123, 124, 125, (140), 147, 151, 152, 156, 157, 158, 159, 162, (190, 191), 207, (217)
Sowa, T., 201, (216)
Sparkes, M. J., 137, 138, (141)
Spencer, R. D., 22, (79)
Speth, C., 91, (95)
Spiegel, A. M., 97, 99, (138)
Spiridinova, S. M., 15, 28, 36, (77, 80, 82)
Sporn, M. B., 168, (192), 223, 224, (250)
Sprecker, M. A., 68, 70, (88)
Spring, F. S., 161, (191)
Sprinzl, M., 30, 37, 42, 67, (80, 82, 88), 161, 167, 169, (191, 192, 193)
Staab, H. A., 214, (218)
Staub, M., 150, (191)
Stenberg, K., 71, (88)
Stening, G., 71, (88)
Sternbach, H., 56, (85), 106, 110, (139), 161, 167, 168, 169, (191, 192, 193), 225, 232, 244, (250, 251, 252)
Stevens, C. L., 168, (192), 223, (250)
Stevens, M. A., 18, 19, (78)
Stevenson, R., 19, (78)
Stewart, J. M. B., 198, 199, (216)
Stockx, J., 4, (12)
Stout, M. G., 66, (87)
Stowring, L., 25, (80)

Streeter, D. G., 66, 70, 71, (87, 88), 130, (141)
Strehlke, P., 30, (80), 171, (193)
Stridh, S., 71, (88)
Strotmann, H., 249, (253)
Stryer, L., 43, 44, (83)
Stütz, A., 113, (139), 197, (215)
Suck, D., 32, 58, 74, (81, 86, 89)
Suda, K., 46, (83)
Suhadolnik, R. J., 14, 65, (77, 87), 161, (191)
Sulston, J. E., 247, (253)
Sund, H., 37, (82), 239, 240, 241, (252)
Sundaralingam, M., 2, 9, (11, 12), 32, 67, (81, 88), 129, (140)
Sunthankar, D. C., 49, (84)
Suzaki, S., 152, 153, (191)
Suzuki, H., 229, (251)
Swerdlov, E. D., 15, (78)
Swierkowski, M., 56, (85)
Switzer, R. L., 23, 49, (79)
Sykes, M. P., 24, (79)
Szabo, L., 53, (85)
Szekeres, G. L., 147, (190)
Szer, W., 17, 56, (78, 85)

Taguchi, Y., 206, (217)
Tahara, K., 46, (83)
Takaku, H., 206, 210, (217)
Takei, S., 186, (194)
Takenishi, T., 18, 42, (78, 83), 201, 202, (216)
Tan, T. C., 24, (79)
Tanaka, S., 172, (193)
Tapiero, C. M., 149, (190)
Tavale, S. S., 36, 39, (82)
Taylor, M. J., 187, (194)
Taylor, W. G., 187, (194)
Tazawa, I., 35, 36, 38, (81, 82)
Tazawa, S., 171, (193)
Tener, G., 203, (217)
Tezuka, T., 67, (88)
Thedford, R., 17, 56, (78, 85)
Thelander, L., 235, 237, (252)
Thewalt, U., 26, (80)
Thiry, L., 52, (85)
Thomas, H. J., 24, 62, (80, 86)
Thornton, E. R., 220, (250)
Toda, J., 42, (83)
Todd, A. R., 14, 15, 16, 19, 42, 56, (77, 83, 85), 120, 122, (140), 207, 211, (217), 233, (251)
Toji, L., 164, 165, (192)
Tole, H. C., 161, (191)
Tolman, G. L., 22, (79)
Tolman, R. L., 65, (87), 144, (190)
Tomasz, M., 95, (95)
Tomaszewski, M., 14, 19, (77)
Torrence, P. F., 55, 65, (85, 87), 168, (192)
Townsend, L. B., 34, 35, 36, (81)
Trentham, D. R., 247, (253)
Tritsch, G. L., 19, 20, (78)
Trowbridge, D. B., 97, 115, (140), 203, (216)
Trueblood, K. N., 6, (12)
Truman, J. T., 162, 163, (192)
Ts'o, P. O. P., 171, (193)
Tsou, K. C., 64, (87), 200, 201, (216)
Turner, A. F., 200, (216)
Tytell, A. A., 167, (192)

Uchic, J. T., 20, (78), 134, (141)
Uchic, M. E., 20, (78)
Uchida, K., 176, (194)
Ueda, T., 25, 29, 59, (80, 86), 202, (216)
Uematsu, T., 65, (87), 161, (191)
Uesugi, S., 36, 37, 38, 39, 40, 67, (82, 88), 149, 176, (190, 194)
Uhlenbeck, O. C., 173, (193)
Usher, D. A., 102, 104, (139)

van Boom, J. H., 213, (218)
van Broeckhoven, C., 54, (85)
van der Haar, F., 37, 42, 67, (82, 88), 110, (139), 161, 184, (191, 194), 225, (250)
van der Kraan, J., 37, (82)
van der Lijn, P., 69, 70, (88)

van Lear, G. E., 142, (189)
van Praag, D., 28, (80)
van Tamelen, E. E., 143, (189)
Vaughan, J. R., 62, (86)
Verheyden, J. P. H., 164, 167, 168, (192)
Vesely, J., 75, (89)
Vienze, A., 19, 20, (78)
Vink, A. B., 213, (218)
Visser, D. W., 50, 52, 60, 61, (84, 85, 86)
Vizsolyi, J. P., 203, (216)
Voet, D., 150, (190)
Von Tigerstrom, R., 134, (141)
Vorbrüggen, H., 30, (80), 171, (193)
Vortruba, I., 77, (89)

Wagenvoord, R. J., 37, (82)
Wagner, D., 168, (192)
Waldek, S., 236, (252)
Walker, G. C., 19, 20, (78), 173, (193)
Wallace, J. L., 235, (252)
Wang, A. H.-J., 22, 23, (79)
Ward, D. C, 43, 44, 53, 54, 63, 67, (83, 85, 87, 88)
Warner, H. R., 150, (191)
Warren, S. G., 195, (215)
Watanabe, K. A., 186, (194)
Waters, J. A., 65, (87), 168, (192)
Watson, D. R., 143, (189)
Way, J. L., 62, (86)
Webb, T. E., 63, (87)
Weber, G., 21, 22, (79)
Webster, D., 137, 138, (141)
Wechter, W. J., 144, 146, 147, 149, (190)
Wei, C. M., 14, 15, (77)
Weimann, G., 203, (216)
Weinhouse, S., 161, (191)
Weissman, S., 62, (87)
Welch, A. D., 49, 50, 72, 73, (84, 88, 89)
Wempen, I., 24, 26, 27, 28, 43, 49, (79, 80, 84)
Werner, D., 66, (87)
Westheimer, F. H., 220, 246, (250, 253)
Wetzel, R., 227, 228, (251)
Whistler, R. L., 174, 175, (193)
Wiegand, G., 44, 45, (83)
Wightman, R., 77, (89)
Wigler, P. W., 133, (141)
Wilchek, M., 221, 225, 230, (250)
Wilkes, J. S., 115, 116, (140), 203, (216)
Willett, R., 98, (138)
Wilson, D. P., 167, (192)
Wilson, H. R., 6, (12)
Wilson, R. G., 65, (87)
Wist, E., 150, (190)
Witkop, B., 65, (87), 168, (192)
Witkowski, J. T., 32, 66, 70, 71, (81, 87, 88), 130, (141)
Wittmann, R., 134, 135, (141)
Witzel, H., 198, (215)
Woenckhaus, C., 66, (87)
Wold, F., 235, (252)
Wolff, M. E., 125, (140)
Wood, D. J., 74, (89)
Woodruff, H. B., 161, (191)
Woodside, G. L., 62, (86)
Wu, J. M., 14, (77)
Wu, S. E., 235, (252)

Yamada, Y., 42, (83)
Yamaji, N., 45, 46, 64, (83, 87)
Yamamoto, D. M., 97, 115, (140), 203, (216)
Yamamoto, M., 59, (86)
Yamane, A., 59, (86)
Yamaguchi, K., 179, 180, (194)
Yamazaki, A., 18, 42, (78, 83), 152, 153, (191)
Yamazaki, Y., 229, (251)
Yano, J., 144, (190)
Yano, M., 33, (81)
Yelisseeva, G. I., 31, 59, 60, (80)
Yengoyan, L., 125, 126, 127, (140)
Yip, K. F., 64, (87)
Ynasa, Y., 45, (83)

York, J. L., 150, (190)
Yoshida, K., 39, 40, (82), 176, (194)
Yoshikawa, M., 201, 202, (216)
Yoshioka, Y., 42, (83)
Yount, R. G., 97, 98, 99, 100, 134, 135, (138, 141)
Yugeoglu, M., 24, (79)
Yung, N. C., 49, (84)
Zarytova, V. F., 136, (141), 225, 237, 239, (251, 252)
Zaychikov, E. F., 136, (141)
Zeile, K., 198, (215)
Zemlicka, J., 174, (193)
Zhdanov, R. I., 200, (216)
Zhenodarova, S. M., 200, (216)
Ziff, E. B., 33, (81)
Zimmermann, S. B., 49, (84)
Zmudzka, B., 171, (193)

Subject Index

N^4-Acetylcytidine 2′,3′-cyclic phosphate, 149
N-Acetyl-5-iodo-3-indolylphosphorodichloridate, 201
Active-site-directed reagents, carbene generating, 220
Active-site-directed reagents, photogenerated, 220
Adenine, 9-S-(2′,3′-dihydroxypropyl), 182
Adenine nucleotides:
  8-azido, reaction with thiols, 243
  8-chloro, 39
Adenine-N^1-oxide derivatives, photoreaction, 44
Adenine-9-(β-D-ribofuranosyl uronic acid):
  mixed anhydride with tripolyphosphate, 226
  mixed anhydride with phosphate, 226
Adenosine:
  2-amino, 44
  2-bromo, 42
  2-chloro, 42
  2-methyl, 42
β-L-Adenosine, biological properties, 159
Adenosine 5′-bis-(dihydroxyphosphinylmethyl)-phosphinate, 97
Adenosine 3′,5′-cyclic phosphate:
  2-aza, 64
  8-azido-1,N^6-etheno, photolysis, 242
  8-azido, photoaffinity labelling of protein kinase, bovine brain, 238, 239
  1,N^6-etheno, 45, 64
  2′-(ethyl-2-diazomalonyl), 247
    photoaffinity labelling of rabbit muscle fructokinase, 247
  2-halogeno, 46
  phosphoramidate analogs, 118
    properties with cAMP-dependent kinase, 118
  thiophosphoramidate analogs, 118
Adenosine 3′,5′-cyclic phosphorothioate, 103
  hydrolysis by phosphodiesterase, 103
Adenosine 5′-deoxy-5′-thiophosphate, 113
Adenosine 5′-deoxy-5′-thiotriphosphate, 113
  properties with hexokinase, 113
  properties with RNA-polymerase, 113
β-L-Adenosine 5′-diphosphate, 159
Adenosine 5′-diphosphate:
  2-amino, 44
  N^6-(6-aminohexyl)carbamoyl, 227
  8-azido, 37
    photoaffinity labelling of glutamate dehydrogenase, 239
    photoaffinity labelling of nucleotide carrier from beef heart mitochondria, 240
    photolysis, 239

8-bromo, 36
2-chloro, 42
dialdehyde, affinity labelling of sulfotransferase, 235
2-dimethylamino, 44
1,N^6-etheno, 22
2′-*O*-ethyl, 170
N^6-fluorobenzoyl, 231
2-methyl, 42
N^1-methyl, 15
Dimroth rearrangement, 15
N^6-methyl, 15
2′-*O*-methyl, 171
8-oxy, 37
oxidation with periodate, 184
N^1-oxide, 18
8,2′-thioanhydro, 176
Adenosine 5′-*N*-diphosphate, 116
properties with polynucleotide phosphorylase, 116
Adenosine 5′-(α-D-glucopyranosyl pyrophosphate:
8-bromo, 36
N^1-methyl, 15
Adenosine 5′-hypophosphate, 137
properties with kinases, 137
Adenosine 5′-methylenediphosphonate, *see* α,β-Methylene ADP
Adenosine 5′-methylenediphosphono P²-phosphate, *see* α,β-Methylene ATP
Adenosine-N^1-oxide, photoreaction, 44
β-L-Adenosine 5′-phosphate, 159
Adenosine 5′-phosphate:
N^6-adamantyl, 19
2-amino, 44
N^6-(6-aminohexyl)carbamoyl, 229
D,L-anantiomers, 154
antibodies, specific for, 39
N^6-benzyl, 19
8-bromo, 36
2-bromo, 42
2-chloro, 42
2-dimethylamino, 44
1,N^6-etheno, 21
conformation of, 22
2′-(ethyl-2-diazomalonyl), 247
N^6-fluorobenzoyl, 231
N^6-furfuryl, 19
N^6-Δ^2-isopentenyl, 19
2-methyl, 42
N^1-methyl, 15
Dimroth rearrangement, 15
N^6-methyl, 15
2′-*O*-methyl, 170
N^1-oxide, 18, 25
reaction with cyanoethylene, 72
8,2′-thioanhydro, 177
Adenosine 5′-phosphoazidate, photolysis, 283
Adenosine 5′-phosphohypophosphate, 137
properties with tRNA synthetase and hexokinase, 137
Adenosine 5′-phosphorothioate, 101
Adenosine 5′-(*O*-1-thiodiphosphate), diastereomers, 106
Adenosine 5′-(*O*-2-thiodiphosphate), 105
Adenosine 5′-(*O*-1-thiotriphosphate):
diastereomer A, 106
enzymatic resolution of diastereomers, 106
Adenosine 5′-(*O*-2-thiotriphosphate), 105
absolute stereochemistry of diastereomers, 110
enzymatic resolution of diastereomers, 106
Adenosine 5′-triphosphate:
N^6-(6-aminohexyl)carbamoyl, 228, 229
2-amino, 44
8-amino, 37
8-azido, 37
photoaffinity labelling of Na^+,K^+-ATPase and Mg^{++}-ATPase from human red cells, 238
photolysis, 243
8-azido-1,N^6-etheno, photolysis, 242

4-azido-2-nitrophenyl caproic acid ester, 249
4-azido-2-nitrophenyl aminobutyric acid ester, 249
8-bromo, 37, 46
2-chloro, 42
dialdehyde, affinity labelling of pyruvate carboxylase, 235
1,N^6-etheno, 22
N^6-fluorobenzoyl, affinity labelling of kinases, 232
8-mercapto, 37
2'-*O*-methyl, 171
properties with RNA polymerase, 172
3'-*O*-methyl, properties with RNA polymerase, 172
oxidation with periodate, 184
N^1-oxide, 18
sepharose bound, substrate for hexokinase, 228, 229
β-L-Adenosine 5'-triphosphate, 159
properties with RNA polymerase from *E. coli*, 159
Adenosine 5'-*N*-triphosphate, 115
properties with hexokinase, 115
Adenylosuccinate AMP-lyase, irreversible inhibition, 226
Adenylyl 5'-chloromethylphosphonate, 133
Adenylyl 5'-imidodiphosphate, *see* β,γ-imido ATP
Adenylyl 5'-methylenediphosphonate, *see* β,γ-methylene ATP
Adenylyl 5'-methylphosphonate, 133
N^6-Alkyladenine derivatives, cytokinine activity, 19
P^3-Alkyl P^1-5'-nucleoside 5'-triphosphates, 135
properties with RNA polymerase, 136
2-Aminoadenine nucleotides, biochemical properties, 44
8-Aminoadenosine 5'-phosphate, conformation in solution, 40
N^6-(ω-Aminoalkyl)adenine nucleotides, coupling to Sepharose, 227
P^3-(ω-Aminoalkyl-1-yl)nucleoside triphosphate esters, 225
1-(3'-Amino-3'-deoxy-β-D-glucopyranosyl)cytosine 6'-phosphate, 186
1-(3'-Amino-3'-deoxy-β-D-glucopyranosyl)uracil 6'-phosphate, 186
9-(6-Aminomethyl-6-deoxy-β-D-allofuranosyl)adenine 5'-phosphate, 188
9-(6-Aminomethyl-6-deoxy-L-talofuranosyl)adenine 5'-phosphate, 188
(4-Amino-5-carboxamidopyrrolo(2,3-d)pyrimidine-7-β-D-ribofuranose, *see* Sangivamycin
2-Amino-6-chloro-9-(β-D-ribofuranosyl)-purine 5'-diphosphate, 20
2-Amino-6-chloro-9-(β-D-ribofuranosyl)-purine 5'-phosphate, 20
(4-Amino-5-cyanopyrrolo(2,3-d)pyrimidine)-7-β-D-ribofuranose, *see* Toyocomycin
P^3-(6-Aminohex-1-yl)2'-deoxyguanosine 5'-triphosphate, coupling to Sepharose, 225
3'-(4-Aminophenylphosphoryl)2'-deoxythymidine, diazotization, 223
5'-(4-Aminophenylphosphoryl)2'-deoxythymidine, diazotization, 224
3'-(Aminophenylphosphoryl)2'-deoxythymidine 5'-phosphate, 221
diazotization, 223
6-Amino-9-(β-D-psicofuranose-enyl)-adenine, 143
6-Amino-9-(β-D-psicofuranosyl)adenine, 142
2-Amino-7-(5-phospho-β-D-ribofuranosyl)-pyrimidine(1,2-c)-pyrimidine 5-ium-6(7H)-on, 72
9-Amino-3-(phospho-β-D-ribofuranosyl)-pyrimido(2,1-i)-purin 5-ium, 72
2-Aminopurine, mutagenic function, 43

2-Aminopurine nucleotides, biochemical properties, 43
2-Aminopurine-9-β-D-ribofuranoside, 27
2-Aminopurine-9-(β-D-ribofuranoside):
 5′-diphosphate, 43
 5′-phosphate, 43
 5′-triphosphate, 43
5-Amino-1-(β-D-ribofuranosyl)imidazole-4-carboxamide 5′-phosphate, 42
5-Amino-1-(β-D-ribofuranosyl)imidazole-4-carboxylic acid 5′-phosphate, 70
2-Amino-8-(β-D-ribofuranosyl)imidazo(1,2-a)-s-triazine-4-on, *see* 5-Aza-7-deazaguanosine
7-Amino-3-(β-D-ribofuranosyl)pyrazolo(4,3-d)pyrimidine, *see* Formycin
4-Amino-1-(β-D-ribofuranosyl)pyrazolo(3,4-d)pyrimidine-3-carboxamide, *see* 6-Azasangivamycin
AMP amino hydrolase, irreversible inhibition, 226
AMP kinase, rabbit muscle, irreversible inhibition, 226
8,2′-*O*-Anhydroadenosine 5′-phosphate, 144
8,5′-Anhydro-8-oxy-adenosine 2′,3′-cyclic phosphate, 178
P^3-Anilido P^1-5′-adenosine 5′-triphosphate, 136
Arabinoadenosine, 143
 biological properties, 150
 crystal structure, 150
Arabinoadenosine 3′,5′-cyclic phosphate, 144
Arabinoadenosine 5′-phosphate, 143
 8,2′-anhydro-8-oxy, 177
  properties with various enzymes, 177
Arabinoadenosine 3′-phosphate,8,2′-anhydro-8-oxy, 176
Arabinoadenosine 5′-phosphate,-N^1-oxide, 25
Arabinoadenosine 5′-triphosphate, 144
 8,2′-anhydro-8-oxy, properties with adenine nucleotide carrier of rat liver mitochondria, 179
 biochemical properties, 150
 enzymatic synthesis, 144
9-(β-D-arabinofuranosyl)adenine, *see* Arabinoadenosine
Arabinocytidine, biological properties, 150
Arabinocytidine 3′,5′-cyclic phosphate, 147
Arabinocytidine 3′-phosphate, 149
Arabinocytidine 2′,(3′)-phosphate, 146
Arabinocytidine 5′-triphosphate, biochemical properties, 151
1-(β-D-Arabinofuranosyl)cytosine 5′-phosphate, 145
9-(β-D-Arabinofuranosyl)hypoxanthine 5′-phosphate, 144
Arabino-4-thiouridine 3′,5′-cyclic phosphate, 147
Arabinouridine, reaction with triethyl phosphite, 146
Arabinouridine 3′,5′-cyclic phosphate, 147
Arabinouridine 5′-phosphate, 146
P^2-Arsonomethyl P^1-5′-adenosine 5′-diphosphate, 137
ATP, interaction with divalent cations, 99
 pK_a-values, 99
8-Azaadenosine 5′-phosphate, conformation, 63
5-Azacytidine:
 alkaline hydrolysis, 76
 biological properties, 76
5-Azacytidine 5′-phosphate, 76
5-Azacytidine 5′-triphosphate, 76
 properties with RNA polymerase from *E. coli*, 76
6-Azacytidine, biological properties, 74
6-Azacytidine 5′-diphosphate, 75

6-Azacytidine 5′-phosphate, 75
6-Azacytidylic acid, function in mRNA, 74
5-Aza-7-deazaguanosine, 67, 68
5-Aza-7-deazaguanosine 5′-phosphate, 68
8-Azaguanine, biological properties, 62
8-Azaguanosine 5′-phosphate, 63
8-Azaguanosine 5′-triphosphate, properties with RNA polymerase from *E. coli*, 62
8-Azaguanylic acid, biochemical properties, 63
6-Azasangivamycin, 66
6-Azasangivamycin 5′-diphosphate, 66
6-Azasangivamycin 5′-phosphate, 66
6-Azauracil, biological properties, 72, 73
β-L-6-Azauridine, biological properties, 159
6-Azauridine 5′-methylphosphonate, 133
6-Azauridine 5′-diphosphate, 73
6-Azauridine 3′-phosphate, 73
6-Azauridine 5′-phosphate, 73
  crystal structure, 74
  inhibitor of orotidylate decarboxylase, 74
6-Azauridine 5′-phosphite, reaction with hexachloroacetone, 124
6-Azauridine 5′-triphosphate, properties with RNA polymerase from *E. coli*, 73
6-Azauridylic acid, function in protein synthesis, 73
P^3-Azido P^1-5′-guanosine 5′-triphosphate, properties with elongation factor G GTPase, 136
3′-*O*-[*N*-(4-Azido-2-nitrophenyl)amino]-butyryladenosine 5′triphosphate, affinity labelling of ATPase, 249
3′-*O*-[*N*-(4-Azido-2-nitrophenyl)amino]-proprionyladenosine 5′-phosphate, 248
3′-*O*-[*N*-(4-Azido-2-nitrophenyl)amino]-proprionyladenosine 5′-triphosphate, photoaffinity labelling of F_1 ATPase, myosin subfragment 1 ATPase and heavy meromyosin, 248
1-(4-(Azidophenyl)-2-(5′-guanylyl)-pyrophosphate, affinity labelling of elongation factor G GTPase, 237

S-Benzoyl-(3′-dephospho-8-azido) coenzyme A, properties with acyl coenzyme A-glycine *N*-acyltransferase, 241
Bis-(2-*t*-butylphenyl)phosphorochloridate, 199
Bis-(2-cyanoethyl)phosphorochloridate, 198
Bis-(4-nitrophenyl)phosphorochloridate, 198
  synthesis of nucleoside 5′-phosphate 4-nitrophenyl esters, 198
Bis-(2,2,2-trichloroethyl)phosphorochloridate, 198
8-Bromoadenosine 5′-phosphate, conformation in solution, 39
P^3-Bromoacetamidoethyl P^1-5′-guanosine 5′-triphosphate, 225
3′-(4-Bromoacetamidophenylphosphoryl)2-deoxythymidine 5′-phosphate, 222
5′-(4-Bromoacetamidophenylphosphoryl)2′-deoxythymidine, irreversible inhibition of Staphylococcal nuclease, 222
8-Bromopurine nucleosides, crystal structures, 36
8-Bromopurine nucleotides, nucleophilic reactions, 37
5-Bromo-2′-deoxyuridine 5′-methylphosphonate, 133
  biological properties, 133

N^6-Carboxymethyladenosine 5′-triphosphate, 228
2-Cyanoethylphosphate, activation by

dicyclohexylcarbodiimide, 203
2-Cyanoethyl,2-pyridiylphosphate, reaction with phosphate anions, 215
6′-Cyanohomoadenosine 6′-phosphonate, 132
  epimeric forms, 132
  properties with AMP kinase and AMP aminohydrolase, 132
2-Chloro-4,5-dimethyl-2-oxo-P^{V}-1,3,2-dioxaphosphole, 205
2-Chloromethyl-4-nitrophenylphosphorochloridate, 200
2-Chloro-2-oxo-P^{V}-1,3,2-benzo-dioxaphosphole, 199
6-Chloropurine-9-(β-D-ribofuranosyl)-5′-phosphate, 20, 227
6-Chloropurine-9-(β-D-ribofuranosyl)-5′-diphosphate, 20
6-Chloropurine-9-(β-D-ribofuranosyl)-5′-triphosphate, reaction with aminohexyl-Sepharose, 227
8,5′-Cycloadenosine 5′-phosphates:
  mixture of epimers, 178
  stereoselective synthesis, 178
β-L-Cytidine, biological properties, 159
Cytidine:
  3,N^4-etheno:
    crystal structure, 23
    fluorescence, 23
  5-hydroxymethyl, 50
  tautomeric forms, 1
β-L-Cytidine 5′-diphosphate, 159
Cytidine 5′-diphosphate:
  5-chloro, 50
  5-methyl, 56
  6-methyl, 58
  N^3-methyl, 15
  2′-*O*-methyl, 171
  2-thio, 30
Cytidine 2′,3′-cyclic phosphate, reaction with trimethylsilyl chloride, 149
β-L-Cytidine 5′-phosphate, 159
Cytidine 5′-phosphate:
  conversion to arabino derivatives, 149, 150
  5-dimethylamino, 52
  3,N^4-etheno, 23
  2′,(3′)-*O*-ethyl, 171
  5-ethyl, 56
  5-fluoro, 49
  5-methyl, 56
  6-methyl, 58
  N^3-methyl, 15
  2′-*O*-methyl, 170
  N^3-oxide, 18
  reaction with cyanoethylene, 72
  2-thio, 30
Cytidine 5′-triphosphate:
  5-fluoro, 49
  6-methyl, 58
  2-thio, 30
Cytosine,-1-*S*-(2′,3′-dihydroxypropyl), 182

7-Deazaadenosine, *see* Tubercidine
3-Deazaadenosine 5′-diphosphate, 66
1-Deazaadenosine 5′-diphosphate, 66
1-Deazaadenosine 5′-phosphate, 66
3-Deazaadenosine 5′-phosphate, 66
3-Deazaguanosine, 66
7-Deazainosine 5′-diphosphate, 65
7-Deazainosine 5′-phosphate, 65
7-Deazanebularin, 67
  biological properties, 67
7-Deazanebularin 5′-diphosphate, 67
7-Deazanebularin 5′-triphosphate, 67
  properties with RNA polymerase from *E. coli*, 67
1-Deazapurine-3-(β-D-ribofuranosyl)-5′-phosphate, 66
3′-Deoxyadenosine, biological properties, 161
3′-Deoxyadenosine 5′-phosphate, 161
2′-Deoxyadenosine 5′-phosphate:
  N^1-oxide, 25
  N^7-oxide, 19
2′-Deoxyadenosine 5′-phosphite, 122
3′-Deoxyadenosine 5′-triphosphate, 161
  biochemical properties, 161

5′-Deoxy-5′-alkylaminoadenosine 3′-phosphoric acid-(4-nitrophenyl-ester)-3′,5′-cycloamide, 117
3′-Deoxy-3′-aminoadenosine, biological properties, 162
3′-Deoxy-3′-aminoadenosine 3′,5′-cyclic phosphate, 119
properties with cAMP-dependent kinase, 119
3′-Deoxy-3′-aminoadenosine 3′,2′-cyclic phosphorothioate, 119
3′-Deoxy-3′-aminoadenosine 3′,5′-cyclic phosphorothioate, 119
3′-Deoxy-3′-aminoadenosine 3′,5′-cyclic phosphorothioate, separation of diastereomers, 119, 120
3′-Deoxy-3′-aminoadenosine 5′-phosphate, 162
3′-Deoxy-3′-aminoadenosine 5′-triphosphate, biochemical properties, 162
2′-Deoxy-2′-aminoadenosine 5′-triphosphate, properties with RNA polymerase from *E. coli*, 169
2′-Deoxy-8-azaguanosine 5′-triphosphate, properties with DNA polymerase I from *E. coli*, 62
2′-Deoxy-2′-azidoadenosine 5′-diphosphate, 169
2′-Deoxy-2′-azidonucleoside 5′-diphosphates, affinity labelling of ribonucleotide reductase, 235
2′-Deoxy-2′-aminocytidine 5′-diphosphate, 169
5′-Deoxy-5′-aminoguanosine 5′-*N*-phosphoric acid di-(4-nitrophenyl)-ester, 117
5′-Deoxy-5′-aminonucleoside 5′-*N*-phosphate esters, 114
properties with phosphodiesterase, 114
3′-Deoxy-3′-aminothymidine 5′-phosphate, 168
5′-Deoxy-5′-aminothymidine 5′-*N*-phosphate, 114
2′-Deoxy-2′-aminouridine 3′-phosphate, 168
2′-Deoxy-2′-aminouridine 5′-phosphate, 168
2′Deoxy-2′-aminouridine 5′-triphosphate, properties with RNA polymerase from *E. coli*, 169
2′-Deoxy-2′-azidoadenosine 5′-phosphate, 169
2′-Deoxy-2′-azidocytidine 5′-diphosphate, 169
2′-Deoxy-2′-azidouridine 5′-diphosphate, 168
2′-Deoxy-2′-azidouridine 5′-phosphate, 168
2′-Deoxy-2′-chloronucleoside 5′-diphosphates, affinity labelling of ribonucleotide reductase, 235
2′-Deoxy-2′-chlorouridine 5′-triphosphate, properties with RNA polymerase from *E. coli*, 169
3′-Deoxycytidine 5′-phosphate, 162
2′-Deoxy-α-D-cytidine 5′-phosphate, 160
2′-Deoxy-α-D-cytidine 5′-phosphate, 5-ethyl, 56
2′-Deoxy-β-D-cytidine 5′-phosphate, 5-ethyl, 56
2′-Deoxycytidine 5′-phosphate, 5-hydroxymethyl, 50
2′-Deoxycytidine 5′-phosphorothioate, 101
3′-Deoxycytidine 5′-triphosphate, 162
5′-Deoxy-5′-(dihydroxyphosphinyl)-hydroxymethyladenosine, 130
3′-Deoxy-3′-(dihydroxyphosphinyl)-methyladenosine, crystal structure, 129
5′-Deoxy-5′-(dihydroxyphosphinyl)-methylnucleosides, 127
3′-Deoxy-3′-(dihydroxyphosphinyl)-methylnucleosides, 129
5′-Deoxy-5′-(dihydroxyphosphinyl)-thymidine, *see* Thymidine 5′-phosphonate
5′-Deoxy-5′-(dihydroxypyrophosphor-

oxyphosphinyl)-cyanomethyl-
adenosine, 132
3′-Deoxyguanosine 5′-triphosphate, 162
properties with DNA polymerase I and RNA polymerase from *E. coli*, 162
6′-Deoxyhomoadenosine 6′-phosphonic acid, 128
6′-Deoxyhomoadenosine 6′-phosphonic acid ethyl ester, 128
6′-Deoxyhomouridine 6′-phosphonic acid, 128
2′-Deoxy-5-iodouridine 5′-triphosphate, photoinactivation of DNA polymerase I and II from *E. coli*, 244
2′-Deoxy-α-D-nucleoside 5′-phosphates, properties with snake venom nucleotidase, 160
1-(2′-Deoxy-α-D-ribopyranosyl)thymine 3′,4′-cyclic phosphate, 187
1-(2′-Deoxy-β-D-ribopyranosyl)thymine 3′,4′-cyclic phosphate, 187
2′-Deoxy-tetrahydrouridine 5′-phosphate, 60
2′-Deoxy-2′-thiocytidine 2′,3′-cyclic phosphorothioate, 112
conformation in solution, 112
5′-Deoxy-5′-thiothymidine 3′,5′-cyclic phosphorothioate, 112
5′-Deoxy-5′-thiouridine 2′,3′-cyclic phosphate, 113
2′-Deoxythymidine 5′,3′-di-(3-bromoacetamidophenyl)-phosphate, irreversible inhibition of RNA-dependent DNA polymerase, 223, 224
2′-Deoxythymidine, haloacetamidophenyl phosphate esters, 223
irreversible inhibition of exoribonuclease, 223
2′-Deoxythymidine 5′-phosphate:
2,4-dithio, 31
2-thio, 31
2′-Deoxythymidine 5′-triphosphate:
2,4-dithio, 31
N^3-methyl, 17
2′-Deoxyuridine, 5-iodo, 49
2′-Deoxy-α-D-uridine 5′-phosphate, 160
2′-Deoxyuridine 5′-phosphate:
5-acetyl, 57
5-allyl, 57
5-azidomethyl, 57
5-benzyloxymethyl, 57
5-formyl, 57
5-fluoro, 49
5-hydroxymethyl, 57
5-mercapto, 53
5-methoxymethyl, 57
5-nitro, 55
5-(2,3-oxypropyl), 57
5-trifluoromethyl, 57
3′-Deoxyuridine 5′-phosphate, 162
2′-Deoxyuridine 5′-phosphonate, 126
3′-Deoxyuridine 5′-triphosphate, 162
2′-Deoxyuridine 5′-triphosphate, 5-trifluoromethyl, 57
2′,3′-Dideoxyadenosine 5′-phosphate, 164
2′,3′-Dideoxyadenosine 5′-triphosphate, 164
properties with DNA polymerase I from *E. coli*, 165
5′,3′-Dideoxy-5′,3′-diaminoadenosine 3′,5′-cyclic phosphorothioate, 120
2′,3′-Dideoxy-3′-chlorothymidine, 165
1-(2′,3′-Dideoxy-2′,3′-didehydro-β-D-glyceropentofuranosyl)5-fluorouracil 5′-triphosphate, 166
2′,3′-Dideoxy-3′-fluorothymidine, biological properties, 165
2′,3′-Dideoxynucleosides, biological properties, 165
1-(5,6-Dideoxy-β-D-ribohexofuranosyl-6-phosphonic acid)-1,2,4-triazole-3-carboxamide, 129, 130
2′,3′-Dideoxythymidine 5′-phosphate, 163
2′,3′-Dideoxythymidine 5′-triphosphate, 164

properties with DNA polymerase I
from *E. coli*, 165
2′,3′-Dideoxyuridine 5′-phosphate, 163
2′,(3′)-*O*-Dihydrocinnamoylnucleoside 5′-diphosphates, 173
5,6-Dihydrouridine 5′-diphosphate, 60
5,6-Dihydrouridine 5′-phosphate, 59
5,6-Dihydrouridine 5′-triphosphate, 60
2′(*R*),3′-Dihydroxyethyl-1′-(uracil-1-yl)propane 2′,3′-cyclic phosphate, 184
R,S-9-(3′,4′-Dihydroxybutyl)adenine 4′-phosphate, 183
R,S,-1-(3′,4′-Dihydroxybutyl)thymine 4′-phosphate, 183
R-9-(2′,3′-Dihydroxypropyl)adenine, 182
3-(2′,3′-Dihydroxypropyl)adenine 3′-phosphate, 179
9-(2′,3′-Dihydroxypropyl)adenine 3′-phosphate, 179
R,S-9-(2′,3′-Dihydroxypropyl)adenine 3′-triphosphate, 184
3-(2′,3′-Dihydroxypropyl)xanthine 3′-phosphate, 179
R,S-1-(2′,3′-Dihydroxypropyl)thymine 2′,3′-cyclic phosphate, 180
R,S-1-(2′,3′-Dihydroxypropyl)thymine 3′-phosphate, 179
R,S,-1-(2′,3′-Dihydroxypropyl)thymine 2′-phosphate, 181
R,S,-1-(2′,3′-Dihydroxypropyl)uracil 2′,3′-cyclic phosphate, 180
properties with ribonucleases, 180
R,S,-1-(2′,3′-Dihydroxypropyl)uracil 3′-phosphate, 179
Diesterphosphorochloridates, 198
8-Dimethylaminoadenosine 5′-phosphate, conformation in solution, 40
2-(*N,N*-Dimethylamino)-4-nitrophenyl phosphate, transesterification with nucleosides, 206
4,5-Dimethyl-2-(1-imidazolyl)-2-oxo-P^{v}-1,2,3-dioxaphosphate, 205
P^1-Diphenyl P^2-(nucleoside 5′-pyrophosphate), reaction with nucleophiles, 211
Diphenyl phosphorochloridate, 198
Dist-benzoadenine, 68
Dithiophosphate, thiophosphorylation of nucleosides, 112
DNA, alkylation, 14
DNA-dependent RNA polymerase from *E. coli*, stereochemistry of polymerization, 107

1,N^6-Ethenoadenine derivatives, fluorescence, 22
1,N^6-Ethenoadenine nucleotides, reaction with divalent mercury, 23
1,N^6-Ethenoadenine nucleotides, biological functions, 22
1,N^6-Ethenoadenosine 3′,5′-cyclic phosphate:
2-halogeno, 46
2-substituted, 45
1,N^6-Etheno-2-azaadenosine 5′-diphosphate, 64
1,N^6 Etheno-2-azaadenosine 5′-phosphate, 64
3,N^4-Ethenocytosine nucleotides, reaction with divalent mercury, 23

Flavine adenine dinucleotide:
8-azido, 39
properties with glucose oxidase and D-amino acid oxidase, 240
1,N^6-etheno, 22
γ-Fluoro ATP, 134
properties with myosin, heavy meromyosin and hexokinase, 135
γ-Fluoro GDP, 135
γ-Fluoro GTP, 135
properties with elongation factor G GTPase, 135
P^3-Fluoro P^1-5′-adenosine triphosphate, *see* γ-Fluoro ATP
P^2-Fluoro P^1-5′-guanosine 5′-diphos-

phate, *see* γ-Fluoro GDP
P^3-Fluoro P^1-5′-guanosine 5′-triphosphate, γ-fluoro GTP, 135
5-Fluoropyrimidine derivatives, antitumor activities, 48
  mode of biological function, 48
4′-Fluorouridine 5′-phosphate, 166
Formycin:
  biological properties, 67
  crystal structure, 67
Formycin 5′-diphosphate, 67
Formycin 5′-phosphate, 67
  biochemical properties, 67
Formycin 5′-triphosphate, 67
  biochemical properties, 67
5-Formyl-1-(α-D-ribofuranosyl)uracil 5′-triphosphate, affinity labelling of RNA polymerase from *E. coli*, 232
5-Formyl-1-(β-D-ribofuranosyl)uracil 5′-triphosphate, properties with RNA polymerase from *E. coli*, 232
5-Formyluridine, 55

2′,(3′)-*O*-Glycyladenosine 5′-diphosphate, 174
2′,(3′)-*O*-Glycyladenosine 5′-triphosphate, 174
9-(β-D-Glucopyranosyl)purine phosphates, 185
9-(β-D-Glucopyranosyl)hypoxanthine phosphates, 185
Guanosine:
  1,N^2-etheno, 22
  6-seleno, 35
  tautomeric forms, 1
  6-thio, 24
    crystal structure, 26
    desulfurization, 27
Guanosine 5′-(3-azidotriphosphate, 238
Guanosine 5′-(3-(4-azidobenzylamido)-triphosphate, photoaffinity labelling of elongation factor G GTPase, 237
Guanosine 5′-(2-azidodiphosphate), 238
Guanosine 3′,5′-bis(methylenediphosphonate), 100
Guanosine 3′,5′-cyclic phosphate:
  8-acetyl, 41
  8-bromo, nucleophilic reactions, 38
  8-(1-hydroxyethyl), 41
Guanosine 2′,3′-cyclic phosphorothioate, 101
  diastereomers, ^{31}P-nmr, 103
  properties with ribonuclease T_1, 103
Guanosine 5′-diphosphate:
  8-amino, 37
  8-aza, 62
  dialdehyde, affinity labelling of elongation factor G GTPase, 234
  8-bromo, 36
  O^6-methyl, 18
  8-methylamino, 37
  8-oxy, 37
  8,2′-thioanhydro, 176
Guanosine 5′-methylenediphosphono P^2-phosphate, *see* α,β-Methylene GTP
Guanosine 5′-methylenephosphonate, *see* α,β-Methylene GDP
Guanosine 2′,(3′)-phosphate, 8-aza, 62
Guanosine 5′-phosphate:
  8-aza, 62
  8-bromo, 36
  3-deaza, 66
  8-methyl, 40
  N^7-methyl, 14
    hydrolysis, 14
  O^6-methyl, 19
  6-thio, 24, 25
  8,2′-thioanhydro, 176
Guanosine 5′-phosphoazidate, photolysis, 283
Guanosine 5′-phosphohypophosphate, 136
  properties with elongation factor G GTPase, 136

Guanosine 5′-triphosphate:
8-aza, 62
8-bromo, 37
dialdehyde, affinity labelling of elongation factor G GTPase, 234
8-oxy, 37
6-thio, 25
Guanosine 5′-(*O*-3-thiotriphosphate), 105
5′-Guanylyl imidodiphosphate, *see* β,γ-Imido GTP
5′-Guanylyl methylenediphosphonate, *see* β,γ-Methylene GTP

5-Halogenopyrimidine derivatives, displacement reactions, 52
Heterocyclic bases:
alkylation at exocyclic groups, 19
alkylation at ring nitrogens, 14
electronic structures, 1
site of proton abstraction, 2
site of protonation, 2
N-oxidation, 18, 19
reaction with formaldehyde, 21
reaction with glyoxal, 21
reaction with ketoxal, 21
tautomeric forms, 1
pK_a-values, 3, 4
Hexokinase, stereospecificity with adenosine 5′-(*O*-2-thiotriphosphate), 111
N^4-Hydroxy-6-azacytidine 3′-phosphate, 75
N^4-Hydroxy-6-azacytidine 5′-phosphate, 75
6′-Hydroxy-6′-homoadenosine 6′-phosphonate, 130
epimeric forms, 130
properties with AMP kinase, pyruvate kinase and AMP aminohydrolase, 131
Hypoxanthine, 9-*S*-(2′,3′-dihydroxypropyl), 182

β,γ-Imido ATP, 97
interactions with divalent cations, 99
pK_a-value, 99
Imidodiphosphate, bond angles and bond distances, 98
β,γ-Imido GTP, 97
Inosine 3′,5′-cyclic phosphate, 8-methyl, 42
Inosine 5′-deoxy-5′-thiophosphate, 113
Inosine 5′-diphosphate:
alkylation, 15
2-methyl, 42
2-methylthio, 42
6-thio, 25
8,2′-thioanhydro, 176
Inosine 5′-phosphate:
2-aza, 63
8-aza, 62
2-chloro, 44
D,L-enantiomers, 154
2-mercapto, 42
2-methoxy, 42
2-methyl, 42
N^7-methyl, hydrolysis, 15
2-methylthio, 42
6-methylthio, 24
N^1-oxide, 18
6-thio, 24
8,2′-thioanhydro, 176
Inosine 5′-triphosphate, 6-thio, 25
5-Iodo-2′-deoxyuridine, chemotherapy of viral diseases, 48
Irreversible enzyme inhibitors, active-site-directed, 220
3-Isoadenine nucleotides, biochemical properties, 91
3-Isoadenosine, 90, 91
3-Isoadenosine 5′-diphosphate, 91
3-Isoadenosine 2′,(3′)- phosphate, 91
3-Isoadenosine 5′-phosphate, 91
3-Isoadenosine 5′-triphosphate, 91
Isoguanosine 5′-diphosphate, 44
Isoguanosine 5′-phosphate, 44
Isoguanosine 5′-triphosphate, 44
2′,3′-Isopropylideneuridine, methyl, 51

2′,3′-Isopropylidene-2-thiouridine, 5-hydroxymethyl, 51
2′,(3′)-*O*-Isovaleroylnucleoside 5′-diphosphates, 175

Levulinate-$O^{2',3'}$-nucleotide ketals, 236
Lin-benzoadenine, 68
Lin-benzoadenosine, fluorescence, 70
Lin-benzoadenosine 3′,5′-cyclic phosphate, 69
Lin-benzoadenosine 5′-diphosphate, 69
Lin-benzoadenine nucleotides, biochemical properties, 69
Lin-benzoadenosine 5′-phosphate, 69
Lin-benzoadenosine 5′-triphosphate, 69
9-(β-D-Lyxofuranosyl)adenine, biological properties, 153
9-(α-D-Lyxofuranosyl)adenine 2′,3′-cyclic phosphate, 157
9-(α-L-Lyxofuranosyl)adenine 2′,3′-cyclic phosphate, 156
9-(α-L-Lyxofuranosyl)adenine 5′-phosphate, 157
9-(α-L-Lyxofuranosyl)hypoxanthine 2′,3′-cyclic phosphate, 157
9-(α-D-Lyxofuranosyl)hypoxanthine 2′,3′-cyclic phosphate, 157
  properties with ribonuclease T_2, 157
1-(β-D-Lyxofuranosyl)uracil 2′,3′-cyclic phosphate, 151
1-(β-D-Lyxofuranosyl)uracil 3′,5′-cyclic phosphate, 152
1-(β-D-Lyxofuranosyl)uracil, reaction with triethyl phosphite, 151

9-(α-D-Mannofuranosyl)adenine 5′-phosphate, 187
  properties with AMP kinase, 187
5-Mercuripyrimidine nucleotides, 53
  biochemical properties, 53
  chemical reactivities, 53
2′-*O*-(α-Methoxyethyl)nucleoside 5′-diphosphates, 172
8-Methylaminoadenosine 5′-phosphates, conformation in solution, 40
5-Methyl-6-azauridine 3′-phosphate, 73
α,β-Methylene ADP, 97
α,β-Methylene ATP, 97
β,γ-Methylene ATP, 97
  interaction with divalent cations, 99
  pK_a-values, 99
α,β-Methylene GDP, 97
α,β-Methylene GTP, 97
β,γ-Methylene GTP, 97
Methylenediphosphonate, 97
  bond angles and bond distances, 98
2′-*O*-Methylnucleotides, 170
Methyl phosphorodichloridate, 200
6-Methylpyrimidine nucleotides, biochemical properties, 58
2-Methylthio-2-oxo-P^V-4H-1,2,3-benzodioxaphosphorin, reaction with nucleosides, 204
Modification of proteins by active-site-directed reagents, 219
8-Monoalkylaminoadenosine 5′-phosphates, conformation in solution, 39
Monoesterphosphates:
  activation by aryl sulfonylchlorides, 204
  activation by dicyclohexylcarbodiimide, 203
  activation by triphenylphosphite, 209, 210
Monoester phosphites, reaction with α-haloketo compounds, 207
Monoester phosphorodichloridates, 200
Monothiophosphate, thiophosphorylation of nucleosides, 101

Nicotinamide adenine dinucleotide:
  8-azido, 39
  1,N^6-etheno, 22
  3′-*O*-*N*-(4-azidophenyl)amino proprionyl, photoaffinity labelling of alcohol dehydrogenase, 249

P^1-(3-Nicotinamide-1-β-D-ribofuranosyl) 5′-P^2-thioinosine) 5′-pyrophosphate, reaction with bromoacetamido groups on solid supports, 230
Nicotinamide-8-azidoadenine dinucleotide, properties with dehydrogenases, 240
Nicotinamide-3,N^4-ethenocytosine dinucleotide, 23
S-(2-Nitro-4-carboxyphenyl)-6-mercaptopurine-9-(β-D-ribofuranosyl) 5′-phosphate, affinity labelling of phosphorylase b, 231
Nitronium tetrafluoroborate, 55
4-Nitrophenylphosphate, transesterification with nucleosides, 205
4-Nitrophenyl,5-(2-pyridyl)phosphorothioate, transesterification with nucleosides, 206
Nucleoside 5′-deoxy-5′-thiophosphates, conformation in solution, 113
Nucleoside 5′-diphosphates, 2′-deoxy-2′-azido, properties with polynucleotide phosphorylase, 169
Nucleoside 5′-diphosphates:
  2′-deoxy-2′-chloro, properties with polynucleotide phosphorylase, 167
  2′-deoxy-2′-fluoro, properties with polynucleotide phosphorylase, 167
  2′-*O*-ethyl, 171
  2′-*O*-methyl, 170
  2′-*O*-substituted, properties with polynucleotide phosphorylase, 172
Nucleoside 5′-fluorophosphates, 134
  properties with phosphatases, nucleotidases and phosphodiesterases, 134
Nucleoside 2′,3′-cyclic phosphates, 207
Nucleoside 5′-phosphoazidates, 238
Nucleoside 5′-*S*-phosphates, 197
Nucleoside phosphites, 120
  oxidation by 2,2′-dipyridylyl disulfide, 208
  oxidation by permanganate, 122
  reaction with α-haloketo compounds, 124
  by transesterification with triethylphosphite, 207
  by transesterification with triphenylphosphite, 207
Nucleoside phosphonates, 125
Nucleoside 5′-phosphoramidates, 212
Nucleoside 5′-phosphoric di-*n*-butylphosphinothioic anhydrides, 212
Nucleoside 5′-phosphorimidazolidates, 214
Nucleoside 5′-phosphormorpholidates, 213
Nucleoside phosphorothioates, 101
  chemical reactions, 112, 204
  hydrolysis by phosphatases, 104
Nucleoside 5′-polyphosphates:
  imido analogs, 99, 100
  methylene analogs, biochemical properties, 99
  phosphorazidate analogs, 136
  thiophosphate analogs, 105
Nucleotides:
  N-alkylation, 14
  conformation, 3, 4, 8, 9
  ultraviolet absorption spectra, 6
Nucleotide 5′-aminophenyl esters, coupling to Sepharose, 225
Nucleotide 5′-aminophenylethyl esters, 224
Nucleotides, thiophosphate analogs, properties with various enzymes, 107, 109

Orotidine 5′-phosphate, 59
2′-*O*-(Ortho-nitrobenzyl)nucleoside 5′-diphosphates, 172
2-Oxypyrimidine-4-sulfonates, 33

Pancreatic ribonuclease A:
  mechanism of hydrolysis, 102

mechanism of transesterification, 102
Phosphitylation of nucleosides, 207
Phosphorous oxychloride, 201
selective phosphorylation, 201
Polyarabinouridylic acid, 150
Polyphosphoric acid, phosphorylation of nucleosides, 202
Prox-benzoadenine, 68
Pseudo-ATP, 138
properties with tRNA synthetase and hexokinase, 138
Pseudouridine, 92
Pseudouridine 5′-diphosphate, 92
Pseudouridine 2′,(3)′-phosphate, 94
Pseudouridine 5′-phosphate, 92
Pseudouridine 3′-phosphate, photochemical reactions, 95
Pseudouridine 5′-triphosphate, 95
Purine nucleotides, 6-seleno, 34
Purine-9-β-D-ribofuranoside, 27
Purine 6-sulfonates, 27
Purine-6-sulfonate-9-β-D-ribofuranosyl 5′-phosphate, 28
N-Purin-6-carbamoyl-9-β-D-ribofuranosyl 5′-phosphates, 28
α-Pyridylphosphate esters, 206
Pyrimidine nucleotides, chlorination by tetrabutylammonium iodotetrachloride, 50
Pyrophosphate, bond angles and bond distances, 98
Pyrophosphoric acid, phosphorylation of nucleosides, 202
Pyrophosphorylchloride, selective phosphorylation, 202
Pyrrolo(4,3-d)pyrimidine-3-β-D-ribofuranoside, *see* 7-Deazanebularin
Pyruvate kinase, stereospecificity with adenosine 5′-(*O*-2-thiotriphosphate), 111

Ribavirin, biological properties, 70
Ribavirin 5′-phosphate, 71
Ribavirin 2′,(3′)-phosphate, 71
Ribavirin 5′-phosphate, conformation, 71
Ribavirin 5′-triphosphate, 71
properties with viral RNA polymerases, 71
3-β-D-Ribofuranosyladenine, *see* 3-Isoadenosine
3-(β-D-Ribofuranosyl)adenine 9,5′-(P)-cyclic phosphonate, 134
9-(β-D-Ribofuranosyl)adenine 5′-phosphate, 156
3-(β-D-Ribofuranosyl)adenine 5′-phosphate, 156
1-(β-D-Ribofuranosyl-4-amino-*s*-triazine-2(1H)-on, *see* 5-Azacytidine
α,D-Ribofuranosylcytosine 5′-phosphate, 155
α,L-Ribofuranosylcytosine 5′-phosphate, 155
β,L-Ribofuranosylcytosine 5′-phosphate, 156
P^1-5′-*O*-(1-β-D-Ribofuranosylnicotinamide) P^2-5′-*O*-(9-(4-thio-β-D-ribofuranosyl)adenine)pyrophosphate, 174
2-(β-D-Ribofuranosyl)-pyridazon-(3) 5′-diphosphate, 77
2-(β-D-Ribofuranosyl)-pyridazon-(3) 5′-phosphate, 77
1-(β-D-Ribofuranosyl)-pyridon-(2) 5′-diphosphate, 77
1-(β-D-Ribofuranosyl)-pyridon-(2) 5′-phosphate, 77
1-(β-D-Ribofuranosyl)-2-pyrimidine, biological properties, 77
7-(β-D-Ribofuranosyl)theophyllin 5′-phosphate, 72
1-(β-D-Ribofuranosyl)thymine 2′,3′-cyclic phosphate, 187
1-(α-L-Ribofuranosyl)thymine 5′-phosphate, 155
1-(β-L-Ribofuranosyl)thymine 5′-phosphate, 155
1-(β-D-Ribofuranosyl)-1,2,4-triazole-3-carboxamide, *see* Ribavirin

3-(β-D-Ribofuranosyl)uracil 2′,3′-cyclic phosphate, 92
5-(β-D-Ribofuranosyl)uracil, *see* Pseudouridine
Ribonuclease T_1:
  mechanism of hydrolysis, 103
  mechanism of transesterification, 103
Ribonucleoside 2′,3′-cyclic phosphate, 125
β-L-Ribonucleoside 2′,3′-cyclic phosphates, properties with ribonuclease A, ribonuclease T_1 and T_2, 158
β-L-Ribonucleoside 5′-diphosphates, properties with polynucleotide phosphorylase, 159
β-L-Ribonucleoside 5′-phosphates, properties with 5′-nucleotidases, 159
Ribonucleoside 2′,(3′)-phosphites, 124
Ribonucleotide dialdehydes, preparation of protein-nucleotide conjugates, 235
Ribonucleotides, oxidation with periodate, 233
P^1-Ribosylnicotineamide-5′ P^2-(3-β-D-ribofuranosyl)adenine 5′-pyrophosphate, 91
3-β-D-Ribosyluric acid, 90
3-β-D-Ribosyluric acid 5′-phosphate, 90
tRNA nucleotidyltransferase from yeast, stereochemistry of internucleotide bond formation, 110

Sangivamycin, 65
Sangivamycin 5′-diphosphate, 65
Sangivamycin 5′-phosphate, 65
Sangivamycin 5′-triphosphate, 65
Seleoxopyrimidine derivatives, 35
  absorption spectra, 35
  pK_a-values, 35
Staphylococcal nuclease, irreversible inhibitors, 222, 223

Tetra-(4-nitrophenyl)pyrophosphate, 203
Tetrazolo(5,1-b)-6-oxopurine-9-(β-D-ribofuranosyl)5′-phosphate, 45
Tetrazolo(5,6-b)-6-oxopurine-9-(β-D-ribofuranosyl)5′-triphosphate, 45
8,3′-Thioanhydroadenosine 5′-phosphate, 176
6-Thioinosine 5′-diphosphate:
  inactivation of polynucleotide phosphorylase, 230
  reaction with bromoacetamido groups at solid supports, 229 230
6-Thioketoguanosine 5′-triphosphate, photoaffinity labelling of RNA polymerase from *E. coli*, 245
6-Thioketoinosine, 24, 27
6-Thioketo purine derivatives:
  alkylation, 27
  photooxidation, 27
Thioketopyrimidine derivatives:
  alkylation, 33
  oxidative hydrolysis, 33
  photooxidation, 33
  reaction with mercurials, 33
Thioketopyrimidine nucleosides, crystal structure, 32
4-Thioketopyrimidine nucleotides:
  interaction with enzymes, 32
  photooxidation, 244
  spectroscopic properties, 32
9-(4-Thio-β-D-ribofuranosyl)adenine 3′,5′-cyclic phosphate, 175
4-Thiouridine 3′-phosphate, 245
2-Thiouridine 4-sulfonates, 30
4-Thiouridine 5′-triphosphate, photoaffinity labelling of RNA polymerase from *E. coli*, 244
Thymidine, tautomeric forms, 1
Thymidine 5′-deoxy-5′thiophosphate, 113

Thymidine 5′-deoxy-5′-triphosphate, 113
properties with DNA polymerase from *E. coli*, 113
Thymine, 1-*S*-(2′,3′-dihydroxypropyl), 182
α-D-Thymidine 5′-diphosphate, 160
α-D-Thymidine 5′-phosphate, 160
β-L-Thymidine 5′-phosphate, 159
Thymidine 3′-phosphite, 122
Thymidine 5′-*N*-phosphate phenyl ester, 116
Thymidine 5′-phosphonate, 125
Thymidine 5′-phosphonylpyrophosphate, 126
properties with DNA polymerase I from *E. coli*, 126
Thymidine 5′-phosphorthioate, 101
Thymine-1-(β-D-ribofuranosyl uronic acid), 227
Thymidylate synthetase inhibitors, 57
Thymidine 5′-(*O*-1-thiotriphosphate), 104
Thymidine 5′-*N*-triphosphate, 115
properties with DNA polymerase I from *E. coli*, 115
α-D-Thymidine 5′-triphosphate, 160
Thymidylyl-(3′-5′)-5′-deoxy-5′-aminothymidine, 115
Toyocomycin, 65
Toyocomycin 5′-phosphate, 65
Trialkylphosphites, reaction with 5′-azido-5′-deoxynucleosides, 208
1,2,4-Triazine-3,5(2,3,4,5-tetrahydro)-dione, *see* 6-Azauracil
2,2,2-Tribromethyl phosphormorpholino chloridate, 213
2,2,2-Trichloroethyl phosphate, 203
Trichloromethyl phosphonodichloridate, 208
2,2,2-Trichloroethyl groups, removal, 199
Triethylphosphite, transesterification with ribonucleosides 124
5-Trifluoromethyl-2′-deoxyuridine 5′-phosphate, 57
Trimetaphosphate, bismethylene analog, 97
Trimetaphosphate:
phosphorylation of 5′-deoxy-5′-aminonucleosides, 203
phosphorylation of nucleosides, 202
2′,(3′)-*O*-(2,4,6-Trinitrophenyl)adenosine 5′-triphosphate, 175
Triphenylphosphite, transesterification with nucleosides, 122
Triphosphopyridine nucleotide, 1,N^6-etheno, 22
Tris-(8-hydroxychinolyl) phosphate, transesterification with nucleosides, 206
Tris-imidazolylphosphinoxide, 28
phosphorylation of nucleosides, 214
Tris-imidazolyl-1-phosphinesulfide, 101
Tubercidine, 64
Tubercidine 2′,(3′)-phosphate, 65
Tubercidine 5′-triphosphate, 65

Uracil:
1-*S*-(2′,3′-dihydroxypropyl), 182
1-(β-D-ribofuranosyl uronic acid), 227
2-seleno, 34
4-seleno, 34
Uridine:
5-hydroxymethyl, 50
5-iodo, 49
4-seleno, 35
tautomeric forms, 1
α-D-Uridine 2′,3′-cyclic phosphate, 160
properties with ribonuclease A, 160
Uridine 2′,3′-cyclic phosphate:
5-iodo, 49
5-substituted, 57
Uridine 2′,3′-cyclic phosphorothioate, 101
diastereomers, crystal structure, 102
properties with ribonuclease A, 102
Uridine 5′-deoxy-5′-thiophosphate, 113

Uridine 5′-deoxy-5′-thiotriphosphate, 113
properties with polynucleotide phosphorylase, 113
Uridine 5′-deoxy-5′-thiotriphosphate, 113
properties with RNA polymerase from *E. coli*, 113
α-D-Uridine 5′-diphosphate, 160
β-L-Uridine 5′-diphosphate, 159
Uridine 5′-diphosphate:
5-bromo, 49, 50
5-carboxyethyl, 56
6-carboxamido, 59
5-chloro, 50
dialdehyde, affinity labelling of galactosyl transferase, 234
2,4-dithio, 29
5-ethyl, 56
2′,(3′)-*O*-ethyl, 171
5-hydroxy, 50
5-hydroxymethyl, 52
5-iodo, 50
5-methyl, 56
N^3-methyl, 17
2′-*O*-methyl, 170
5-methylsulfonyl, 56
2-thio, 30
4-thio, 28
Uridine 5′-(α-D-glucopyranosyl pyrophosphate):
analogs, 59, 60, 61
catalytic hydrogenation, 59
N^3-methyl, 17
2-thio, 31
4-thio, 31
Uridine nucleotides, bromination, 49
α,D-Uridine 5′-phosphate, 160
α,L-Uridine 5′-phosphate, 155
β,L-Uridine 5′-phosphate, 155, 159
Uridine 5′-phosphate:
5-amino, 52
N^3-(3-L-amino-3-carboypropyl), 17
5-bromo, 59
5-chloro, 49
6-cyano, 59
2,4-dithio, 29
5-ethoxycarbonyl, 56
5-ethyl, 56
2′,(3′)-*O*-ethyl, 171
5-fluoro, 49
5-formyl, 56
5-hydroxy, 50
5-hydroxymethyl, 50, 51
5-iodo, 49
5-methyl, 56
6-methyl, 58
N^3-methyl, 17
5-methylsulfonyl, 56
5-nitro, 55
4-thio, 28
2-thio, 30
6-thiocarboxamido, 59
Uridine 3′-phosphate, N^3-methyl, 17
Uridine 2′,(3′)-phosphate, 4-thio, 28
Uridine 5′-phosphonate, 126
Uridine 5′-phosphorothioate, 4-thioate, 29
Uridine 5′-(*O*-3-thiotriphosphate), 105
α-D-Uridine 5′-triphosphate, 160
Uridine 5′-triphosphate:
5-amino, 52
2,4-dithio, 29
5-fluoro, 49
5-formyl, 56
5-hydroxy, 50
2-thio, 30
4-thio, 29

Xyloadenosine 5′-phosphate:
8,3′-anhydro-8-oxy, 177
properties with enzymes, 177
Xyloadenosine 5′-triphosphate:
8,3′-anhydro-8-oxy, 179
properties with nucleotide carrier from rat liver mitochondria, 179
9-(β-D-Xylofuranosyl)adenine, 153
9-(β-D-Xylofuranosyl)adenine 3′,5′-cyclic phosphate, 152
9-(β-D-Xylofuranosyl)adenine 5′-phosphate, 152